Guidelines for Inherently Safer Chemical Processes:
A Life Cycle Approach
3rd Edition

Guidelines for Inherently Safer Chemical Processes: A Life Cycle Approach
3rd Edition

CENTER FOR CHEMICAL PROCESS SAFETY
of the
AMERICAN INSTITUTE OF CHEMICAL ENGINEERS
New York, NY

An AIChE Technology Alliance
CCPS
Center for Chemical Process Safety

WILEY

This edition first published 2020

A Joint Publication of the American Institute of Chemical Engineers and John Wiley & Sons, Inc.

Registered Office
John Wiley & Sons, Inc., 111 River Street, Hoboken, NJ 07030, USA

Editorial Office
111 River Street, Hoboken, NJ 07030, USA

For details of our global editorial offices, customer services, and more information about Wiley products visit us at www.wiley.com.

Wiley also publishes its books in a variety of electronic formats and by print-on-demand. Some content that appears in standard print versions of this book may not be available in other formats.

Library of Congress Cataloging-in-Publication Data is available

Hardback ISBN: 9781119529163

Cover Design: Wiley
Cover Images: Silhouette, oil refinery © manyx31/iStock.com; Stainless steel © Creativ Studio Heinemann/Getty Images; Dow Chemical Operations, Stade, Germany/Courtesy of The Dow Chemical Company

Printed in the United States of America.

V10014469_100719

Preface

The American Institute of Chemical Engineers (AIChE) has been involved with process safety and loss prevention issues in the chemical, petrochemical, hydrocarbon process and related industries and facilities for more than 50 years. AIChE publications and symposia are information resources for chemical engineers and other professionals on the causes of process incidents and the means of preventing their occurrences and mitigating their consequences.

In 1985, the Center for Chemical Process Safety (CCPS), a Technology Alliance of the AIChE, was established specifically to develop and disseminate technical information for use in the prevention of major chemical process incidents. With the support and direction of the CCPS Advisory and Managing Boards, a multifaceted program was initiated to address the need for Process Safety and Risk Management systems capable of reducing potential exposures to the public, the environment, personnel, and facilities. This program includes:

- developing and publishing Guidelines and Concept Books relating to specific areas of Process Safety and Risk Management
- publishing a monthly newsletter, Process Safety Beacon
- organizing, convening and conducting seminars, symposia, training programs, and meetings on process safety-related matters
- cooperation with other organizations, both internationally and domestically, to promote process safety.

CCPS activities are supported by more than 200 corporations that provide funding and professional expertise. Several government agencies and academic institutions also participate in CCPS endeavors.

In 1989, CCPS published Guidelines for Technical Management of Chemical Process Safety, which presented a model for process safety management characterized by twelve distinct, essential and interrelated elements. These Guidelines were refined over the ensuing years and updated in the Guidelines for Risk Based Process Safety (2007), which expanded the definition of process safety and risk management into twenty distinct elements.

The previous editions of this book, which were part of the "Concept Series," supported many of those twenty key elements of process safety, as identified in the Guidelines for Risk Based Process Safety, including process safety competency, workforce involvement, hazard identification and risk analysis, auditing, and management review and continuous improvement. The purpose of this guideline series book is to update the previous 2nd edition concept book in order to demonstrate the on-going and improved application of inherently safer strategies throughout all the stages of the chemical process life cycle, including additional case studies and examples of inherently safer design (ISD) implementation.

Inherently safer (IS) concepts continue to be well-received by industry and there has been significant advancement in the concept since the original version of this book was published in 1996. Both the original 1996 concept book and the 2009 update are frequently cited as authoritative sources on inherent safety, and, as such, it is important that an updated edition be published. The Third Edition will reflect not only the most current knowledge on the subject but will harness the concepts introduced in the first two editions, provide additional examples of IS and provide a guideline for the practical application of inherent safety techniques.

This book encourages engineers, process safety experts, and others involved in analyzing and reducing the risks from chemical processes to make conservative choices that apply the principles and spirit of inherent safety to the extent feasible. The discussion and examples range from the basic process chemistry through the details of the design of hardware and procedures. Inherently safer thinking can be applied at all levels of the process design, from the overall concept through the detailed equipment design and procedure development.

Acknowledgements

The American Institute of Chemical Engineers (AIChE) and its Center for Chemical Process Safety (CCPS) thank the subcommittee members and their CCPS member companies for their generous efforts and technical contributions to this book. CCPS also thanks the members of the CCPS Technical Steering Committee for their advice and support.

Guidelines for Inherently Safer Chemical Processes
A Life Cycle Approach, 3rd edition

The Chair of the subcommittee was John Wincek, originally of Croda, Inc. and currently of Dekra. The CCPS staff consultant was Dan Sliva.

The Subcommittee members were:

Steve	Arendt	*ABS Group*
Susan	Bayley	*Linde Process Plants, Inc.*
Prashanti	Bhupathi	*Reliance*
Wayne	Chastain	*Eastman*
Oliva	Cheng	*Chevron Corporation*
Robert	Coover	*Praxair*
Carolina	Del Din	*PSRG*
Jonas	Duarte	*Lanxess*
Emmanuelle	Hagey	*Nova chemicals*
Scott	Haney	*Marathon Petroleum Company*
Dennis	Hendershot	*CCPS*
Reyyan	Koc Karabocek	*ExxonMobll*
Nicole	Loontjens	*Americas Styrenics*
Dan	Miller	*CCPS Emeritus*
Jitesh	Patel	*New Jersey DEP*
Katherine	Perm	*Arkema*
Sabrina	Petruele	*Pluspetrol*
Morgan	Reed	*MMI Engineering*
Sonny	Sachdeva	*PSRG*
Randy	Sawyer	*Contra Costa Health Services*
Dallas L.	Singleton	*Jacobs*
Scott	Wallace	*Olin*
Bob	Weber	*PSRG*

This manuscript was written by AcuTech Group team of Rich Santo, Dave Moore, Cara Hamel, Mike Hazzan, Dave Heller, Marty Rose. It was copy edited by Megan Fennell under the direction of Anil Gokhale of CCPS.

Peer Reviewers

Before publication, all CCPS books receive a thorough peer review. CCPS gratefully acknowledges the thoughtful comments and suggestions of the following peer reviewers. Their work enhanced the accuracy and clarity of these guidelines.

Denise	Chastain-Knight	*Exida*
Alan	Evankovich	*AVAN Engineering*
Graffar	Keshavarz	*NovaChem*
Beverly	Perozzo	*NovaChem*
Gill	Sigmon	*Advansix*
Ken	Tague	*ADM*
Rob	Savarese	*EmeraldMaterials*

Contents

Figures

Tables

1. Introduction

1.1 OBJECTIVES, INTENDED AUDIENCE, AND SCOPE OF THIS BOOK

1.1.1 Objectives

The primary objective of this book is to provide a tool that can be used by any industrial company that handles hazardous chemicals to better understand inherent safety concepts and provide guidelines on how to implement them. The goal of this book is to provide practical guidelines for illustrating and emphasizing the merits of integrating research, development, and design into a comprehensive approach that balances safety, capital, and environmental concerns throughout the life cycle of the process. The authors hope that this book will help influence the next generation of engineers, chemists, and current practitioners and managers in the field of chemical processing.

Inherent safety is a powerful and effective means of reducing hazards and risk, versus managing risk by adding layers of protection. A hazard eliminated is one that then doesn't need to be addressed, which may have many direct and indirect benefits. Responsible companies understand the concept and apply it wherever and whenever it may be useful throughout the life cycle of the process. Companies who have an internal motivation to apply inherent safety routinely to the fullest capacity "as a way of doing business" recognize that doing so is beneficial.

In 1996, the Center for Chemical Process Safety (CCPS) published the first edition of its inherent safety concept book. Lessons learned in the ensuing years, combined with the fact that inherently safer design (ISD) was becoming more widely accepted, prompted CCPS to update the concept book in 2009. In the subsequent years, inherent safety has been of greater interest to industry, government, and the public as a concept to reduce hazards. In fact, several governmental entities have mandated consideration of inherently safer design for certain facilities. In the United States, for example, specific regulations exist now for Contra

Costa County in California, the State of New Jersey, and the U.S. Environmental Protection Agency.

Clearly, there is an ongoing need for more guidance, especially in practical, step-wise approaches to conduct inherently safer design studies. This edition builds on the philosophy provided in the first two concept books. It also provides guidelines and approaches to using these concepts with recent research, practitioner observations, new examples and industry methods.

1.1.2 Intended Audience

This book is written for those interested in the application of inherent safety strategies to process safety in industry. It is intended for facility site managers, process safety experts, engineers, chemists, regulators, research and development groups, process development teams, engineering educators and others responsible for workplace safety.

1.1.3 Scope

This book includes key principles on inherent safety and guidelines for how to implement them. It also includes guidance on how to conduct inherent safety studies as well as how to apply inherent safety to an organization's process safety management processes. It also briefly covers the history, research, and basic concepts of inherent safety. The methods described in this book may be widely applicable to inherent safety as it relates to process safety, environmental, and security issues.

1.2 INTEGRATION OF THIS GUIDANCE WITH OTHER CCPS GUIDANCE

Inherent safety is an integral component of process risk management. It is a foundation topic to managing chemical risks and has been referenced in other CCPS guideline books including:

- *Guidelines for Safe Storage and Handling of Reactive Materials*, 1995.
- *Layer of Protection Analysis, Simplified Process Risk Assessment*, 2001.
- Guidelines for Identifying and Analyzing the Security Vulnerabilities of Fixed Chemical Sites, 2003.
- *Guidelines for Risk-Based Process Safety*, 2007.

- *Guidelines for Hazard Evaluation Procedures, Third Edition, with Worked Examples*, 2007.
- *Guidelines for Engineering Design for Process Safety, Second Edition*, 2012, includes a section (Chapter 5.2) on inherently safer design.
- *Guidelines for Implementing Process Safety Management Systems, Second Edition*, 2016.
- *Guidelines for Siting and Layout of Facilities, Second Edition*, 2018.
- *Guidelines for Integrating Process Safety into Engineering Projects*, 2019.

1.3 ORGANIZATION OF THE BOOK

The book is written with the key principles for inherent safety in the body of the book. Tools for implementing the approach, as well as indicative examples and checklists, are included in the appendices. The guidelines begin with an explanation of the concept of Inherent Safety (IS) and then explain its role in Process Risk Management.

Chapter 2 introduces the topic of inherent safety. The key terms and the philosophy behind inherent safety are described. The different ways in which inherent safety is applied can be categorized into "strategies." These strategies—**minimize, substitute, moderate, and simplify**—are discussed in detail in Chapters 3 through 6. Human factors are an extremely important subset of the **simplify** strategy, and Chapter 6.11 of this book presents a detailed discussion of human factors as related to inherently safer design for the human-machine interface. Chapter 7 outlines how each of the four IS strategies can be applied to traditional protection layers.

"Inherently safer" is a way of thinking and, to successfully implement it, this thinking must be well-understood and continually employed wherever possible. Improved understanding of the process may result in a better, more reliable, and even more productive and profitable process that produces higher quality products. Processes should be reviewed for hazards and risks periodically throughout their life cycle. Chapters 8 discusses review methods to do this. Chapter 9 discusses the role of inherent safety in chemical process security. Chapter 10 presents available methods for implementing inherently safer strategies. These can either be independent, special studies, done periodically or before a major project or change is undertaken, or opportunistically applied into day-to-day process risk management strategies. Chapter 11 describes the relationship between the four main IS strategies, i.e., *Substitution, Minimization, Moderation, and Simplification* and each element of a PSM/Risk-Based Process Safety (RBPS) program. Chapter 12 addresses tools available to assist with implementing these IS strategies.

Chapter 13, entitled "Inherently Safer Design Conflicts," describes the conflicts that often develop between the various attributes of safety, operability, cost, as well as other risk parameters and the ways to understand and make decisions considering those constraints. With the advent of regulations requiring inherent safety consideration or implementation, Chapter 14 was written to help guide regulators and industry through the various considerations and challenges of IS.

Chapter 15 contains worked examples of IS study methods and case studies to show a step-wise process that can be followed for an IS evaluation. It also gives practical examples of successful implementation.

Lastly, Chapter 16 describes potential future IS initiatives, including needs, research, expected practice issues, and regulatory issues.

1.4 HISTORY OF INHERENT SAFETY

Inherent Safety is a modern term for an age-old concept: to progress towards eliminating or reducing hazards rather than accepting and managing them. This concept goes back to prehistoric times. For example, building villages near a river on high ground, rather than managing flood risk with dikes and walls, is an inherently safer design concept.

There are many examples of milestones in the application of inherently safer design. For example, in the United States back in 1866, following a series of explosions involving the handling of nitroglycerine, which was being shipped to California for use in mines and construction, state authorities quickly passed laws forbidding its transportation through San Francisco and Sacramento. This action made it virtually impossible to use the material in the construction of the Central Pacific Railroad.

The railroad desperately needed the explosive to maintain its construction schedule in the mountains. Fortunately, a British chemist, James Howden, approached Central Pacific and offered to manufacture nitroglycerine at the construction site. This is an early example of an inherently safer design principle – *minimize* the transport of a hazardous material by in-situ manufacture at the point of use. While nitroglycerine still represented a significant hazard to the workers who manufactured, transported, and used it at the construction site, the hazard to the general public from nitroglycerine transport was eliminated. At one time, Howden was manufacturing 100 pounds of nitroglycerine per day at railroad construction sites in the Sierra Nevada Mountains. The Central Pacific Railroad's experience with the use of nitroglycerine was quite good, with no further fatalities directly attributed to use of the explosive during the Sierra Nevada construction (Ref 1.22 Rolt), (Ref 1.2 Bain).

By today's standards, little about 19th Century railroad construction would qualify as safe, but the in-situ manufacture of nitroglycerine by the Central Pacific Railroad did represent an advance in inherent safety for its time. A further, and probably more important advance occurred in 1867, when Alfred Nobel invented dynamite by absorbing nitroglycerine on a carrier, greatly enhancing its stability. This is an application of another principle of inherently safer design – *moderate*, by using a hazardous material in a less hazardous form (Ref 1.9 Henderson).

A milestone event in process safety and inherent safety was the 1974 Flixborough explosion in the United Kingdom that caused twenty-eight deaths. On December 14, 1977, inspired by this tragic event, Dr. Trevor Kletz, who was at that time safety advisor for the ICI Petrochemicals Division, presented the annual Jubilee Lecture to the Society of Chemical Industry in Widnes, England. His topic was "What You Don't Have Can't Leak," and this lecture was the first clear and concise discussion of the concept of inherently safer chemical processes and plants.

Following the Flixborough explosion, interest in chemical process industry (CPI) safety increased, from within the industry, as well as from government regulators and the general public. Much of the focus of this interest was on controlling the hazards associated with chemical processes and plants through improved procedures, additional safety instrumented systems and improved emergency response. Kletz proposed a different approach—to change the process to either eliminate the hazard completely or sufficiently reduce its magnitude or likelihood of occurrence to eliminate the need for elaborate safety systems and procedures. Furthermore, this hazard elimination or reduction would be accomplished by means that were inherent in the process, and thus, permanent and inseparable from it.

Kletz repeated the Jubilee Lecture two times in early 1978, and it was subsequently published (Ref 1.14 Kletz 1978). In 1985, Kletz brought the concept of inherent safety to North America. His paper, "Inherently Safer Plants" (Ref 1.13 Kletz 1985), won the Bill Doyle Award for the best paper presented at the 19th Annual Loss Prevention Symposium, sponsored by the Safety and Health Division of the American Institute of Chemical Engineers (AIChE).

Since 1978, interest in inherently safer chemical processes and plants has grown, and that growth was particularly rapid in the 1990s (Ref 1.15, Kletz, 1996). Inherent Safety received additional attention from the chemical industry because of several major industrial accidents. Some examples are:

- Mexico City (1984) – a series of explosions in an LPG terminal killed approximately 500 people and destroyed the facility. A contributing factor was the very high quantity of LPG stored onsite at the terminal, which was in an urban area. Applying IS strategies may have reduced the LPG quantity onsite, making the propagation of the event and the impacts less severe.
- Bhopal, India (1984) – the worst disaster in the history of the chemical industry occurred when water entered a storage tank containing methyl isocyanate (MIC). The resulting reaction generated heat and pressure, lifting the tank relief valve and releasing highly toxic MIC vapors into the city. Exact casualty figures are disputed, but tens to hundreds of thousands of people were exposed. In 1994, the official Indian government estimate of fatalities was 4,000 (Ref 1.8 Hendershot). The two-step process required MIC to be stored onsite. Applying IS strategies by interconnecting the process would have eliminated the need to store MIC altogether.

- Charleston, West Virginia (2008) – debris from a fire and explosion in a pesticide manufacturing facility narrowly missed damaging a neighboring storage tank containing methyl isocyanate (MIC). The company embarked on a redesign of the MIC storage and handling processes at the plant, incorporating IS techniques, eliminating above-ground storage, greatly reducing overall stored volume, and simplifying the internal logistics of MIC handling. (Ref 1.20 NAS). In this case, the MIC tank was not damaged and there was no release of MIC. However, applying IS strategies may have reduced the total quantity onsite.
- West, Texas (2013) - An ammonium nitrate explosion occurred at fertilizer storage and distribution facility while emergency services personnel were responding to a fire at the facility. Fifteen people were killed, more than 160 were injured, and more than 150 buildings in the nearby town of West were damaged or destroyed, raising questions about the siting of the facility relative to populated area (Ref 1.24 CSB). This event highlighted the debate about the explosiveness of Fertilizer Grade Ammonium Nitrate (FGAN) and whether an inherently safer option could be developed and used.

In addition, a reflection on these incidents using Kletz's seminal 1978 paper over thirty years later provides a glimpse what has been learned from these incidents and how effectively managing their hazards and risks can be addressed (Ref 1.25 Vaughen 2012a).

Inherent safety has received considerable research attention from academia, industry, government and regulatory organizations (Ref 1.1 Ashford) (Ref 1.17 Lin) (Ref 1.18 Mansfield), joint industry-government working groups, such as the INSIDE (INherent SHE In DEsign) Project in Europe (Ref 1.21 Rogers) (Ref 1.19 Mansfield), and environmental and public interest organizations (Ref 1.23 Tickner).

In a similar manner, the work of Etowa et al. (Ref 1.5 Etowa) sought to clearly identify the inherent safety features using the Dow Fire & Explosion Index and the Dow Chemical Exposure Index – two commonly used tools in process safety management for relative risk ranking. Goraya, et al. (Ref 1.6 Goraya) developed and validated a methodology for incident investigation that makes explicit use of the principles of inherent safety in determining the root causes of an incident. By means of an illustrative example, Gupta et al. (Ref 1.7 Gupta) provided evidence of the linkages between inherent safety and the cost of process safety. Their work helps to establish a clear business case for the use of inherent safety principles in management efforts directed at enhancing process safety.

Further motivation for the current research is found in the comments of employees who have reviewed the field of inherent safety and inherently safer design, including (Ref 1.3 Bollinger), (Ref 1.7 Gupta), (Ref 1.12 Kletz), (Ref 1.10 Khan) and (Ref 1.26 Vaughen 2012b). For example, Khan and Amyotte (Ref 1.10 Khan) have remarked that the various elements of process safety management can be seen to have at least a partial basis in inherent safety. This fact has been recognized by companies that have incorporated inherent safety as a "named feature" in their safety management documentation and have developed internal standards for the use of inherent safety principles. Yet the term *inherent safety* is typically not named as such in the general description of process safety management systems. Per Bollinger, et al. (Ref 1.3 Bollinger), explicit use of inherent safety terminology within such management systems is a possible means of furthering the adoption of inherent safety principles in industry.

Over the years, the interest in inherent safety from government, industry, and the public has increased. Inherent safety's promise has produced heightened expectations and it is seen almost as a panacea to reducing risks in the chemical process industries as the public becomes aware of the concept. Inherent safety is incorporated into safety and security regulations in parts of the United States at the local, state, and federal levels. Inherent safety has been proposed as a leading requirement for chemical security regulations in the U.S. Congress. To better clarify and more precisely define the terminology, the US Chemical Security Analysis Center (CSAC) of the Department of Homeland Security (DHS) contracted with CCPS to provide a technology-based definition of Inherently Safer Technology (IST) (Ref 1.4 CCPS/DHS).

Some of the best compilations of information on IS can be found in the works of (Ref 1.14 Kletz 1978), (Ref 1.15 Kletz 1996), (Ref 1.16 Lees) and the final report of the INSIDE (INherent SHE in DEsign) project in

Europe (1997) (Ref 1.21 Rogers). Khan and Amyotte (Ref 1.10 Khan) provide an excellent summary of the applications of IS. The second edition of Process Plants: A Handbook for Inherently Safer Design (Ref 1.11 Kletz, 2010) is a practical guide for plant engineers for incorporating IS techniques in their operations. The aforementioned 2012 National Academy of Sciences report on the "Use and Storage of Methyl Isocyanate (MIC)" at Bayer CropScience (Ref 1.20 NAS) is an excellent case study of IS principles applied to an existing chemical manufacturing facility.

1.5 REFERENCES

1.1 Ashford, NIA. The Encouragement of Technological Change for Preventing Chemical Accidents: Moving Firms From Secondary Prevention and Mitigation to Primary Prevention, Massachusetts Institute of Technology, 1993.

1.2 Bain, D. H. *Empire Express: Building the First Trans-continental Railroad*. New York: Viking, 1999.

1.3 Bollinger, R. E., Clark, D. G., Dowell, A.M., Ewbank, R. M., Hendershot, D.C., Lutz, W. K., et al. (1996). Inherently Safer Chemical Processes: A Life Cycle Approach (D. A. Crowl, Ed.), American Institute of Chemical Engineers, 1996.

1.4 Center for Chemical Process Safety, Report: Definition for Inherently Safer Technology in Production, Transportation, Storage, and Use, Chemical Security Analysis Center, Science & Technology Directorate, U.S. Department of Homeland Security (DHS), 2010.

1.5 Etowa, C. B., Amyotte, P.R., Pegg, M. J. and Khan, F.I., Quantification of inherent safety aspects of the Dow indices. Journal of Loss Prevention, 15, 477-487, 2002.

1.6 Goraya, A., Amyotte, P.R. and Khan, F. I, An inherent safety-based incident investigation methodology. Process Safety Progress, 23 (3), 197-205, 2004.

1.7 Gupta, J. and Edwards D. W., A simple graphical method for measuring inherent safety, Journal of Hazardous Materials, 104, 15-30, 2003.

1.8 Hendershot, D.C. An overview of inherently safer design. Risk Management: The Path Forward: 20th Annual CCPS International Conference, Atlanta, GA, 2005.

1.9 Henderson, D. C, and Post, R.L. (2000). Inherent safety and reliability in plant design. In Beyond Regulatory Compliance, Making Safety Second Nature. Mary Kay O'Connor Process Safety Center 2000 Annual Symposium. College Station, TX: Texas A&M University.

1.10 Khan F.I. and Amyotte, P.R., How to make inherent safety practice a reality. The Canadian Journal of Chemical Engineering, 81, 2-16, 2003.

1.11 Kletz, T., Amyotte, P., Process Plants – A Handbook for Inherently Safer Design, 2nd Ed., CRC Press, 2010.

1.12 Kletz, T.A., Inherently safer design—Its scope and future. Process Safety and Environmental Protection, 81(B6), 401-405, 2003.

1.13 Kletz, T.A. Inherently safer plants. Plant/Operations Progress 4 (3), 164-167, 1985.

1.14 Kletz, T.A. What you don't have, can't leak. Chemistry and Industry, 287-292, 1978.

1.15 Kletz, T.A., Inherently safer design: Achievements and prospects, International Conference and Workshop on Process Safety Management and Inherently Safer Processes, October 8-11, 1996, Orlando, FL (pp. 197-206), American Institute of Chemical Engineers, 1996.

1.16 Lees, F.P., Loss Prevention in the Process Industries, 2nd Ed., Butterworth-Heinemann, 1996.

1.17 Lin, D., Mittelman, A., Halpin, V. and Cannon, D., Inherently Safer Chemistry: A Guide to Current Industrial Processes to Address High Risk Chemicals, U. S. Environmental Protection Agency, 1994.

1.18 Mansfield, D.P., Inherently Safer Approaches to Plant Design, United Kingdom Atomic Energy Authority, 1994.

1.19 Mansfield, D.P., Viewpoints on implementing inherent safety. Chemical Engineering, 103 (3), 78-80, 1996.

1.20 National Academy of Sciences (NAS), The Use and Storage of Methyl Isocyanate (MIC) at Bayer CropScience, National Academies Press, 2012.

1.21 Rogers, R.L., Mansfield, D.P., Malmen, Y., Turney, R.D., and Verwoerd, M. (1995). The INSIDE Project: Integrating inherent safety in chemical process development and plant design. In G.A. Melhem and H.G. Fisher (Eds.). International Symposium on Runaway Reactions and Pressure Relief Design, August 2-4, 1995, Boston, MA (pp. 668-689). American Institute of Chemical Engineers, 1995.

1.22 Rolt, L.T.C, The Railway Revolution: George and Robert Stevenson (pg.147). New York: St. Martin's Press, 1960.

1.23 Tickner, J. The case for inherent safety. Chemistry and Industry, 796, 1994.

1.24 U.S. Chemical and Hazard Investigation Board (CSB), West Fertilizer Company Fire and Explosion, Final Report, 2013.

1.25 Vaughen, B. K., and Klein, J. A., What you don't manage will leak: A tribute to Trevor Kletz. Process Safety and Environmental Protection, 90, 411-418, 2012a.

1.26 Vaughen, B. K., and Kletz, T. A., Continuing our process safety management journey. Process Safety Progress, 31(4), 337-342, 2012b.

2. The Concept of Inherent Safety

2.1 INHERENT SAFETY AND PROCESS RISK MANAGEMENT

The modern design and implementation of chemical process safety programs incorporates risk-based approaches. This includes recognition of the hazards posed by the process, and a continual effort to analyze the risks, and to reduce or control them to the lowest levels practical, while balancing other objectives. A hazard is defined as "An inherent chemical or physical characteristic that has the potential for causing damage to people, property, or the environment" (Ref 2.12 CCPS Glossary). Risk is defined as "A measure of human health effects, environmental damage, or economic loss in terms of both the incident likelihood and the magnitude of the loss or injury. A simplified version of this relationship expresses risk as the product of the likelihood and the consequences (i.e., Risk = Consequence x Likelihood) of an incident" (Ref. 2.12 CCPS Glossary). In other words, *risk* is a function of both *consequence* (hazard) and *likelihood* (frequency).

Safety can be defined as a condition of tolerable risk compared to the benefit of the activity. CCPS formally defines safety as: "The expectation that a system does not, under defined conditions, lead to a state in which human life, economics or environment are endangered" (Ref 2.12 CCPS Glossary). This considers who receives the benefit when compared to who bears the risk. Inherent safety is one of the tools available to improve safety; it is a preferable method, if feasible, but other approaches are valid and successful depending on the specific situation. Inherent safety is used where it meets overall safety objectives, where it is sensible given various alternatives to manage risks, and where there is an application for the inherent safety principles. In general, the application of inherent safety principles has reached a higher level of maturity in situations where high hazards exist and the risk is also high, i.e., in chemical process safety and in the security of such facilities.

Process risk management is the term given to collective efforts to manage process risks through a wide variety of strategies, techniques, procedures, policies, and systems that can reduce the hazard of a process, the probability of an accident, or both. In general, the strategy for reducing process risk, whether directed toward reducing the frequency or the consequences of potential accidents, can be classified into five categories:

1. *Inherent* - Minimizing and possibly eliminating the hazard by using materials and process conditions that are non-hazardous; i.e., substituting water for a flammable solvent;

2. *Spatial* - The use of distance to reduce the effects of hazards, such as siting the process as far as possible from occupied buildings and outside receptors.

3. *Passive* - Minimizing the hazard through process and equipment design features that reduce either the frequency or consequence of the hazard without the active functioning of any device or requiring human input; i.e., providing a dike wall around a storage tank of flammable liquids;

4. *Active* - Using SCAI (safety controls, alarms, and interlocks), and mitigation systems to detect and respond to process deviation from normal operation; e.g., a pump which is shut off by a high-level switch in the downstream tank when the tank is 90% full. These systems are commonly referred to as engineering controls or safeguards.

5. *Procedural* - Using policies, operating procedures, training, administrative checks, emergency response, and other management approaches to *prevent* incidents, or to *minimize* the effects of an incident; i.e., hot work procedures and permits. These approaches are commonly referred to as administrative controls and require humans that interface with the process (primarily operators) to recognize and react to abnormal or adverse events in a prompt and correct manner as directed by the approved procedures and the training on those procedures.

All five categories can contribute to the overall safety of a process. Ideally, the steps of analyzing, reducing, and managing risk will be considered in a hierarchical manner as shown in Figure 2.1 (Ref 2.21 Kletz 2010).

Inherent safety uses the properties of a material or process to eliminate or reduce the hazard and are preferable to passive, active, and procedural measures. Passive protective layers are preferable to active layers, and passive and active measures are preferable to procedural ones, which involve direct human interface. The fundamental difference between inherent safety and the other four categories is that inherent safety seeks to remove or at least reduce the hazard at the source, as opposed to accepting the hazard and attempting to control it in some manner. If implementing inherently safer approaches alone to meet project risk goals is feasible, other layers of protection - and their associated costs in time, capital, and ongoing expenses - may not be required.

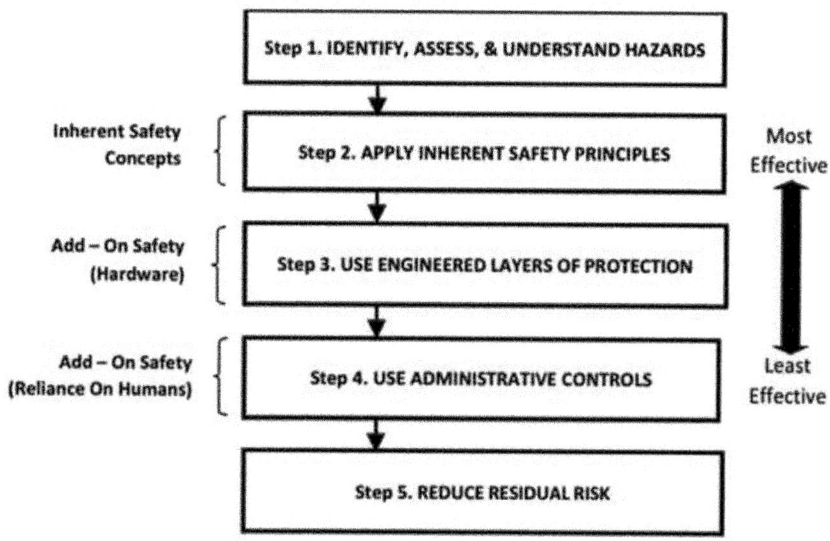

Figure 2.1 Hierarchy of Controls (adapted from Ref 2.4 Amyotte 2006)

2.2 INHERENT SAFETY DEFINED

> **in·her·ent**: Adjective. Existing as an essential constituent or characteristic; intrinsic. From the Latin *inharens*, *inhaerent-*, present participle of *inhaerere*, to inhere.

What do we mean when we speak of "inherent safety" or "inherently safer?" "Inherent" has been defined as "existing in something as a permanent and inseparable element, quality, or attribute" (Ref 2.1 American Heritage). Inherent safety is a concept, an approach to safety that focuses on eliminating or reducing the hazards associated with a set of conditions. A chemical manufacturing process is inherently safer if it reduces or eliminates the hazards associated with materials and operations used in the process *and* this reduction or elimination is permanent and inseparable. The process of identifying and implementing inherent safety in a specific context is called inherently safer design (ISD). A process with reduced hazards is described as inherently safer compared to a process with only passive, active, and procedural controls.

Since the 2nd Edition of this book was published in 2009, additional definitions of inherent safety, inherently safer, inherently safer technologies (IST), and inherently safer design (ISD) have appeared in the technical literature.

However, for the purposes of this book, a general definition of inherent safety is adapted from the 2010 CCPS project for the U.S. Department of Homeland Security to define "inherently safer technologies." (Ref 2.13 CCPS DHS) This definition is as follows:

> *"The application of inherent safety concepts permanently eliminates or reduces hazards to avoid or reduce the consequences of incidents. Inherent safety is a philosophy, applied to the entire life cycle of chemical processes, including design, construction, operation, maintenance, and decommissioning, as well as all modes of operation of these processes including manufacture, transport, storage, use, and disposal. It is an iterative process that considers options, including eliminating a hazard, reducing a hazard by having less of the hazardous materials,*

substituting less hazardous materials, using less hazardous process conditions, and designing processes to reduce the potential for, or consequences of, human error, equipment failure, or intentional harm. Overall safe design and operation options cover a spectrum from inherent through passive, active and procedural risk control measures. There is no clear boundary between inherent safety and other strategies."

This is consistent with the definition of inherent safety and its related terms found in the current CCPS Process Safety Glossary (Ref 2.12 CCPS Glossary). In North America, the State of New Jersey, the State of California, Contra Costa County, California, and the federal Environmental Protection Agency are currently the only regulators with mandatory IS-related requirements. In Europe the Belgian Chemical Risks Directorate has published guidance for the implementation of the Seveso III Directive (Ref 2.14 Seveso) in Belgium which includes guidance on inherent safety, although this guidance is not mandatory (Ref 2.5 Belgian). Similarly, the UK HSE has published extensive non-mandatory guidance on the subject of inherent safety (Ref 2.26 UK HSE HSG-143) (Ref 2.27 UK HSE HSL 2005). The definition provided above is consistent with the definitions used by these two regulators. Both jurisdictions have used the definitions provided in the previous edition of this book to formulate their Inherent Safety (IS) requirements, and both require mandatory IS reviews/analysis but not mandatory process changes. See Chapter 14 for a detailed discussion of inherent safety regulatory initiatives implemented in the United States.

2.3 SHARED CHARACTERISTICS

While the above definition succinctly captures what inherent safety includes, its application encompasses a broader set of shared characteristics:

- It is a concept - Inherent safety (and related terms such as inherently safer design, inherently safer technologies, etc.) represents a set of concepts rather than a distinct set of rules. In this book, the terms and acronyms Inherent Safety (IS), Inherently Safer Technology (IST), and Inherently Safer Design (ISD) will be used interchangeably. All three terms are used in the literature and in the field, but actually have different connotations. IST is more related to a change of technology to achieve hazard reduction. ISD is more related to a change in design. Additionally, "inherently safer," as it is used for procedures or SCAI, will be used (e.g.,an inherently safer SCAI).

- It applies throughout the life cycle of the process - Inherent safety applies throughout the life cycle of the process and is not something to consider only during the design phase of a process.
- It focuses on hazard reduction which may then reduce risks - In application, inherent safety seeks to avoid hazards to begin with rather than attempt to control them. If complete avoidance cannot be achieved, then reducing the severity and likelihood of the risk are attempted.
- It attempts to find more permanent solutions to hazard reduction rather than adding layers of protection - The avoidance of risk or the reduction of risk using inherently safer solutions/options are intrinsic to the process, i.e., they are permanent and inseparable.
- It is not uniformly applicable and requires case by case analysis - The application of inherent safety is relative, i.e., the same inherently safer solution applied in one process, given its materials, conditions, of operation, location, and other factors, may result in a different level of risk reduction in another process, or may not even be appropriate.
- It is involved, and inherently safer decisions must be taken in careful consideration of the context - The decision process to apply inherently safer solutions must consider the entire life cycle of the process, the full spectrum of hazards and risks, and the potential for transfer of risk from one impacted population to another, as well as the technical and economic feasibility of the options being considered.
- It can reduce hazards but not necessarily all risks - In most cases an inherently safer process should not be considered "inherently safe" or "absolutely safe." While implementing inherent safety concepts will move a process in the direction of reduced risk (inherently safer), it will not typically remove all risks.

Inherently safer concepts will enhance overall risk management programs, whether directed toward reducing the frequency or consequences of potential incidents.

2.4 INHERENTLY SAFER STRATEGIES

Approaches to achieving inherently safer processes and plants have been grouped into four major inherent safety strategies. The names or titles of these strategies were established in the 1st and 2nd editions of this book. These groupings follow the original treatment of the topic of inherent safety as published by IChemE and IPSG (Ref 2.17 IChemE) and Kletz (Ref 2.20 Kletz 1984, Ref 2.19 Kletz 1991). However, in this earlier, seminal work, different names/titles were used. In general, these may be considered subsets of the four core strategies. These different names/titles and how some strategies have been incorporated into the four used herein are presented in this section for clarity and as a reference in interpreting older work on inherent safety:

- *Minimize* Using or having smaller quantities of hazardous substances (also called intensification).
- *Substitute* Replace a chemical/material with a benign or less hazardous substance; or replace a process or processing technology with one that is benign or is less hazardous.
- *Moderate* Use less hazardous or energetic processing or storage conditions, a less hazardous form of a material, or facilities that minimize the impact of a release of hazardous material or energy (also called *Attenuation and Limitation of Effects*).
- *Simplify* Design facilities which eliminate unnecessary complexity and make operating errors less likely, and which are forgiving of errors that are made (also called *Error Tolerance*).

These inherently safer design strategies are discussed in more detail in Chapters 3-6. Examples can be found in Chapter 8, which discusses inherently safer solution opportunities throughout the life cycle of a process.

Table 2.1 presents a mapping of the original concepts to the CCPS ISD strategies as restated by Kletz in 1998, as well as others who have published technical work on inherent safety. (Ref 2.3 Amyotte, Ref 2.18 Khan, Ref 2.25 Overton, Ref 2.22 Lutz).

Table 2.1: Strategies for Inherently Safer Design

Kletz (1998)	CCPS (2008)
Inherent Safety Strategies	Inherent Safety Strategies
Intensification	Minimize
Substitution	Substitute
Attenuation	Moderate
Limitation of Effects	Moderate and Simplify
"Friendly Plant Design" Strategies regarded as "add on" features	Mapping to Four Core CCPS Strategies
Simplification	Simplify
Segregation	Moderate
Avoiding knock-on effects	Moderate
Making incorrect assembly impossible	Simplify
Making status clear	Simplify
Tolerance of misuse (error tolerance)	Simplify
Ease of control	Simplify
Understandable software for computer control	Simplify
Instructions and procedures	Simplify

Life cycle friendliness (construction and demolition)	Consider all strategies with regard to all phases of a plant life cycle
Passive safety	Moderate
Others	Mapping of to Four Core CCPS Strategies
Inherent robustness	Moderate or elimination

These four strategies form a protocol by which the risks associated with the loss of containment of hazardous materials or energy can be significantly reduced, and in some cases eliminated. The elimination of risk due to loss of primary containment is very difficult, if not impossible to achieve using other risk reduction measures, i.e., active or passive safeguards. These measures, while effective if installed and maintained properly, generally reduce the likelihood of release, and sometimes will mitigate the consequences of a release. However, they cannot reduce the risk to zero. Truly, Trevor Kletz's statement "What you don't have can't leak" embodies the ultimate goal of inherent safety and embodies the elimination of the risk of hazardous material releases. However, while they are highly effective techniques, it is usually not possible to eliminate all process-related risks using inherently safer strategies.

As previously noted, some researchers and practitioners include other strategies within the inherent safety definition. In general, these may be considered subsets of the four core strategies. However, in the literature, they are sometimes described as separate strategies (Ref 2.3 Amyotte; Ref 2.18 Khan; Ref 2.25 Overton and King; Ref 2.22 Lutz). These inherent safety strategies include:

- Avoid incorrect assembly
- Making status clear
- Inherent robustness

Chapters 3 through 6 provide more detailed definitions and application guidance for these strategies, as well as examples of their application. How IS strategies can be applied to the various stages of the process life cycle is discussed in Chapter 8, while additional examples of IS concepts are presented in Chapters 9, 11, 13, and 15.

While inherently safer strategies offer potential risk reduction measures that are not possible with active or passive safeguards, the application of inherently safer designs and measures is not without its own problems. This is because hazards are interrelated in a process, and the elimination or reduction of one risk may introduce new risks or cause existing ones to become more prominent or even worse. Both components of risk, i.e., both the frequency of occurrence, and its consequences, must be evaluated and compared for each risk affected by the implementation of the inherently safer measure, and a decision must be made as to which path forward is most appropriate. It may be that the risk without implementing certain inherently safer measure(s) is actually *lower* than with it implemented. For example, replacing a flammable solvent with a non-flammable, but highly corrosive one.

2.5 INHERENT SAFETY THROUGHOUT THE PROCESS LIFE CYCLE

The industry has recognized that a process goes through various stages of evolution. In this book, these stages are called life cycle stages as shown in Figure 2.2. At various stages of the life cycle formal hazard/risk analyses (e.g., PHAs) are generally performed, particularly during the early stages when the process is being conceived and designed, but also during other stages, e.g., management of change. During these various hazard/risk analyses the opportunity exists to also formally analyze the process and its design and operations for possible inherent safety improvements and changes. Formal inherent safety assessments can also be performed separately.

Exploring inherently safer alternatives may require more resources during the early stages of development than at other points. However, the resulting understanding and application will, in many cases, minimize or eliminate the need for appended safety mitigation devices and the costs of maintaining them, as well as reduce the possibility of incidents. Inherently safer considerations may reduce the life cycle cost of the process. In general, the earlier inherently safer thinking is applied to a process, the greater the economic benefits and, in some cases, the feasibility. However, inherently safer strategies may also be (and often are) applied at existing facilities, as it is likely that some positive risk reduction can be achieved even after the facility is operational.

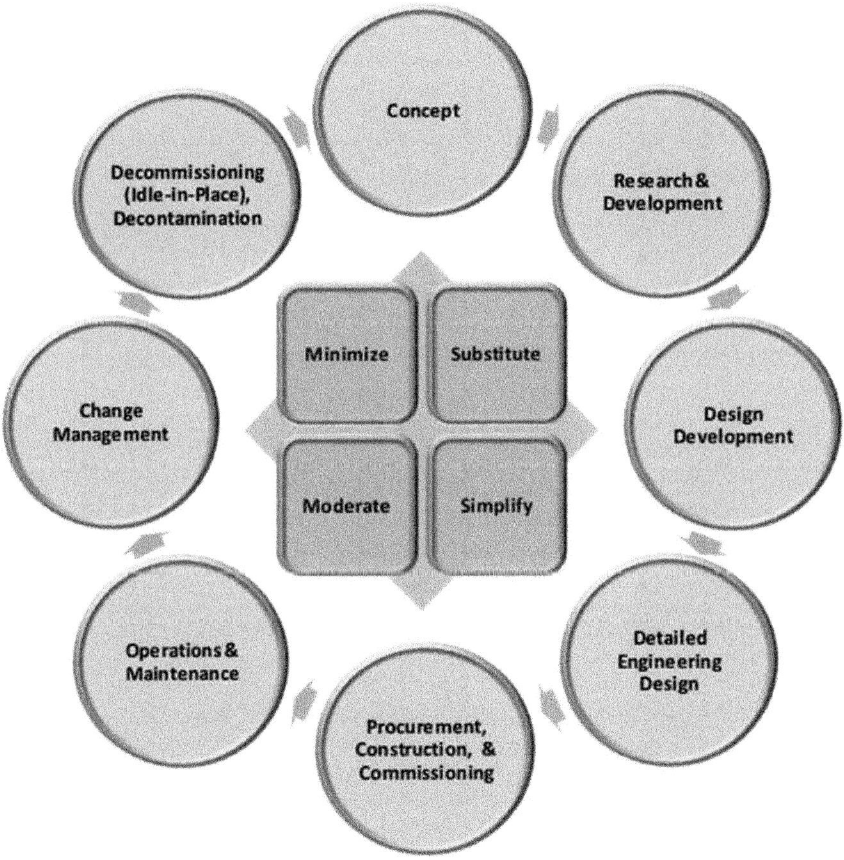

Figure 2.2. Application of Inherent Safety Throughout the Process Life cycle.

2.6 INHERENTLY SAFER APPROACHES

Where feasible, reducing or eliminating hazards through changes in materials, chemistry or process variables is preferable to adding layers of protection to a process. While layers of passive, active or procedural controls *will* reduce the risk, they will do nothing to reduce the nature or presence of the hazard itself.

The CCPS Process Safety Glossary defines a hazard as an inherent chemical or physical characteristic that has the potential for causing damage to people, property, or the environment. Hazards cannot be changed - they are intrinsic to the material or to its conditions of storage or use. They can, however, be removed or mitigated.

Chemical process hazards come from two sources:

1. Hazards that are *characteristic of the materials and chemistry used*, and

2. Hazards that are *characteristic of the process variables*, or the way the chemistry works within the process.

Examples of Chemical Process Hazards

- Phosgene is toxic if inhaled
- Concentrated sulfuric acid is corrosive to the skin
- Ethylene is flammable
- Steam confined in a drum at 600 psig (4137 KPa) contains a significant amount of potential energy and may explode
- Acrylic acid monomer can polymerize, releasing large amounts of heat

Inherent safety approaches can be implemented to address any of the four stages that most incidents follow: (Ref 2.10 CCPS HEP)

- *Initiating Cause*: The operational error, mechanical failure, or external event that is the first event in an incident sequence and marks the transition from a normal situation to an abnormal situation. (synonymous with *Initiating Event*)
- *Intermediate Events*: the events that maintain or expand the incident;

- *Loss Event*: Point of time in an abnormal situation when an irreversible physical event occurs that has the potential for loss and harm impacts. Examples include release of a hazardous material, ignition of flammable vapors or an ignitable dust cloud, and over pressurization rupture of a tank or vessel. An incident might involve more than one loss event, such as a flammable liquid spill (first loss event) followed by ignition of a flash fire and pool fire (second loss event) that heats up an adjacent vessel and its contents to the point of rupture (third loss event).
- *Impact*: A measure of the ultimate loss and harm of a loss event. *Impact* may be expressed in terms of numbers of injuries and/or fatalities, extent of environmental damage, and/or magnitude of losses such as property damage, material loss, lost production, market share loss, and recovery costs.

The most effective strategies will eliminate a hazard. Inherently safer design can also reduce the potential for propagating an incident sequence before there are major impacts on people, property, or the envlronment.

Inherently safer concepts should be an essential aspect of any process safety program. If the hazards can be eliminated or reduced, the extensive layers of protection to control those hazards may not be required or may be less robust. However, inherent safety is not the only process risk management strategy available and may not always be the most effective solution. For example, capital expenditure limitations may preclude the wholesale adoption of a new, inherently safer process in an established, depreciated chemical plant. Targeted application of inherently safer design techniques to parts of the existing process, coupled with additional layers of protection may be the only economically feasible option. A system of strategies that includes both inherently safer design and additional layers of protection may be needed to reduce risks to an acceptable level.

The application of inherent safety concepts in a hierarchy of controls is shown in Figure 2.3. The steps of managing risk should ideally be executed in a hierarchical manner and iteratively. The ability to apply IS strategies following the logic of the hierarchy of controls may be restrained.

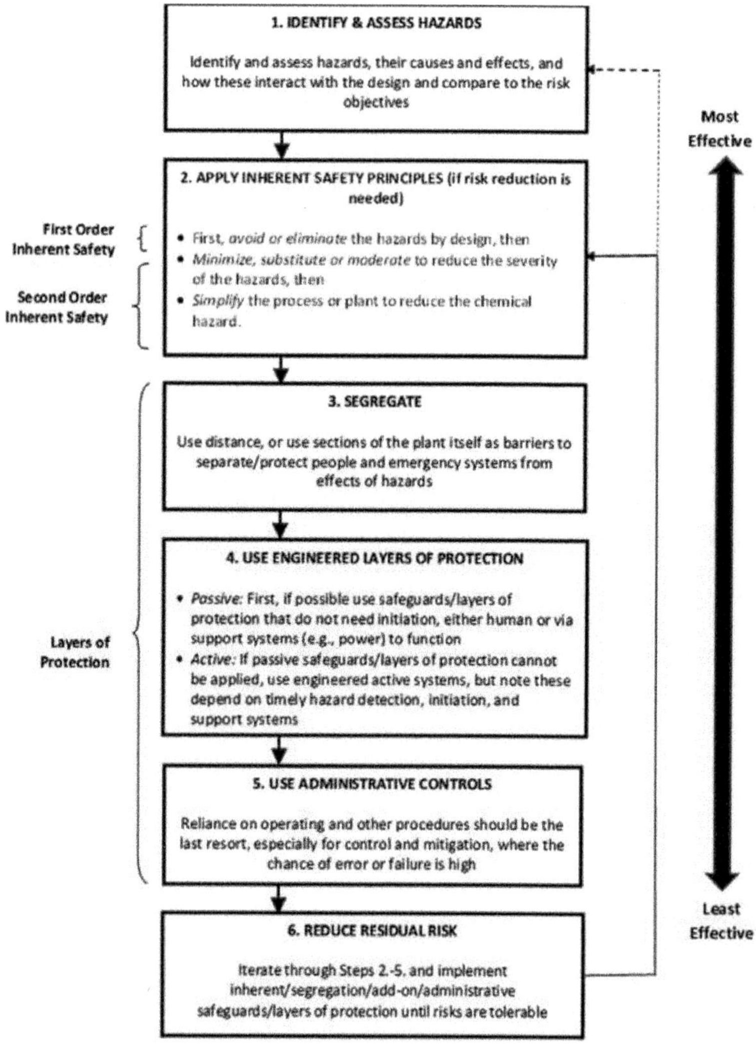

Figure 2.3 Inherent Safety Considerations in Hierarchy of Controls (adapted from Ref 2.6 Amyotte 2006)

In step 2 of Figure 2.3, one of the inherently safer strategies is to minimize the inventory of hazardous chemicals that can cause or contribute to process safety incidents. Some facilities will also employ minimization to escape formal applicability of process safety regulations by reducing the inventories of the chemicals covered by those regulations below the threshold quantities which trigger formal coverage by the regulations. Substitution can be used in the same way, i.e., to replace a chemical included in a process safety regulation with one that is not covered by regulation. Employment of the inherently safer strategies with the expressed goal of escaping formal regulatory coverage is a valid application of the strategies.

The feedback loop in Figure 2.3 contains two possible paths: for a given hazard analysis, the return path to Step 2 (the solid line arrow) would be appropriate to iterate the consideration of inherently safer solutions, segregation, use of layers of protection, and administrative controls after the basic hazards have been identified. However, periodically, either following fixed revalidation schedules of hazard analyses, or during the MOC process, the identification of basic hazards is repeated or revalidated, and the return path to Step 1 would be appropriate (the dotted-line arrow).

Whether or not a particular safety feature or option in a chemical process is "inherent" can be debated. Such discussion may arise, in part, because different people have different perspectives on risk, or view the process at different levels of resolution, ranging from a global view of the entire process to a very detailed view of one specific feature. One could also be referring to addressing the hazards of the process vs. addressing other aspects of managing the risk of the process hazards. However, if the design or operational solution option fits the definition in Section 2.2 then it can be considered as "inherent."

2.6.1 Orders of Inherent Safety

Incorporating inherently safer concepts can either change the conditions under which a process using hazardous chemicals is operated (i.e., change the characteristics of the process parameters) or completely eliminate/reduce the hazardous characteristic (e.g., change the material and chemistry used in the process). Changes in the chemistry of a process that reduce the hazard of the chemicals used or produced completely can be considered First Order Inherent Safety approaches. Changes in process variables or other aspects that do not eliminate a hazard but reduce it can be considered Second Order Inherent Safety approaches. First and second order inherently safer approaches are described in more detail below.

In the strictest sense, or the First Order of Inherent Safety, one could argue that the definition of inherently safer applies only to the complete elimination of a hazard. Elimination or complete avoidance of the hazard is a priority of a First Order solution and certainly fits Trevor Kletz's original dictum that "What you don't have can't leak." Inherently safer strategies that absolutely eliminate a hazard are an optimum solution, while hopefully not introducing another hazard of concern as a result, nor transferring the risk to another part of the value chain for the product or material of concern. Examples of first order IS measures would be shut down or removal of the process presenting the basic hazard(s) or substituting a hazardous material with one that is totally non-hazardous from a process safety viewpoint.

Alternatively, inherently safer approaches can also address the hazard by making it less intense or by virtually eliminating it. These approaches can be labeled as Second Order of Inherent Safety. Such approaches are clearly in line with the philosophy of inherent safety but may not be as powerful as a First Order change. In the Second Order of inherent safety, the hazard is only reduced through the application of IS principles, e.g.,minimization or substitution. It could be that Second Order inherently safer design or operational options result in an acceptable reduction of hazard and, therefore, the risk is adequately addressed. Examples of second order IS measures would be substituting a hazardous material with one that is still hazardous from a process safety viewpoint but less hazardous (e.g., less volatile, or less toxic), or reducing the inventory of a hazardous material but not completely eliminating its use. This is what is meant by "virtually" eliminated.

In the broadest sense, the overall hazard is not completely eliminated or reduced by way of Second Order inherently safer strategies but instead, sublevel hazards are minimized and the likelihood of the event occurring is reduced by adding layers of protection. The strength and reliability of a layer of protection can vary, with some layers designed to be more "robust" than others. For example, "independent" protection layers provide more strength and reliability than protection layers that are not independent of each other. This could mean it is more reliable, effective, simpler, or any other positive safety attributes achievable by comparison, but the fundamental hazard may still exist. This is the difference between the inherently safer design concepts being applied to the hazard, and layers of protection being applied to reduce the overall risk. Layers of protection are more thoroughly discussed in Section 2.7 and in Chapter 7, as well as in the CCPS books *Layer of Protection Analysis: Simplified Process Risk Assessment* and *Guidelines for Initiating Events and Independent Protection Layers in Layer of Protection Analysis* (Ref 2.11 CCPS LOPA, Ref 2.9 CCPS IPL).

In evaluating process safety alternatives, it is important to consider the total life cycle costs and risks. There is an increasing interest in this concept in the environmental area, with recognition of the need to incorporate waste treatment, waste disposal, regulatory compliance, potential liability for environmental damage, and other long-term environmental costs into the economic evaluation for projects. Similarly, we must consider life cycle *safety and process safety* costs. Some examples of the process safety factors that should be considered when evaluating costs include:

- Capital cost of safety and process safety related equipment, including active barriers and systems such as instrumented safety systems and mitigation systems.
- Capital cost of passive barriers, such as containment dikes, and vacant land to provide spacing, required by codes, regulations, and insurers.
- Operating and maintenance costs for instrumented safety systems, fire protection systems, personal protective equipment, and other safety and process safety equipment.
- Increased maintenance cost for process equipment due to safety and process safety requirements, such as safety permits, cleaning and purging equipment, personal protective equipment, training, and restricted access to process areas.
- Operator safety training costs for hazardous materials or processes.
- Regulatory compliance costs.
- Insurance costs.
- Potential property damage, product loss, and business interruption costs if a process safety incident occurs.

- Potential impact to reputation and market value
- Potential liability if a process safety incident occurs.

In addition, the cost differential of the physical process technology, licensing, operating costs, and equipment must be considered in the trade-off analysis. These must be weighed against the change in risk from adopting one alternative or the other.

2.7 LAYERS OF PROTECTION

Layers of Protection is "a concept whereby a device, system, or human action is provided to *reduce the likelihood and/or severity* of a specific loss event." (Ref 2.12 CCPS Glossary). This strategy works on the premise that multiple layers of protective features and countermeasures reduce the potential for the event to occur. Although the reduction of the likelihood of occurrence by employing layers of protection can be substantial, there is still residual risk that has not been eliminated after their use. This is different from IS measures, where the hazard(s) have been eliminated or virtually eliminated (also see Chapter 7). Ideally each layer should be independent of the other layers, such that each device, system, or action can prevent the event from proceeding to the undesired consequence regardless of the initiating event or the action of any other protection layer. In addition to independence, each layer should be effective and auditable as defined by CCPS (Ref 2.11 CCPS LOPA, Ref 2.9 CCPS IPL). Independent layers of protection are inherently safer than layers of protection that share sources of common cause failure. This is a general assumption and the actual reliability and robustness may vary.

The passive, active, and procedural process risk management strategies mentioned earlier are considered layers of protection, as they involve the add-on of safety or protective devices and systems or work processes that are intended to reduce the risk. Passive layers do not require any human input, nor do they rely on any support or utility systems to function, e.g., power. Sometimes this is referred to as "fail safe." Active layers of protection do require human or support system input to function. Procedural layers, or administrative controls, utilize safe work practices and procedures to reduce risk and rely on humans to recognize, diagnose, and react correctly to detect, prevent, or mitigate incidents.

Independent protection layers (IPL) that meet the prerequisites of independence, effectiveness, and auditability may be credited qualitatively or semi-quantitatively in formal Layers of Protection Analysis (LOPA) to determine if the collective protective features (i.e., the layers) are adequate to reduce the risk of an undesired hazard scenario to a tolerable level. These are generally the most reliable and robust layers and may include safety instrumented systems (SIS), basic process control systems (BPCS), relief devices, operator response to alarms, certain mitigation systems, and certain key design features related to preventing process safety incidents. Creating multiple layers of protection, and those that qualify as IPLs, between a hazard and potentially impacted people, property and the environment can be highly effective. Their application has significantly improved the safety and process safety performance of the chemical/processing industry. However, such an approach may have significant disadvantages:

- *The basic hazard(s) remains*, and some combination of failures of the layers of protection may result in an incident, thereby allowing the consequences to be realized. Every active or passive layer has a certain likelihood of failure, due either to mechanical means or management systems failures, such as not maintaining or keeping administrative controls active. The outcome of the event may be limited by whatever passive or inherent layers have been applied.
- *Potential impacts could be realized by some unanticipated route or mechanism*. Hazardous event can occur by means beyond what were anticipated by process safety engineers. Accidents can occur by mechanisms that were unanticipated or that were poorly understood. Complicated and overlapping layers of protection increase the possibility of unanticipated failure routes, particularly when common cause failures are shared by some of the layers, i.e., they are not independent. Therefore, the actual risk may be the same or even increase after the application of additional layers of protection. For example, a complicated shutdown system may cause an inadvertent sudden shutdown, which presents an overpressure, leading to a release.

- *The layers of protection can be expensive to build and maintain throughout the life of the process*. Initial capital expense, and the costs of operation, training, and maintenance, along with diversion of scarce and valuable technical resources into maintenance and operation of the active layers of protection can add significantly to plant operating budgets. In particular, SISs are difficult and expensive to design, procure, and maintain. Also, to maintain the calculated level of reliability of SISs, i.e., the safety integrity level (SIL), interval testing schedules and procedures must be diligently followed.

For these reasons, it is more advantageous to design or modify the process to be able to inherently withstand the hazard(s) if possible, rather than, as the old adage says, add more "belts and suspenders." For example, it is inherently safer to design the maximum allowable working pressure in a process to be higher than the maximum pressure that can be achieved in the process rather than to add additional active overpressure protection devices such as relief valves (although relief valves may be required by applicable regulations and/or engineering codes and standards). Figure 2.4 shows the relationship between inherently safer designs and layers of protection. This figure depicts the concept of layers of protection (including the inherently safer solutions in the "design" layer) representing how the layers of an onion build upon each other to form the whole (adapted from Ref 2.29 CCPS Metrics)

2.8 INTEGRATING INHERENT SAFETY IN PROCESS RISK MANAGEMENT SYSTEMS

How does inherent safety fit into an overall process risk management program? As discussed above, "risk" is defined as "the likelihood that a defined consequence will occur." Efforts to reduce the risk arising from the operation of a chemical processing facility can be directed toward reducing the likelihood of incidents (incident frequency), reducing the magnitude of the loss, injury or damage should an incident occur (incident consequences), or eliminating the potential consequences altogether.

A key engineering risk tool is a management system appropriate for the risks being addressed (i.e., health, occupational safety, process safety, equipment reliability, etc.). Process safety management (PSM) is the application of management systems that are focused on prevention of, preparedness for, mitigation of, response to, and restoration from catastrophic releases of chemicals or energy from a process associated with a facility (Ref 2.12 CCPS Glossary). Process safety management systems are comprehensive sets of policies, procedures, and practices designed to ensure that barriers to episodic incidents are in place, in use, are effective (Ref 2.12 CCPS Glossary) and are comprised of elements—management of change, process hazards analysis, mechanical integrity, and others. These systems are recognized and accepted worldwide as best-practice methods for managing process safety risks. PSM programs typically consist of 10 to 20 program elements that must be effectively carried out to manage the risks in an acceptable way.

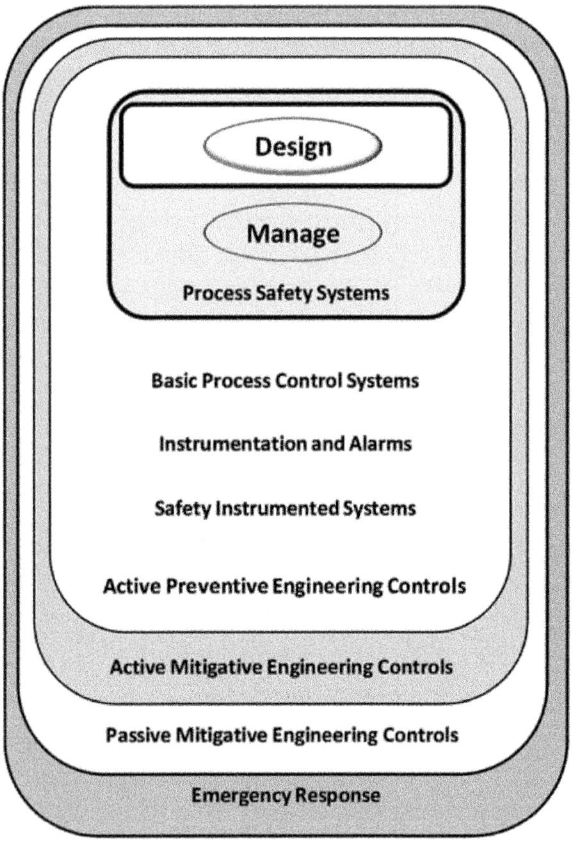

Figure 2.4 Layers of Protection (Ref 2.29 CCPS Metrics)

Inherent safety strategies can be applied to these process safety management elements. As an example, by including consideration of inherent safety strategies in a management of change (MOC) program technical or hazard review, the MOC procedure recognizes inherent safety as both a driving force, as well as an opportunity during the review and approval of prospective changes. See section 2.6 and Chapters 3-6 for a detailed description of inherent safety strategies and see Chapter 10 for a more detailed discussion of the application of inherently safer strategies in the various elements of a PSM program.

Inherently safer design is a way of thinking involving a holistic approach, considering the process as a system with interacting concerns, such as toxicity, flammability, reactivity, stability, quality, capital costs, operating costs, transportation risks, pollution prevention and site factors. These considerations make it important to understand how to manage tradeoffs and definition of acceptable risks - be they safety, environmental, security, or financial in nature. Containment of energy, pressure, temperature, or chemicals will be an important consideration in many inherently safer suggestions, but many constraints will be placed on the decision-making process due to economic limitations or perceived risks associated with new approaches. It must be remembered that there is no single answer to the question of what the best way is to build or operate a safe industrial facility that manufactures, stores, processes or uses chemicals. Inherently safer does not always equal "safe," but is one tool for reducing risk. Inherent safety must be used in conjunction with other safety methodologies.

Section 2.1 introduced five categories of process risk. Inherent safety can be integrated into to the implementation of any of these five categories - inherent, spatial, passive, active, and procedural, as shown in Table 2.2. It should be noted that there is not a clear boundary between IST and other strategies (Ref 2.13 CCPS DHS).

An inherently safer process can offer greater safety potential, often at a lower cost. However, selection of an inherently safer approach does not guarantee that the actual implementation of those approaches will result in a safer operation compared to an alternate process that is safer due to multiple layers of protection. For example, implementing a theoretically inherently safer (but completely new) technology carries the risks of unfamiliarity and inexperience. The traditional strategy of providing layers of protection for a hazardous process can be quite effective, although the expenditure of resources to install, operate and maintain the layers of protection may be very large. In some cases, the benefits of the inherently more hazardous technology will be sufficient to justify the costs needed to provide the layers of protection required to reduce its risk to a tolerable level.

Table 2.2: Examples of Process Risk Management Categories (Ref 2.8 CCPS GED)

Process Risk Management Category	Example	Comments
1. Inherent	An atmospheric pressure reaction using non-volatile solvents that are incapable of generating any pressure in the event of a run-away reaction.	There is no potential for reactor overpressure because of the chemistry and physical properties of the materials.
2. Spatial	A rocket propellant mixing process, capable of explosion, carried out in a remote area and operated remotely, beyond the hazardous overpressure footprint.	The process could experience an explosion, but there would be no hazard to occupied buildings or operations personnel.
2. Passive	A reaction capable of generating 150 psig (1034 KPa) pressure in case of a runaway, carried out in a reactor design for 250 psig (1725 KPa)	The reactor can contain the runaway reaction. However, if 150 psig (1034 KPa) pressure is generated, the reactor could still fail due to a defect, corrosion, physical

		damage or other cause.
3. Active	A reaction capable of generating 150 psig (1034 KPa) pressure in case of a runaway, carried out in a 15 psig (103 KPa) design reactor with a 5 psig (34 KPa) high pressure SIS interlock to stop reactant feeds, together with a properly sized 15 psig (103 KPa) rupture disk discharging to an effluent treatment system.	The SIS interlock could fail to stop the reaction in time, and the rupture disk could be plugged or improperly installed, resulting in reactor failure in case of a runaway reaction. The effluent treatment system could fail to prevent a hazardous release.
4. Procedural	The same reactor described in Example 3 above, but without the 5 psig (34 KPa) high pressure SIS interlock. Instead, the operator is instructed to monitor the reactor pressure and stop the reactant feeds if the pressure exceeds 5 psig (34 KPa).	There is a potential for human error, the operator failing to monitor the reactor pressure, or failing to stop the reactant feeds in time to prevent a runaway reaction.

Note: These examples apply specifically to categorization of risk management strategies with respect to high-pressure hazards caused by a runaway reaction. The processes described may involve trade-offs with other risks arising from other hazards. For example, the non-volatile solvent in the first example may be extremely toxic, and the solvent in the remaining examples may be water. Decisions on process design must be based on a thorough evaluation of all hazards involved.

Hendershot (Ref 2.15 Hendershot 1995) compares the inherent safety characteristics of air and automobile transportation, and concludes that automobile transportation is inherently safer for reasons such as:

- The automobile, on the ground, will coast to a stop in case of engine failure, while the airplane will rapidly descend and may not be able to land safely.
- The automobile travels at a lower speed.
- The automobile contains a smaller inventory of passengers.
- The control of an automobile is simpler (in two dimensions) compared to the airplane, which must be controlled in three dimensions.

However, the benefits of air transportation, primarily speed, make it an attractive alternative for longer trips. These benefits have justified the expenditure of large amounts of money to provide extensive layers of protection to overcome the inherent hazards of air travel, including extensive redundancy in aircraft design and construction, the management of airspace by the air traffic control system, rigorous maintenance of aircraft and equipment, and many other systems. The result is that air travel, while inherently more hazardous, is, in fact, safer than automobile travel for long trips, even though the rate of fatal auto accidents dropped from1982-2005. Similar situations can be expected to occur in the chemical/process industry (Ref 2.16 Hendershot 2006).

In developing the definition of IST for the U.S. Department of Homeland Security that has been adapted as the definition of inherent safety for this book above, CCPS included the following caveats (Ref 2.13 CCPS DHS):

ISTs are relative: A technology can only be described as inherently safer when compared to a different technology, including a description of the hazard or set of hazards being considered, their location, and the potentially affected population. A technology may be inherently safer than another with respect to some hazards but inherently less safe with respect to others and may not be safe enough to meet societal expectations. IST options can be location and release scenario dependent, and different potentially exposed populations may not agree on the relative inherent safety characteristics of the same set of options.

ISTs are based on an informed decision process: Because an option may be inherently safer with respect to some hazards and inherently less safe compared to others, decisions about the optimum strategy for managing risks from all hazards are required. The decision process must consider the entire life cycle, the full spectrum of hazards and risks, and the potential for transfer of risk from one impacted population to another. Technical and economic feasibility of options must also be considered. Risk reduction criteria will be determined by the nature of the hazards or threats and will require consideration of conflicts among multiple hazards and threats.

Inherently safer options should also be considered for the entire supply chain, including manufacturing, use, storage, transportation, and disposal. Tradeoffs are involved in terms of moving risk from one location in the supply chain to another (Ref 2.6 Berger, Ref 2.2 ACS). For example, reducing onsite inventory of a hazardous raw material may require more frequent shipments and subsequent risk along the transportation route, or increased inventory at the supplier location.

Marshall (Ref 2.23 Marshall 1990, Ref 2.24 Marshall 1992) discusses accident prevention, control of occupational disease, and environmental protection in terms of strategic and tactical approaches. Strategic approaches have broad significance and represent "once and for all" decisions. The inherent and passive categories of risk management would usually be classified as strategic approaches. In general, strategic approaches are best implemented at an early stage in the process or plant design. Tactical approaches, which include active and procedural risk management categories, tend to be implemented much later in the plant design process, or even after the plant is operating, and often involve much repetition, increasing costs, as well as the potential for failure. However, it is never too late to consider inherently safer alternatives. Major enhancements to inherent safety have also been reported in plants that have been operating for many years (Ref 2.8 CCPS GED, Ref 2.28 Wade, Ref 2.7 Carrithers).

2.9 SUMMARY

Inherent safety represents a fundamentally different way of thinking about the design and operation of chemical processes and plants. It focuses on the elimination or reduction of the hazards, rather than on risk management and control via additional layers of protection. The demarcation line between an IS measure and a layer of protection can be defined as follows: IS measures eliminate or reduce the hazard(s), whereas layers of protection do not eliminate the hazard(s) but reduce their likelihood of occurrence. Sometimes the reduction in likelihood is substantial, but there is still a residual (and measurable) likelihood that the hazard can occur, i.e., the risk of the incident of interest has not been reduced to zero.

This approach will result in safer and more robust processes, and it is likely that these inherently safer processes will also be more economical in the long run (Ref 2.20 Kletz, 1984, Ref 2.21 Kletz, 2010). However, the cost of changing an existing design to a more inherently safer technology may be unjustified or difficult to justify from a strictly investment standpoint. For this reason, the options must be holistically weighed, and the total life cycle costs and risks analyzed for completeness. Indirect costs of implementing inherently safer technologies, such as inadvertent risk transfer, must also be weighed before a final decision is made to implement them.

Eliminating or reducing the hazard through the application of inherently safer design is the first element in the process risk management hierarchy. Other strategies—passive, active or procedural—constitute layers of protection for the hazard.

2.10 REFERENCES

2.1 *The American Heritage® Dictionary of the English Language, Fourth Edition, Houghton Mifflin Company, 2000.*

2.2 American Chemical Society (ACS) Public Policy Statement 2015-2018, Inherently Safer Technology for Chemical and Related Industrial Operations, 2015.

2.3 Amyotte, P, et. al., Reduction of Dust Explosion Hazard by Fuel Substitution in Power Plants, Trans IChemE (81), Institution of Chemical Engineers, *2003.*

2.4 Amyotte, P.R. Goraya, A.U, Hendershot, D.C., and Khan, F.I. (2006). Incorporation of inherent safety principles in process safety management. In Proceedings of 21st Annual International Conference–Process Safety Challenges in a Global Economy, World Dolphin Hotel, Orlando, Florida, April 23-27, 2006 (pp.175-207) New York: American Institute of Chemical Engineers.

2.5 Belgian Administration Of Labour Safety, Technical Inspectorate, Chemical Risks Directorate, Process Safety Study: Practical guideline for analysing and managing chemical process risks, 2001.

2.6 Berger, S, Hendershot, D., Famini, G., Emmett, G., Defining Inherently Safer Technology to Focus Process Safety and Security Improvements, Chemical News, August 2010.

2.7 Carrithers, G., Dowell, A., Hendershot, D., It's Never Too Late for Inherent Safety, International Conference and Workshop on Process Safety Management and Inherently Safer Processes, American Institute of Chemical Engineers, 1996.

2.8 Center for Chemical Process Safety, Guidelines for Engineering Design for Process Safety, American Institute of Chemical Engineers, 1993.

2.9 Center for Chemical Process Safety, Guidelines For Initiating Events And Independent Protection Layers In Layer Of Protection Analysis, American Institute of Chemical Engineers, 2014.

2.10 Center for Chemical Process Safety, Hazard Evaluation Procedures, 3rd Ed., American Institute of Chemical Engineers, 2008.

2.11 Center for Chemical Process Safety, Layer of Protection Analysis -Simplified Process Risk Assessment, American Institute of Chemical Engineers, 2001.

2.12 Center for Chemical Process Safety, Process Safety Glossary, American Institute of Chemical Engineers, www.aiche.org/ccps/resources/glossary.

2.13 Center for Chemical Process Safety, Report: Definition for Inherently Safer Technology in Production, Transportation, Storage, and Use (for U.S. Department of Homeland Security), 2010.

2.14 Council of the European Union, Council Directive, Control of Major-Accident Hazards Involving Dangerous Substances (Seveso III), 2012/18/EU, June 19, 2012.

2.15 Hendershot, D.C. , Some thoughts on the difference between inherent safety and safety. Process Safety Progress 14 (4), 227-228, 1995.

2.16 Hendershot, D.C., Implementing Inherently Safer Design in an Existing Plant, Process Safety Progress 25(1), American Institute of Chemical Engineers, 2006.

2.17 The Institution of Chemical Engineers & The International Process Safety Group, Inherently Safer Process Design, 1995.

2.18 Khan, F., *Evaluation of Available Indices for Inherently Safer Design Options, Process Safety Progress (22) 2, American Institute of Chemical Engineers, 2003.*

2.19 Kletz, T.A., Plant Design for Safety, Rugby, Warwickshire, England: The Institution of Chemical Engineers, 1991.

2.20 Kletz, T.A., Cheaper, Safer Plants, or Wealth and Safety at Work. Rugby, Warwickshire, England: The Institution of Chemical Engineers, 1984.

2.21Kletz, T., Amyotte, P., Process Plants – A Handbook for Inherently Safer Design, 2nd Ed., CRC Press, 2010.

2.22 Lutz, W., *Take Chemistry and Physics into Consideration in All Phases of Chemical Plant Design, Process Safety Progress (14) 3, American Institute of Chemical Engineers, 1995.*

2.23 Marshall, V., *The Social Acceptability of the Chemical and Process industries, IChemE, Part B, 68, Institute of Chemical Engineers, 1990.*

2.24 Marshall, V., *The Management of Hazard and Risk, Applied Energy, 42, 1992.*

2.25 Overton, T., King, G., *Inherently Safer Technology – An Evolutionary Approach, Process Safety Progress (25) 2, American Institute of Chemical Engineers, 2006.*

2.26 UK Health and Safety Executive, Designing and operating safe chemical reaction processes, HSG-143, 2000.

2.27 UK Health and Safety Executive, Review Of Hazard Identification Techniques, HSL/2005/58, 2000.

2.28 Wade, D., *Reduction of Risks By Reduction of Toxic Material Inventory, Proceedings of the International Symposium on Preventing Major Chemical Accidents, American Institute of Chemical Engineers, 1987.*

2.29 Center for Chemical Process Safety, *Process Safety Metrics: Guide for Selecting Leading and Lagging Indicators, Version 3.1,* American Institute of Chemical Engineers (2019).

3. Minimize – An Inherently Safer Strategy

3.1 MINIMIZE

In the context of IS, *minimization* means to reduce the quantity of material, energy, or energy density contained in a manufacturing process or plant. Process minimization is often thought of as resulting from the application of innovative new technology to a chemical process, for example, tubular reactors with static mixing elements, centrifugal distillation techniques, or innovative, high surface area heat exchangers. However, much can be accomplished in process inventory reduction simply by applying good engineering design principles to more conventional technology, as well as by adjusting operational practices where possible.

Often, the inventory of hazardous materials on-site is driven by operational and business considerations, particularly in the number of transportation containers that are stored or used on-site at any given time. Railroad dispatching schedules, trucking schedules, and other logistical issues, most of which are established independently of safety considerations, often influence the amount of hazardous materials present on-site. Sometimes inventories are determined by on-site purchasing considerations, such as timing of incoming or outgoing shipments related to price, or onsite vendor-managed inventories (VMI). In some cases, the inventories were established based on transportation scheduling practices that are no longer applicable. Careful and continuing coordination with these stakeholders is required to minimize inventories.

Changing operational and maintenance practices can also help minimize inventories. For example, the application of reliability-centered (risk-based) maintenance techniques can also increase the inherent safety of a plant by reducing plant downtime, thus reducing the need for intermediate inventory and storage. This in-process storage or surge capacity may be required to allow portions of the plant to continue to operate while other parts are shut down for equipment maintenance.

Improving the reliability of critical pieces of equipment may eliminate or significantly reduce the need for in-process storage of hazardous chemical intermediates.

When designing a process facility or unit, the dimensions of every item of process equipment should be specified as large enough to accomplish its intended purpose, and no larger. Required surge capacities, either for normal operations or for emergency situations, sometimes demand larger equipment. They are part of the intended purpose of a process design and must be maintained. Utilization of this extra space should be kept to a minimum, although the process may be modified in the future to take advantage of additional process capacity. Raw material and in-process intermediate storage tanks should be minimized, if feasible. The need for all in-process inventories should be periodically reviewed and evaluated, particularly those of hazardous materials.

Minimizing the size of equipment not only enhances inherent process safety, but it can often save money. If equipment can be eliminated from a manufacturing process, it will eliminate the need for associated design, engineering, purchasing, operating, and maintenance costs. Equipment which is eliminated also cannot release hazardous material or energy into the surrounding environment. The true engineering art is to determine how to accomplish a given task with a minimum of equipment, and with the required equipment of the smallest size. Siirola (Ref 3.19 Siirola) discusses process synthesis strategies that are helpful in designing and optimizing a process route to minimize the required equipment and operations.

The term "process intensification" is used synonymously with "minimization," though the former is often used more specifically to describe new technologies that reduce the size of unit operations equipment, particularly reactors. It has been defined as "any chemical engineering development that leads to a substantially smaller, cleaner, safer, and more energy efficient technology (Ref 3.17 Reay). The European Federation of Chemical Engineering has held biennial conferences on process intensification since 2007, and other international conferences have taken place as well. These conferences presented several interesting possibilities for a range of unit operations, including reaction, gas-liquid contacting, liquid-liquid separation, heat exchange, distillation, and separation. Also, more recently, AIChE has sponsored the RAPID (Rapid Advancement in Process Intensification Deployment) Institute. (Ref 3.10 RAPID) The institute's purpose is to create a dynamic network of partners who collectively build a sustainable ecosystem, through developing and commercializing new process intensification technology. While the focus of process intensification in general, is on improving process efficiencies and economics, many of the technologies described can also improve the inherent safety of processes by reducing in-process inventories resulting from their application.

Process intensification includes the following novel techniques and designs to minimize the size, inventory, and energy consumption of process equipment (Ref 3.17 Reay; Ref 3.20 Stankiewicz):

- Equipment:
    - Spinning disk reactors, oscillatory-baffled reactors, membrane reactors, microreactors.
    - Compact and micro heat exchangers
    - Non-reaction equipment, such as packed bed contactors and centrifugal absorbers.
- Methods:
    - Intensified mixing providing higher shear and mass transfer rates (spinning disk, induction-heated, inline mixers)
    - Intensified separation techniques (centrifugal, membrane, adsorption)
    - Alternative energy sources: centrifugal fluids, ultrasound, solar energy, microwave, electric fields, plasma technology.
    - Other methods: supercritical fluids, nano fluids, process synthesis, electrolysis

A few examples of process minimization will be presented here. See Kletz (Ref. 3.12 Kletz 1984; Ref 3.13 Kletz 1991), Englund (Ref. 3.5 Englund 1990; Ref 3.6, Englund 1991a; Ref 3.7, Englund 1991b; Ref 3.8 Englund 1993), IChemE and IPSG (Ref. 3.1 IChemE), Lutz (Ref 3.15 Lutz 1995a; Ref 3.16 Lutz 1995b), CCPS (Ref 3.2 CCPS), Stankiewicz (Ref 3.20 Stankiewicz), and Reay, Ramshaw and Harvey (Ref. 3.17 Reay) for more examples.

3.2 REACTORS

Reactors can represent a large portion of the risk in a chemical process. A complete understanding of reaction mechanism and kinetics is essential to the optimal design of a reactor system. This includes the chemical reactions and mechanisms, as well as physical factors, such as mass transfer, heat transfer, and mixing. A reactor may be large because the chemical reaction is slow. However, in many cases, the chemical reaction occurs very quickly, but it appears to be slow due to inadequate mixing and contacting of the reactants. Innovative reactor designs that improve mixing may result in much smaller reactors. Such designs are usually cheaper to build and operate, as well as being safer due to smaller inventory. In many cases, improved product quality and yield also result from better and more uniform contacting of reactants. With a thorough understanding of the reaction, the designer can identify reactor configurations that maximize yield and minimize size, resulting in a more economical process that generates fewer by-products and waste, and increases inherent safety by reducing the reactor size and inventories of all materials.

A relatively new development in reactor design is the spinning disk reactor. In this novel design, the reactions take place in an imposed acceleration field, in this case, centrifugal motion. The fluid acceleration greatly enhances both the mass and heat transfer processes, thereby allowing the same reaction rates to occur in a much smaller volume. The rotating surface of revolution (i.e., the spinning disc) creates an ideal environment for the rapid transmission of mass, heat, and momentum because the thin liquid films generated on the disc are highly sheared. This facilitates rapid physical or chemical processes involving liquids (even viscous liquids), such as polymerization, precipitation, and rapid exothermic organic reactions. For example, the manufacture of a pharmaceutical product may require a 2000-liter conventional stirred batch reactor vessel. Using a 30-cm disc reactor 1000 tons/year of the same pharmaceutical product can be produced at a continuous rate of 30 gram/sec. (Ref 3.20 Stankiewicz).

Another recent development in reaction technology is the micro reaction. The miniaturization of plant mixing, and heat transfer equipment can generate very high mass and heat transfer rates. Firstly, the gradients driving mass and heat transfer, i.e., concentration and temperature, will be increased when the device dimensions are reduced, although viscous losses will increase. Secondly, the surface area-to-volume ratio of the system increases as dimensions decrease, thereby increasing the interface area per unit mass or volume. This increases the mass and heat transfer rates. Both factors together can create extremely efficient mixing devices and heat exchangers. Since volume is a three-dimensional property, the volume of the micro-device, and hence the amount of material in it, is reduced by the inverse of the third power of its characteristic dimensions. In doing so, material inventory is greatly reduced, as are concerns about large temperature and concentration differences because of the greatly reduced response time of the micro-device (Ref 3.20 Stankiewicz).

3.3 CONTINUOUS STIRRED TANK REACTORS

A continuous stirred tank reactor (CSTR) is usually much smaller than a batch reactor for a specific production rate. In addition to reduced inventory, using a CSTR usually results in other benefits that enhance safety, reduce costs, and improve the product quality. For example:

- Mixing in the smaller reactor is generally better. Improved mixing may improve product uniformity and reduce by-product formation.
- Controlling temperature is easier and the risk of thermal runaway is reduced. A smaller reactor provides greater heat transfer surface per unit of reactor volume.
- Containing a runaway reaction is more practical by building a smaller but stronger reactor rated for higher pressure.

In considering the relative safety of batch and continuous processing, it is important to fully understand any differences in the chemistry and processing conditions that may outweigh the benefits of a continuous reactor's reduced size. Englund (Ref 3.7 Englund 1991b) describes continuous latex processes that have enough unreacted monomer in the continuous reactor to be less safe than a well-designed batch process.

3.4 TUBULAR REACTORS

Tubular reactors often offer the greatest potential among reaction devices for inventory reduction. They are usually extremely simple in design, containing no moving parts and a minimum number of joints and connections. A relatively slow reaction can be completed in a long tubular reactor if mixing is adequate. There are many devices available for providing mixing in tubular reactors, including jet mixers, eductors, and static mixers.

It is generally desirable to minimize the diameter of a tubular reactor because the leak rate in case of a tube failure is proportional to its cross-sectional area. For exothermic reactions, heat transfer will also be more efficient with a smaller tubular reactor. However, these advantages must be balanced against the higher-pressure drop due to flow through smaller reactor tubes.

3.5 LOOP REACTORS

A loop reactor is a continuous tube or pipe that connects the outlet of a circulation pump to its inlet (Figure 3.1). Reactants are fed into the loop, where the reaction occurs, and product is then withdrawn. Loop reactors have been used in place of batch stirred tank reactors in a variety of applications, including chlorination, ethoxylation, hydrogenation, and polymerization. A loop reactor is typically much smaller than a batch reactor producing the same amount of product. Wilkinson and Geddes (Ref 3.23 Wilkinson) describe a 50-liter loop reactor for a polymerization process that has a capacity equal to that of a 5000-liter batch reactor. Mass transfer is often the rate limiting step in gas-liquid reactions, and a loop reactor design increases mass transfer, while reducing reactor size and improving process yields. As an example, an organic material was originally chlorinated in a glass-lined batch stirred tank reactor, with chlorine fed through a dip pipe. Replacing the stirred tank reactor with a loop reactor that fed chlorine to the recirculating liquid stream through an eductor reduced reactor size, increased productivity, and reduced chlorine usage. These results are summarized in Table 3.1 (Ref. 3.2 CCPS).

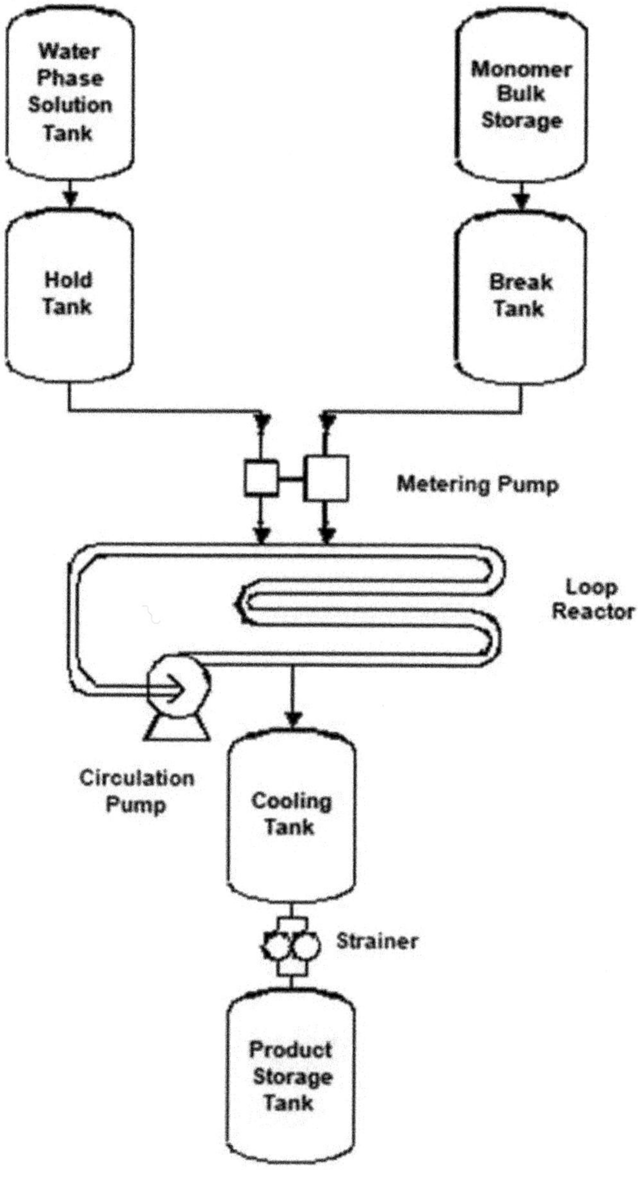

Figure 3.1: A Loop Reactor Production System

Table 3.1: Effect of Reactor Design on Size and Productivity for a Gas-Liquid Reaction (Ref 3.2 CCPS)

Reactor Type	Batch Stirred Tank Reactor	Loop Reactor
Reactor Size (l)	8000	2500
Chlorination Time (hr)	16	4
Productivity (kg/hr)	370	530
Chlorine Usage (kg per 100 kg product)	33	22
Caustic Usage in Vent Scrubber (kg per 100 kg product)	31	5

3.6 REACTIVE DISTILLATION

The combination of several unit operations into a single piece of equipment can eliminate equipment, minimize the inventory of reactants, and simplify a process. However, there may be inherent safety conflicts resulting from this strategy (see Chapter 12). Combining several operations into a single device increases the complexity of that device, but it also reduces the number of vessels or other pieces of equipment required for the process. Careful evaluation of the options with respect to all hazards is necessary to select the inherently safer overall option.

Reactive distillation is a technique for combining several process operations in a single device. One company has developed a reactive distillation process for the manufacture of methyl acetate that reduces the number of distillation columns from eight to three, while also eliminating an extraction column and a separate reactor (Ref 3.1 Agreda; Ref 3.4 Doherty; Ref 3.19 Siirola). Inventory is reduced and auxiliary equipment, such as reboilers, condensers, pumps, and heat exchangers are eliminated. Figure 3.2 shows the conventional design, and Figure 3.3 shows the reactive distillation design. For this process, Siirola (Ref 3.19 Siirola) reports significant reductions in both capital investment and operating cost for the reactive distillation process.

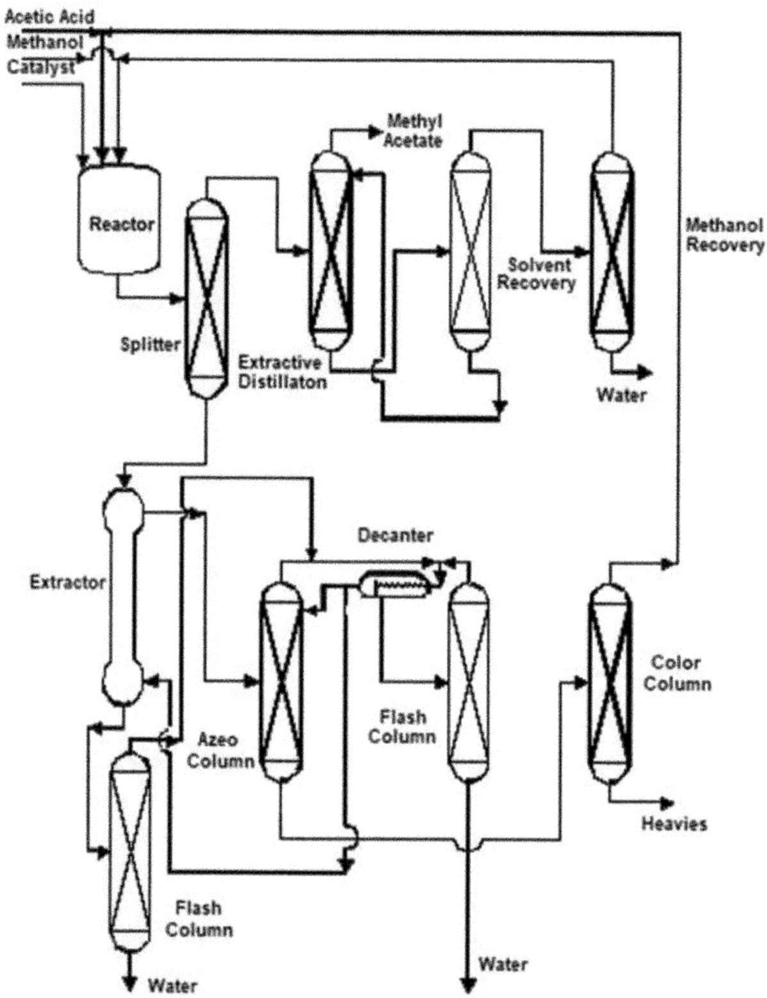

Figure 3.2: Conventional process for methyl acetate (based on Ref 3.19 Siirola).

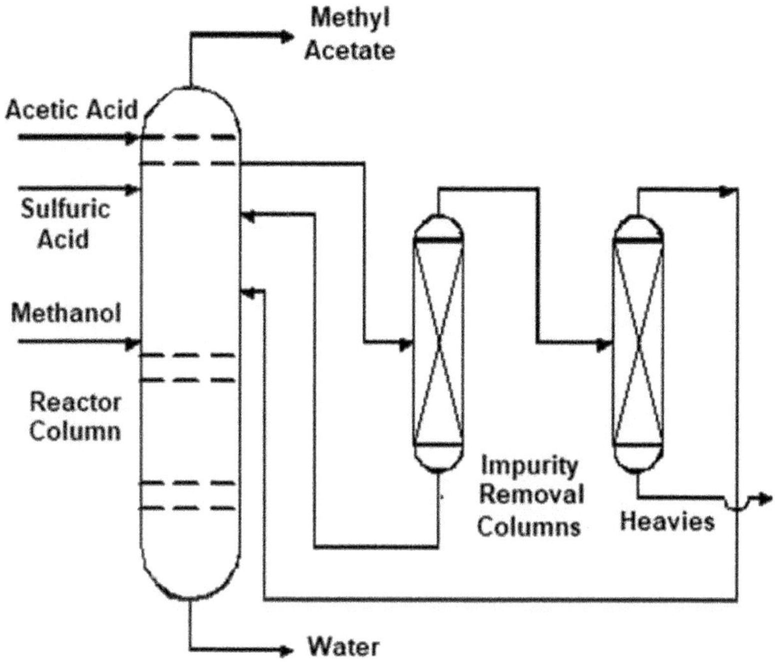

Figure 3.3: Reactive distillation methyl acetate process (based on Ref 3.1 Agreda).

Englund (Ref 3.5 Englund 1990) and Stankiewicz (Ref 3.20 Stankiewicz) describe several other types of minimization:

- Reducing the inventory by changing the process chemistry. The production of latex using styrene and butadiene can be accomplished by the continuous addition of the reactants so that the reaction occurs as the batch reactor is being filled. The maximum inventory of unreacted monomer is greatly reduced, and the temperature control is made easier. Scheffler (Ref 3.18 Scheffler) also addresses the application of IS principles to the production of latex.
- Reducing the inventory by changing the mixing intensity. Some reactions that occur very quickly can be accomplished in piping systems using inline mixers or nozzles. This reduces the holdup

volume of the reactants significantly. Static mixer reactors that incorporate heat transfer also reduce the holdup volume of reactants and solve the heat transfer problems associated with many static mixers. For gas-liquid reactions, which are an integral part of many processes, such as chlorinations, oxidations, sulfonations, nitrations, and hydrogenations, high intensity motionless static mixers can increase the reaction rate, particularly when the gaseous reactant must be transferred to the liquid phase (Ref 3.20 Stankiewicz). For example, sodium hypochlorite (bleach), an inherently safer substitute for elemental chlorine in the treatment of potable and wastewater, is produced by reacting gaseous chlorine with liquid sodium hydroxide in a small static mixer pipe reactor.

3.7 STORAGE OF HAZARDOUS MATERIALS

Raw material and in-process storage tanks and piping/pipelines often represent a significant portion of the risk of a processing facility. Attention to the design of storage and transfer equipment can reduce hazardous material inventory.

Storage tanks for raw materials and intermediates are often much larger than necessary, usually because this makes it "easier" to operate the plant. The operating staff can pay less attention to ordering raw materials on time or can accept downtime in a downstream processing unit because upstream production can be kept in storage until the downstream unit is back on line. This convenience in operation can come at a significant cost in the risk of loss of containment of the hazardous materials being stored. The process design engineers and operating staff must jointly determine the need for all intermediate hazardous material storage and minimize quantities where appropriate. Hendershot (Ref 3.9 Hendershot) describes a detailed risk analysis that demonstrated that changing from the bulk storage of bromine to cylinders reduced the calculated off-site consequences by three orders of magnitude. This minimization change also had the advantage of relieving the facility of having to comply with EPA's Risk Management Program (40 CFR Part 68). In this case, the usage rate of bromine was low, thereby not significantly increasing the number of deliveries nor transferring the risk to the transportation sector. The existing bulk storage tank and its supports were also in need of replacement due to corrosion. Therefore, the proposed inherently safer modification was economically viable as well.

Similarly, hazardous raw material storage should also be minimized, with greater attention being given to "just in time" supply. Inventory reduction lowers inventory costs, while increasing inherent safety. However, in determining appropriate raw material inventories, the entire supply chain should be considered, as follows:

- Will the originating facility for the materials, distribution facilities in the value chain, or both have to increase inventories to provide "just-in-time" service, and will this represent a greater risk than a larger inventory at the end-user facility?
- How much additional burden will "just-in-time" delivery place on operating staff? Will increased number of "just-in-time" deliveries increase the potential for human errors in loading/offloading operations?
- Will the additional time working in transient operating modes, due to unplanned shutdowns and the resulting restarts caused by running out of raw materials, increase the risks?
- Will transportation and temporary storage of raw material, either in parked railroad cars, tank trucks, barges, or other transportation containers along transportation routes or in transportation facilities, present perhaps an even greater risk than on-site storage in a well-designed end-user facility?
- Does the increased number of deliveries increase the risk in the mobile portions of the value chain? Typically, increased rail, truck, or barge shipments increases exposure of populations, property, and the environment along the transportation routes to potential loss of containment events along these routes. Chapter 8 discusses inherently safer options in hazardous materials transportation in more detail.

The reduction in inventory resulting from greater attention to plant operations and design of unit interactions can be substantial. Several excellent examples are provided here:

- An acrylonitrile plant eliminated 500,000 pounds of in-process storage of hydrogen cyanide by shutting down an entire unit when the product purification area shut down. This forced the plant staff to solve the problems that caused the purification area shutdowns. (Ref 3.22 Wade).

- Another acrylonitrile plant that supplied by-product hydrogen cyanide to various other units eliminated an inventory of 350,000 pounds by having the other units draw directly from the plant. This required considerable work to resolve many issues related to acrylonitrile purity and unit scheduling. (Ref 3.22 Wade).
- A central bulk chlorine system with large storage tanks and extensive piping was replaced with several small cylinder facilities local to the individual chlorine users. The total inventory of chlorine was reduced by more than 100,000 pounds. This is another example of conflicting inherent safety strategies. The use of a central bulk chlorine system reduces the need for operators to frequently connect and disconnect chlorine cylinders, a step where increased rates of human error and equipment wear, and failure are possible. But, the disadvantage is a large inventory that could be released if a leak occurs. The use of several local cylinder facilities results in a greater likelihood of a leak because of the necessity to connect and disconnect the cylinders more frequently, but the maximum size of the leak will be limited to the inventory of one cylinder. In such cases the total risk must be analyzed and considered; the increased frequency of failure, combined with the reduced consequences per event when using small chlorine cylinders, must be compared to the combined frequency and consequences of using large bulk chlorine storage and delivery systems. (Ref 3.22 Wade).
- Debris from a fire and explosion in a pesticide manufacturing facility narrowly missed damaging a neighboring storage tank containing methyl isocyanate (MIC). Although the explosion was not related to MIC manufacturing, the company recognized the hazards associated with storing large quantities of MIC in above-ground storage tanks throughout the plant. The company embarked on a redesign of the MIC storage and handling processes at the plant, incorporating IST/ISD techniques, eliminating above-ground storage, reducing overall stored

volume by 80%, and confining the MIC storage to one area of the plant, which simplified the internal logistics of MIC handling (Ref 3.3 Committee).

An additional method to reduce raw material inventory is to manufacture the hazardous component in a just-in-time fashion, rather than purchasing it in bulk and storing it. This potentially achieves similar inventory reductions as a just-in-time purchasing strategy, with the added benefit of eliminating the additional material handling (i.e., transportation container unloading). Consider a chlorination process which uses railcar quantities of purchased chlorine on a daily basis. A chloralkali plant can generate low pressure chlorine as needed for the process, with very little in-process inventory, and whatever inventory is required can be at low pressure (another ISD strategy, i.e., Moderation, discussed in Chapter 5). This has the added benefit of reducing the offsite consequences of a loss of containment (similar to the bromine example above) and provides a supply of caustic soda for either outside sale or reuse/recycle onsite. This example was considered for an actual organic chlorination process which generated hydrogen chloride as a low market value byproduct. The co-generated caustic soda would be used to neutralize and recycle the chlorine essentially lost as byproduct to high-value-added chlorinated organic. The worst-case scenario of a release incident for a co-generated facility is nearly an order of magnitude less severe than the original variation.

3.8 PROCESS PIPING

All process equipment and units must be connected by piping systems, making the layout of piping within a plant a significant influence on the inventories of materials on-site. In turn, the layout and placement of units and equipment will influence the length of inter-unit and intra-unit process piping and transfer piping systems which link these units. Since the volume of piping is a function of the square of the piping diameter, each additional linear foot of piping length can represent a measurable increase in site material inventories. Each 100 feet (30 meters) of piping, with an inner diameter of 2 inches (50 millimeters), adds approximately 16 gallons (60 liters) of liquid to site inventory, and each 100 feet (30 meters) of 6-inch (150 mm) piping adds approximately 150 gallons (568 liters). For the same reason, process piping diameter should also be minimized to that required for the desired flow rate and allowable pressure drop. Although the length of piping systems is generally fixed during initial plant design, the decommissioning of units and equipment during the life of the facility will render some piping obsolete. This piping should be physically removed from the system. However, if this is not possible, then decommissioned piping should never be allowed to contain any hazardous materials, or to receive flow from "live" process equipment and should be appropriately cleaned and physically disconnected at all connection points to live equipment. In addition to representing unneeded inventory, these piping dead legs also represent a source of increased galvanic corrosion and possible leakage. A corroded piping dead leg on a detergent alcohol tank in a New Jersey chemical plant failed when an operator attempted to operate a manual valve on the line, resulting in a multi-thousand-pound spill to containment. A Valero refinery in Texas experienced a significant process safety incident traced to water leaking into and freezing within an out-of-service propane line. The line then failed, and the ensuing fire and equipment damage resulted in two months of downtime and approximately a year of reduced refinery capacity. (Ref 3.21 CSB)

3.9 PROCESS EQUIPMENT

Process systems contain a variety of equipment that perform many necessary functions, including heat exchangers, filters, strainers, dryers, heaters, rotating equipment, and valves. This equipment must be sized to adequately perform its required function but should not have unnecessary additional volume. Inventory reductions in any one device may not have a substantial effect on the total amount of materials on-site, however, the cumulative effect of minimizing these components during an initial design or re-design (i.e., a de-bottlenecking project) can result in significant inventory reductions. Innovative designs that minimize the inventory of materials in the device at any given time, particularly for heat exchangers, can substantially reduce the volume of materials present in a process. Heat exchangers involve several variables whose optimization offer the opportunity to reduce the materials inventory, including:

- Heating/cooling medium

- Heating/cooling medium and process material flow characteristics
- Heating/cooling temperature
- Heat transfer surface area
- Heat transfer geometry
- Location of hazardous material in the heat exchanger (i.e., shell or tube side)

Section 8.4.2 describes different types of compact heat exchangers that achieve a significant degree of process minimization/intensification. These novel heat exchanger designs are characterized by high compactness ratios (heat transfer area-to-volume) and small hydraulic diameters.

In addition to reactors, the use of high gravity or centrifugal forces has also been developed for packed bed applications. A possible equivalent to a large packed-bed column to perform liquid/liquid extractions, gas/liquid interactions, and other similar operations, is a compact rotating packed bed contactor. The heavier component, in this case, the heavier liquid, is introduced at the eye of the packed rotating bed and moves outward, while the lighter component, such as a lighter liquid or gas, is introduced at the periphery and moves inward. The use of an accelerated fluid greatly reduces the size of the packed bed (Ref 3.20, Stankiewicz).

Another development is the potential for desktop manufacturing. Where annual production rates are relatively small, such as for certain pharmaceuticals, replacement of a large batch process that operates infrequently to satisfy desired production volume with a much smaller continuously operating lab or pilot scale process that operates at a very low rate results in a large degree of process minimization. For example, an annual production amount of 500 tons corresponds to a continuous rate of 70 mL/sec. This demand can be met with a desktop process. Scale-up design problems are minimized, and process loads, such as power demand and heat load, are distributed over much wider times, resulting in much smaller equipment (Ref 3.20 Stankiewicz).

Section 8.4.2 on Unit Operations includes additional examples of opportunities for process minimization or intensification for other unit operations and types of equipment.

3.10 LIMITATION OF EFFECTS

An inherently safer way to limit the effects of equipment or control systems failures, or from human error, is to minimize the consequences of these failures through equipment design or process/reaction conditions. (Ref 3.14 Kletz 2010). Contrast this approach with that of adding multiple layers of protection to a process, which typically have discrete probabilities of failure on demand (inferring that they are not foolproof and will eventually fail). Another way to envision this concept is to compare it with how consequence severities are developed in a process hazards analysis without safeguards. A failure scenario in which a reaction cannot run away, or in which a loss of containment results in a much lower amount of material released, is inherently safer than one which relies on protective equipment to mitigate the failure effects.

Some examples of Limitation of Effects are as follows:

- Use of spiral wound gaskets, which leak less than fiber gaskets if bolt torque is lost or not correct
- Limiting the area of storage tank dikes, which limits the evaporation rate, and in the case of flammable liquids, lowers the heat flux from any ensuing fires.
- Limiting the diameter and length of piping systems, which reduces the contained volume, and limiting drain valve and sample point sizes to the absolute minimum, to limit release rates in the event of inadvertent opening or failure to close.
- Decreasing the number of flanges, valves, nozzles, and other piping system appurtenances. Remember, what you don't have can't leak!

Separating the stages of reactions into separate vessels, so that incorrect order of addition of raw materials cannot occur.

Considering the use of continuous processes vs. semi-batch processes, and semi-batch processes vs. traditional batch processes, to limit the amount of unreacted raw materials which could participate in a runaway reaction if cooling or mixing failed during the reaction.

Changing reaction temperatures, concentrations, or other parameters, to limit runaway potential if mixing or cooling are lost.

Limiting the amount of energy available, by reducing the temperature of heating media. The runaway reaction in Seveso, Italy in 1976, which resulted in dioxin contamination of the surrounding farmland, resulted from overheating the reactor vessel with a steam supply whose temperature was both far above that necessary for effective heat transfer, and far above the initiation temperature of the runaway reaction.

3.11 REFERENCES

3.1 Agreda, V. H., Partin, L.R. and Heise, W.H., *High-purity methyl acetate via reactive distillation.* Chemical Engineering Progress, 86 (2), 40-46, 1990.

3.2 Center for Chemical Process Safety (CCPS), *Guidelines for Engineering Design for Process Safety.* New York: American Institute of Chemical Engineers, 1993.

3.3 Committee on Inherently Safer Chemical Processes: The Use of Methyl Isocyanate (MIC) at Bayer CropScience, National Academy of Sciences, 2012

3.4 Doherty, M., and Buzad, G. *Reactive distillation by design.* The Chemical Engineer, s17-s19, 27 August 1992.

3.5 Englund, S. M. *"The design and operation of inherently safer chemical plants."* Presented at the American Institute of Chemical Engineers 1990 Summer National Meeting, August 20, 1990, San Diego, CA, Session 43.

3.6 Englund, S. M. *Design and operate plants for inherent safety - Part 1.* Chemical Engineering Progress, 87 (3), 85-91, 1991a.

3.7 Englund, S.M. *Design and operate plants for inherent safety - Part 2.* Chemical Engineering Progress, 87 (5), 79-86, 1991b.

3.8 Englund, S.M. *Process and design options for inherently safer plants.* In V. M. Fthenakis (ed.). Prevention and Control of Accidental Releases of Hazardous Gases (9-62). New York: Van Nostrand Reinhold, 1993.

3.9 Hendershot, D.C., et al. *Implementing inherently safer design in an existing plant.* Process Safety Progress, 25 (1), 52-57, 2006.

310, www.aiche.org/rapid/about-rapid

3.11 The Institution of Chemical Engineers and The International Process Safety Group. *Inherently Safer Process Design*. Rugby, England: The Institution of Chemical Engineers, 1995.

3.12 Kletz, T.A., *Cheaper, Safer Plants, or Wealth and Safety at Work*. Rugby, Warwickshire, England: The Institution of Chemical Engineers, 1984.

3.13 Kletz, T.A., *Plant Design for Safety,* Rugby, Warwickshire, England: The Institution of Chemical Engineers, 1991.

3.14 Kletz, T., Amyotte, P., *Process Plants – A Handbook for Inherently Safer Design*, 2nd Ed., CRC Press, 2010.

3.15 Lutz, W.K., Take chemistry and physics into consideration in all phases of chemical plant design. Process Safety Progress 14 (3), 153-162, 1995a.

3.16 Lutz, W.K. *Putting safety into chemical plant design*. Chemical Health and Safety, *2* (6), 12-15, 1995b.

3.17 Reay, D, Ramshaw, C., and Harvey, A., *Process Intensification, 2nd. Ed.*, Elsevier, Ltd., 2013

3.18, *Scheffler*, N. E, *"Inherently Safer Latex Plants,"* Process Safety Progress 15, no. 1 (*1996*): 11–17.

3.19 Siirola, J.J. *"An industrial perspective on process synthesis."* In Foundations of Computer-Aided Process Design, American Institute of Chemical Engineers Symposium Series, **91**(304):222-223, 1995.

3.20 Stankiewicz, A. and Moulijn, J., editors, Reengineering the Chemical Processing Plant: Process Intensification, CRC Press, 2004

3.21 United States Chemical Safety and Hazard Investigation Board (CSB), *FINAL REPORT, Valero Refinery Propane Fire*, 7/9/2008

3.22 Wade, D.E. *Reduction of risks by reduction of toxic material inventory.* In J.L. Woodward (Ed.). Proceedings of the International Symposium on Preventing Major Chemical Accidents, February 3-5, 1987, Washington, D.C (2.1-2.8.) New York: American Institute of Chemical Engineers, 1987.

3.23 Wilkinson, M., and Geddes, K., *An award-winning process.* Chemistry in Britain, 1050-1052, December 1993.

4. Substitute – An Inherently Safer Strategy

In the context of IS, "substitution" means the replacement of a hazardous material or process with an alternative that reduces or eliminates the hazard. Process designers, operations managers, and plant technical staff should monitor the state-of-the-art and periodically determine if less hazardous alternatives can be effectively substituted for hazardous materials used in a manufacturing process. However, the substitution concept of inherent safety is best applied during the initial design of a process. Substituting raw materials and intermediates after the process has been built, while possible in some cases, is usually very difficult at best and infeasible at worst.

The most commonly accepted example of Substitution for existing processes is the widespread replacement of elemental chlorine with sodium hypochlorite as a water disinfectant in both potable and cooling water treatment systems. However, this example represents somewhat of the outlier in that the chemical reaction still involves chlorine; the delivery of the chlorine molecules to the water or wastewater being disinfected can be achieved in several different ways. Not all chemical manufacturing processes have this flexibility.

Examples of substitution can be noted in five categories: reaction chemistry, solvent usage, refrigerants, firefighting agents, and heat transfer media are discussed in the following sections. Substitution of chemicals represents a First Order Inherent Safety solution in that it eliminates of the chemicals used or produced as discussed in Chapter 2.

4.1 REACTION CHEMISTRY

Basic process chemistry that uses less hazardous materials and chemical reactions offers the greatest potential for improving inherent safety in the chemical/processing industry. Alternate chemistry may use less hazardous raw materials or intermediates or result in reduced inventories of hazardous materials (minimization), or less severe processing conditions (moderation). Identification of catalysts that can enhance reaction selectivity or allow desired reactions to be carried out at a lower temperature or pressure, is often the key to developing inherently safer chemical synthesis routes. Some specific examples of innovations in process chemistry that result in inherently safer processes include:

- The insecticide carbaryl can be produced by several routes, some of which do not use methyl isocyanate (MIC) (substitution), or which generate only small quantities of this toxic material as an in-process intermediate (minimization) (Ref 4.15 Kletz 1991). One company has developed a proprietary process for manufacturing carbamate insecticides that generates methyl isocyanate as an in-situ intermediate. Total methyl isocyanate inventory in the process is no more than 10 kilograms (Ref 4.14 Kharbanda; Ref 4.20 Manzer). An Israeli company has changed the order of the three materials (α-naphthol, methylamine, and phosgene) used to produce carbaryl so that no MIC is generated. Rather than react α-naphthol with methylamine to produce MIC, α-naphthol and phosgene are reacted to produce a chloroformate ester that is then reacted with methylamine to produce carbaryl. No MIC is produced. Neither of these reactions is ideal because they involve the use of phosgene, which is a more toxic material than chlorine (Ref 4.16 Kletz 1998). However, process designs to produce the phosgene in situ and in "just-in-time" amounts, with little or no interim storage, are common in many applications where phosgene is a required material.
- Production of the carbamate pesticide methomyl commonly uses MIC which is manufactured separately and stored in large quantities under pressure. A major chemical company implemented a new technology for producing methomyl which involves conversion of the less acutely toxic chemical methylformamide into MIC as a vapor-phase (rather than liquid) on an as-needed basis, which is immediately consumed in a subsequent reaction, leaving very little MIC in the system at one time and eliminating phosgene from the production process entirely. (Ref 4.40 US OSHA)
- Acrylonitrile can be manufactured by reacting acetylene with hydrogen cyanide:

$$CH \equiv CH + HCN \rightarrow CH_2 = CHCN$$

Acetylene Hydrogen Cyanide Acrylonitrile

- A new ammoxidation process uses propylene and ammonia, which are less hazardous raw materials (Ref 4.3 Dale; Ref 4.26 Puranik).

  $$CH \equiv CH + HCN \rightarrow CH_2 = CHCN$$

  Acetylene Hydrogen Cyanide Acrylonitrile

  This process does produce HCN as a by-product in small quantities. Puranik, et al. (Ref 4.26 Puranik) report on work to develop an improved, more selective catalyst, and to couple the ammoxidation process with a second reactor in which a subsequent oxycyanation reaction would convert the by-product HCN to acrylonitrile.

- The Reppe process for manufacturing acrylic esters reacts an alcohol with acetylene and carbon monoxide, using a nickel carbonyl catalyst having both high acute and long-term toxicity, to make the corresponding acrylic ester:

  $$CH_2 = CHCH_3 + NH_3 + \frac{3}{2}O_2 \rightarrow CH_2 = CHCN + 3H_2O$$

  Propylene Ammonia Acrylonitrile

  The alternative propylene oxidation process uses less hazardous materials to manufacture acrylic acid, followed by esterification with the appropriate alcohol (Ref 4.12 Hochheiser).

$$CH_2{=}CHCH_3 + \frac{3}{2}O_2 \xrightarrow{catalyst} CH_2{=}CHCO_2H + H_2O$$

Propylene Acrylic acid

$$CH_2{=}CHCO_2H + ROH \xrightarrow{H+} CH_2{=}CHCO_2R + H_2O$$

Acrylic Acid Alcohol Acrylic ester

- Polymer-supported reagents, catalysts, protecting groups, and mediators can be used in place of the corresponding small molecule materials (Ref 4.29 Sherrington; Ref 4.33 Sundell). The reactive species is tightly bound to a macromolecular support that immobilizes it. This generally makes toxic, noxious, or corrosive materials much safer. Although it has been phased out as a motor fuel additive for environmental reasons, methyl tert-butyl ether (MTBE) was produced commercially by the use of a polystyrene sulfonic acid catalyst with methanol and isobutene (isobutylene).

$$CH_3OH + CH_2{=}C(CH_3)_2 \xrightarrow{\text{Polystyrene sulfonic acid}} CH_3{=}OC(CH_3)3$$

Methanol + Isobutene MTBE

- The chemistry of side reactions and by-products may also offer opportunities for increasing the inherent safety of a process. For example, a process involving a caustic hydrolysis step uses ethylene dichloride (EDC; 1,2-dichloroethane) as a solvent. Under reaction conditions, a side reaction between sodium hydroxide and EDC produces small, but hazardous, quantities of vinyl chloride:

$$C_2H_4Cl_2 + NaOH \rightarrow C_2H_3Cl + NaCl + H_2O$$

Ethylene Dichloride Vinyl chloride

An alternative non-reactive solvent has been identified which eliminates this hazard (Ref 4.10 Hendershot 1987).

- Hydrofluoric acid (HF) has a number of industrial uses, including petroleum refining as an alkylation catalyst, detergent component manufacturing, metals processing, glass etching, organofluoride manufacture, semiconductor manufacturing, and uranium enrichment. Recent process developments have resulted in possible substitutes for liquid HF acid:
    - Solid HF catalysts have been demonstrated as a substitute for liquid HF alkylation processes. Alkylation units in petroleum refineries are used to produce high octane gasoline blending components that have no sulfur or aromatic content. One such process, currently under development, uses a liquid phase riser reactor with a solid catalyst similar in concept to a Fluidized Catalytic Cracking Unit (FCCU). The reactor operates at about 350 psia (24 bar) and refrigerated temperatures of 50-100°F (10 - 38°C). These operating conditions are similar to existing liquid HF Alkylation units, as are the feed input ratios (isobutane-to-olefin). Catalyst reactivation and regeneration are key steps in these processes (Ref 4.21 McCarthy).
    - Solid bed HF catalysts are used in the production of linear alkylbenzenes, a material used in the generation of sulfonation compounds. These are components used in the manufacture of bio-degradable detergents.
    - The use of sulfuric acid as an alkylation catalyst in refineries, although less efficient as a catalyst, presents a demonstrated opportunity for replacement of HF with a less hazardous alternative. Several refineries have performed this conversion as a way to reduce risk. HF often presents the potential for a large off-site (as well as on-site) impact in the event of a catastrophic release, whereas sulfuric acid normally presents a hazard to personnel in the immediate release area only, and as a liquid, is easier to contain. However, there is a risk transfer issue associated with using sulfuric acid in refinery alkylation units. Unlike HF alkylation units, where the HF acid can be regenerated and re-used, the

use of sulfuric acid as an alkylation catalyst is a once-through process, resulting in large quantities of spent sulfuric acid (i.e., sulfuric acid - hydrocarbon mixtures) which are flammable and have been involved in explosions and fires. Unless re-processed onsite, the spent sulfuric acid must be transported to a separate facility for re-processing to recover the acid, resulting in increased transportation risks, as well as the risk associated with the reprocessing facilities. Also, increased sulfuric acid production is required on the front-end of the value chain for this process, which involves the increased use of sulfur dioxide and sulfur trioxide, which also have significant process safety risks. Overall, however, sulfuric acid has a clear advantage over HF for alkylation from a safety and environmental standpoint (Ref 4.22 Norton).

- Phase transfer catalysis processes (Ref 4.31 Starks 1978; Ref 4.32 Starks 1987) for the synthesis of many organic materials use less, or sometimes no, organic solvent. They may use a less toxic solvent, or less hazardous raw materials, such as aqueous HCl instead of anhydrous HCl, or they may operate at milder conditions. Some types of reactions where phase transfer catalysis has been applied include:
    - esterification
    - nucleophilic aromatic substitution
    - dehydrohalogenation
    - oxidations
    - alkylation
    - aldol condensations

Rogers and Hallam (Ref 4.27 Rogers) provide other examples of chemical approaches to inherent safety, involving synthesis routes, reagents, catalysts and solvents, for example:

- Use of air instead of peroxide as a source of oxygen for oxidation reactions
- Use of catalysts which become deactivated when specified operating limits (e.g.,temperature) are exceeded, limiting the degree of excess reactions

- Use of a reaction solvent with a high enough boiling point to prevent vaporization in case of excess reaction

Innovative chemical synthesis procedures have been proposed as offering economical and environmentally-friendly routes to manufacturing a variety of chemicals. These novel chemical reactions may also potentially offer increased inherent process safety by eliminating hazardous materials or chemical intermediates, or by allowing less severe operating conditions. Some examples of interesting and potentially inherently safer chemistries include:

- Electrochemical techniques, proposed for the synthesis of naphthaquinone, anisaldehyde, and benzaldehyde (Ref 4.41 Walsh).
- Extremozymes, or enzymes which can tolerate relatively harsh conditions, suggested as catalysts for complex organic synthesis of fine chemicals and pharmaceuticals (Ref 4.8 Govardhan).
- Domino reactions, in which a series of carefully planned reactions occur in a single vessel, used to prepare complex biologically active organic compounds (Ref 4.9 Hall; Ref 4.34 Tietze).
- Laser light "micromanaged" reactions directed at the production of desired products (Ref 4.7 Flam).
- Supercritical processing, which allows the use in chemical reactions of less hazardous solvents like liquid carbon dioxide or water. This benefit must be balanced against the high temperatures and pressures required for handling supercritical fluids. Johnston (Ref 4.13 Johnston), DeSimone, et al. (Ref 4.5 DeSimone), and Savage (Ref 4.28 Savage) review some potential applications of supercritical processing.
- The use of glucose in lieu of benzene (a toxic and flammable hydrocarbon) for the production of adipic acid. It may be possible to produce glucose from biological residue materials, such as plant husks and straw (Ref 4.16 Kletz 1998).
- The substitution of toxic/flammable gases, such as phosphine, diborane, and silane, in the manufacture of semiconductors with less hazardous liquids, such as trimethyl phosphite, trimethyl borate, and tetraethyl-*o*-silicate (Ref 4.16 Kletz 1998).

- Phosphorous oxychloride can be used in place of phosgene, and 1,1,1-trichloroethane can be used in place of hydrochloric acid gas (Ref 4.16 Kletz 1998). There are some tradeoffs associated with these substitutions. Trimethyl phosphite and phosphorous oxychloride are considered potential chemical weapon precursors and are subject to specific restrictions in some countries. Phosphorous oxychloride is also a toxic material from an airborne release viewpoint, although it is less toxic than phosgene. Phosphorous oxychloride also water-reactive and generates large quantities of HCl on contact with water including atmospheric humidity. In addition, 1,1,1-trichloroethane is a suspected human carcinogen and an ozone depleting chemical.
- The use of membrane cells vs. mercury or diaphragm cells for production of chlorine to avoid the use of mercury and asbestos in the older technology (Ref 4.17 Kletz 2010).
- The use of carbon dioxide instead of mineral acid (sulfuric acid, hydrochloric acid or nitric acid) for pH control of drinking water to avoid the potential for overdosing due to the solubility limit of carbon dioxide. (Ref 4.17 Kletz 2010).
- The use of acid solutions (e.g.,hydrofluoric and hydrochloric acid) instead of the anhydrous gas forms stored under pressure (e.g.,hydrogen fluoride and hydrogen chloride) (Ref 4.36 US EPA & US OSHA).
- The use of onsite sulfur burning equipment instead of the bulk storage of oleum and sulfur trioxide. (Ref 4.36 US EPA & US OSHA).
- The use of supercritical carbon dioxide instead of chlorofluorocarbons (CFC), other ozone-depleting chemicals, or flammable or toxic hydrocarbons as a blowing agent for polystyrene foam production (Ref 4.38 US EPA – Dow).

Many of these alternative chemistry options may still be undergoing research, and there are few, if any, current commercial applications of these techniques. However, the potential environmental and safety benefits of these and other innovative chemical synthesis techniques will encourage further research and development.

4.2 GREEN CHEMISTRY

A United States Environmental Protection Agency (EPA) report (Ref 4.18 Lin) contains an extensive review of inherently safer process chemistry options that have been discussed in the literature. This report includes chemistry options that have been investigated in the laboratory, as well as some that have advanced to pilot plant and even to production scale.

- In the 1990's, the US EPA established the Green Chemistry Program with the goal of promoting innovative chemical technologies that reduce or eliminate the use or generation of hazardous substances in the design, manufacture, and use of chemical products. Green chemistry is a highly effective approach to pollution prevention because it applies innovative scientific solutions to real-world environmental situations. For these reasons, the Green Chemistry Program covered activities that are broader in scope than the IS concepts described in this book. Hendershot (Ref 4.11 Hendershot 2006) provides a clear description of the relationship between inherently safer technologies and green chemistry. However, among the twelve principles of Green Chemistry, as published by EPA, are:
    - to design chemical products to be fully effective, yet have little or no toxicity, and
    - to design syntheses to use and generate substances with little or no toxicity to humans and the environment.

These principles, as well as the others that define the Green Chemistry Program are compatible with the goals and objectives of IS. An example of an award given to companies that have redesigned their processes under the Green Chemistry Program include:

- The use of transition metals for catalyzing reactions is of growing importance in modern organic chemistry. These catalysts are widely used in the synthesis of pharmaceuticals, fine chemicals, petrochemicals, agricultural chemicals, polymers, and plastics. Of particular importance is the formation of C–C, C–O, C–N, and C–H bonds. Traditionally, the use of an inert gas atmosphere and the exclusion of moisture have been essential in both organometallic chemistry and transition-metal catalysis. The research of McGill University has focused on the development of numerous transition-metal-catalyzed reactions, both in air and water. Specifically, Professor Li has developed a novel [3+2] cycloaddition reaction to generate 5-membered carbocycles in water; a synthesis of beta-hydroxyl esters in water; a chemoselective alkylation and pinacol coupling reaction mediated by manganese in water; and a novel alkylation of 1,3-dicarbonyl-type compounds in water. His work has enabled rhodium-catalyzed carbonyl addition and conjugate addition reactions to be carried out in air and water. A highly efficient, zinc-mediated Ullman-type coupling reaction (a metal-catalyzed Nucleophilic Aromatic Substitution between various nucleophiles with aryl halides), catalyzed by palladium in water, has also been designed. This reaction is conducted at room temperature under an atmosphere of air. In addition, a number of Barbier-Grignard-type reactions (metal-catalyzed methods for forming carbon–carbon bonds) in water have been developed. These novel synthetic methodologies are applicable to the synthesis of a variety of useful chemicals and compounds. Transition-metal catalyzed reactions in water and air offer many advantages. Water is readily available and inexpensive, and is not flammable, explosive, or toxic. Using water as a reaction solvent can save synthesis steps by avoiding protection and deprotection processes that affect overall synthesis efficiency and contribute to solvent emission (i.e., simplification). Product isolation may be facilitated by simple phase separation, rather than energy-intensive and organic-emitting processes involving distillation of organic solvent. The temperature of reactions performed in aqueous media is also easier to control since water has such a high heat capacity (i.e., moderation). (Ref 4.38 U.S. EPA – Dow)

4.3 SOLVENTS

The replacement of volatile organic solvents with aqueous systems or less hazardous organic materials improves the safety of many processing operations and final products. In evaluating the hazards of a solvent, or any other process chemical, it is essential to consider the properties of the material at the processing conditions. For example, a combustible solvent is a major fire hazard if handled above its flash point or boiling point. Some examples of solvent substitutions include:

- Water-based paints and adhesives, replacing solvent-based products.
- Less volatile solvents with a higher flash point, used for agricultural formulations (Ref 4.2 Catanach). In many cases, aqueous or dry flowable formulations for agricultural chemicals can be used instead of organic formulations.
- Aqueous and semi-aqueous cleaning systems used for printed circuit boards and other industrial degreasing operations (Ref 4.19 Mandich; Ref 4.4 Davis).
- Abrasive media cleaning systems, replacing hazardous organic solvents for paint stripping (Ref 4.4 Davis).
- N-Methyl pyrrolidone, dibasic ethers, and organic esters, substituting for more hazardous paint removers (Ref 4.25 Consumer Reports; Ref 4.4 Davis).

The effort to substitute inherently safer and more environmentally-friendly solvents in many industries continues to be active. Scores of solvent substitutions have been identified in a variety of industries, including food processing, textile, wood and furniture, printing and casting/metals processing. The United States EPA developed a guide, *Aqueous and Semi-Aqueous Solvent Chemicals: Environmentally Preferable Choices*, (nepis.epa.gov/) to provide companies with information to help them make responsible environmental choices for chemicals used in metal cleaning. Slater and Salveski developed method to calculate the overall "greenness" of a pharmaceutical process that uses multiple solvents, including the ability to compare the greenness of two different process alternatives. A solvent selection table containing over 60 solvents and associated chemicals common in the pharmaceutical and chemical industries (Ref 4.31 Slater). Although this methodology is intended to help minimize environmental impacts, it has the added benefit of reducing safety risk through the use of less volatile organic solvents.

4.4 REFRIGERANTS

Refrigeration systems require the use of a material that has a high vapor pressure, low flash point, and high heat capacity in order to function successfully as a cooling medium. Unfortunately, most of the materials that have desirable properties as refrigerants also have other properties that introduce undesirable risks. The three major refrigerants used in industrial refrigeration systems are light hydrocarbons (i.e., ethylene, propane, propylene, etc.), anhydrous ammonia, and chlorofluorocarbons (CFC). Each of these materials presents individual hazards that may preclude their use in refrigeration systems. Hydrocarbons, such as propane and propylene, are highly flammable and represent significant fire and explosion hazards if released due to leakage. Anhydrous ammonia is a toxic inhalation hazard, and CFCs, first developed in the 1920's as a safer alternative to ammonia and hydrocarbon-based refrigerants, deplete the earth's protective ozone layer when released to the atmosphere. The choice of refrigerant will depend on availability of the material on-site, ease of maintenance of the refrigeration system, and amount of refrigeration tonnage required for the application. Due to the international banning of CFC usage, many facilities have changed back to the use of hydrocarbon refrigerants. If a hydrocarbon-based refrigeration system only incrementally increases the total amount of flammable hydrocarbons on-site, it may be the best choice. Anhydrous ammonia refrigeration systems have been used extensively in the food processing industry because ammonia is an environmentally-compatible refrigerant with no ozone depletion potential (ODP). Ammonia also has advantageous thermodynamic properties at the temperatures and pressures common to refrigeration applications, resulting in smaller, more compact systems and less energy consumption than other refrigerants when used in large industrial systems.

There has been research in recent years on alternative refrigeration materials that have lower ODP levels, or lower flammability, such as hydrofluorocarbons, hydrochlorofluorocarbons (HCFC), and hydrofluoroethers. These materials can be used on their own or as carriers for flammable materials, resulting in a refrigerant that is less flammable than a pure hydrocarbon, such as propane (i.e., substitution). HCFCs are scheduled for phase-out in the first third of the 21st century under the Montreal Protocol and Section 605 of the US Environmental Protection Agency Clean Air Act and are to be phased out completely by 2020. Liquid nitrogen and liquid carbon dioxide also offer possibilities as refrigerants. These materials do not have the same safety or environment hazards as the materials discussed above. However, they are not risk free, either. Both of these materials are colorless and odorless, unlike ammonia, which has a very low odor threshold in humans and, if released in an enclosed space, would represent a significant asphyxiation hazard (Ref 4.16 Kletz 1998).

4.5 FIREFIGHTING AGENTS

Some of the same issues associated with CFC refrigerants affect some traditional firefighting agents. Bromofluorochlorocarbons (BFC), or Halon (Halomethanes) have been used extensively as firefighting agents, particularly in spaces containing electronic or electrical equipment where other fire suppression agents (i.e., water) would cause serious equipment damage. BFCs have the same adverse effects on the ozone layer as CFCs, and their production has been terminated and use is being phased out. Newer halogenated hydrocarbon materials have been developed, however, they are not as efficient firefighting agents as BFCs, and larger systems, with larger inventories, are required to extinguish the same size fire (i.e., substitution). Carbon dioxide is a good firefighting alternative, however, the concentration needed to extinguish most fires is high enough to cause asphyxiation if someone is in the space and not properly equipped with emergency breathing capability. Halon and other clean agents can be used in concentrations that will not cause asphyxiation (Ref 4.16 Kletz 1998). As of 2010, use of CFCs and Halons for fire protection have been phased out in accordance with the Montreal Protocol. Clean agents with low ozone-depletion potential have been introduced as alternatives, such as 3M's Novec-1230 or FM-200, which are both safe to be used in occupied spaces and have low global-warming potential.

4.6 HEAT TRANSFER MEDIA

Heat transfer oils have commonly been used for process heating at temperatures above their flash point, resulting in the potential for a fire if released. Modern heat transfer fluids with high boiling points are now commercially available which present an inherently safer option (i.e., moderation). Heat transfer oils with boiling points between 275oC and 400oC and working ranges of -50oC to 500oC are commercially available. For temperatures between 200-2000oC, molten salts can be used (i.e., substitution). (Ref 4.17 Kletz 2010).

Water has been used in place of kerosene as a heat transfer fluid in the production of ethylene oxide; higher pressures must be accommodated but the potential hazard associated with a release is significantly less (Ref 4.17 Kletz 2010). In many cases, the use of steam as a heating medium also presents an inherently safer option, eliminating the potential for combustion if released (i.e., substitution). However, the risks associated with high pressure or superheated steam are not insignificant with respect to burns and other severe injuries to individuals in the direct vicinity of a steam leak.

Another example is substitution of a combustible heat transfer fluid (operated above its flash point) with an alternative heating medium is the use of a nonflammable mixture of propylene glycol and water for use in a polymer extrusion process. (Ref 4.6 Edwards).

4.7 INFORMED SUBSTITUTION

The United States Occupational Safety and Health Administration (US OSHA) has published resources for Informed Substitution as guidance for replacement of hazardous materials with safer alternatives, which involves identifying alternatives and evaluating their health and safety hazards, potential trade-offs, and technical and economic feasibility. The result may include using chemical substitutes or product or process redesigns that completely eliminate the need for specific hazardous chemicals. It involves the following a 7-step process within the two major stages, as described below:

1. *Substitution Planning*. Process to systematically set goals and priorities to reduce hazards, develop a chemical use inventory, evaluate alternatives, identify preferred alternatives, and implement alternatives.
2. *Alternatives Assessment*. Process to identify, compare, and select safer alternatives for hazardous chemicals on the basis of their hazards, performance, and economic viability. An alternatives

assessment is a key component of a substitution planning process and is used in the evaluation and comparison of alternatives.

A number of tools and methods are available for evaluating hazards, performance, and economic viability as part of an alternatives assessment. For example, there are **comparative chemical hazard assessment** tools which provide a method to compare chemical alternatives on the basis of their chemical hazards. (Ref 4.40 US OSHA)

The steps for conducting an Informed Substitution process are as follows:

1. Form a team to develop a plan
2. Examine current chemical use
3. Identify alternatives
4. Assess and compare alternatives
5. Select a safer alternative
6. Pilot the alternative
7. Implement and evaluate the alternative

There are a number of resources for researching alternatives to hazardous chemicals. These resources include chemical substitution case studies and alternatives, including:

- SUBSPORT - Substitution Support Portal – Moving Towards Safer Alternatives (www.subsport.eu/case-stories-database)
- Intergovernmental Forum on Chemical Safety (IFCS) – Substitution and Alternatives Case Studies, Examples and Tools (www.who.int/ifcs/documents/standingcommittee/substitution/en/)
- University of Massachusetts Lowell Toxics Use Reduction Institute (TURI) Case Studies (www.turi.org/TURI_Publications/Case_Studies)
- ISTAS Risctox (in Spanish only) (www.istas.net/web/abreenlace.asp?idenlace=3911)
- U.S. Environmental Protection Agency – Safer Chemical Ingredients List (www.epa.gov/saferchoice/safer-ingredients)

Completed alternatives assessments (Interstate Chemicals Clearinghouse, www.theic2.org/aa-wiki-archive) are also included. Performing a single search of multiple online resources related to substitution is another rapid way to identify possible alternatives. Guidance is also provided for assessing hazards, performance and cost (Ref 4.40 US OSHA). Comparative Assessment Tools include the following:

- European Commission, *Minimising Chemical Risk to Workers' Health and Safety Through Substitution*, 2012
- Ontario Toxics Reduction Program, *Reference Tool for Assessing Safer Chemical Alternatives*, 2012

Although the focus is on reducing occupational (and household) exposure to hazardous chemicals as well as environmental protection, many of these tools are applicable to chemical processes and process safety as well.

The Ontario Toxics Reduction Program is intended to promote green chemistry and engineering and has developed a valuable reference tool for evaluating chemical substitution using the framework shown in Figure 4.1. This tool is intended to be general for use in industries and companies of various sizes.

It should be noted that several processes for assessing the use of alternative chemicals have been developed, however there is currently no standard method established. Defining metrics for selection of the most attractive alternative is a major challenge; this is particularly important given the tradeoffs that are often necessary when substituting a hazardous chemical with a safer alternative. (Ref 4.23 Ontario)

The U.S. Environmental Protection Agency (EPA) has also developed a process for evaluating safer chemical alternatives, known as Design for the Environment (DfE) Alternatives Assessments. (Ref 4.35 U.S. EPA) This effort is focused primarily on product safety and consists of the following steps:

1. Determine the feasibility of an alternatives assessment
2. Collect information on chemical alternatives
3. Convene stakeholders
4. Identify viable alternatives
5. Conduct the hazard assessment
6. Apply economic and life cycle context
7. Apply the results in decision-making for safer chemical substitutes

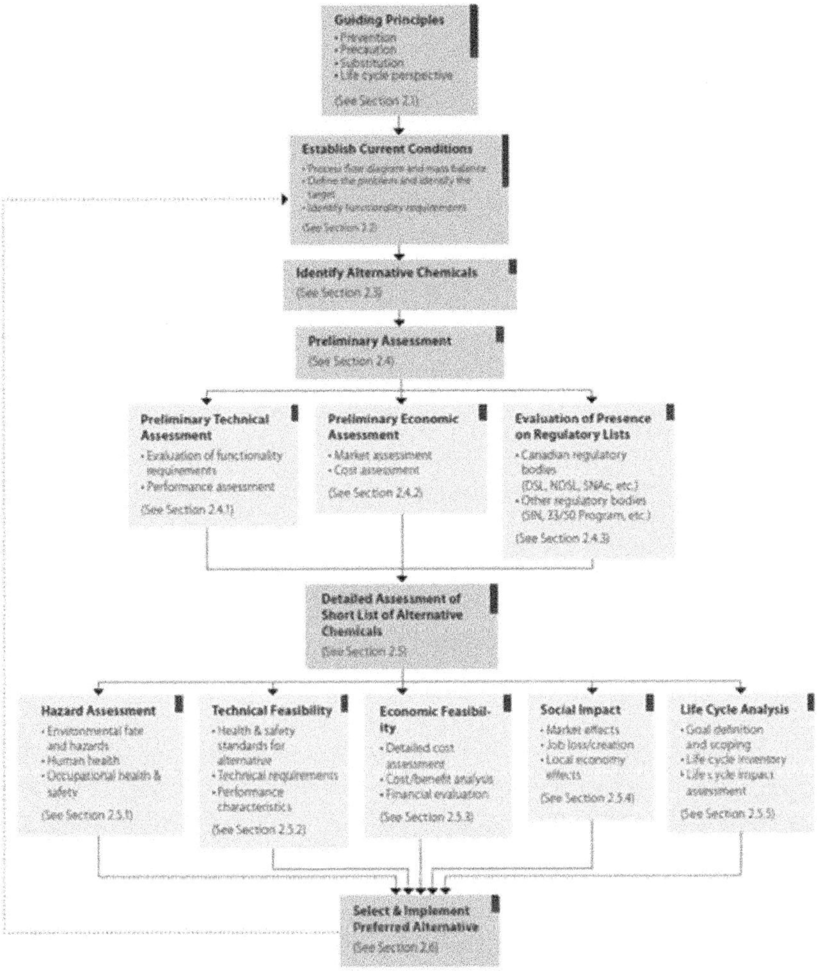

Note: Adapted from Rossi et al. (2006)

Figure 4.1 Framework for Assessing Safer Chemical Alternatives (Ref 4.23 Ontario)

A separate Assessment Criteria for Hazard Evaluation has also been developed which includes toxicological criteria for both human health effects and environmental toxicity and fate. (Ref 4.37 U.S. EPA, Office of Pollution Prevention & Toxics) Completed assessments include those for flame retardants used in flexible polyurethane foam and printed circuit boards. The website of the Canadian Centre for Occupational Health and Safety (CCOHS), Canada's national center for occupational health and safety information, provides guidance for selection of alternate chemicals. This includes hazard assessments for alternative chemicals which should include consideration of the following issues in order to minimize risk (Ref 4.1 CCOHS):

- Vapor pressure
- Short-term health effects
- Long-term health effects
- Skin toxicity
- Sensitization of the respiratory system
- Cancer-causing potential and reproductive effects
- Physical hazards (e.g.,flammability)

The Organization for Economic Cooperation and Development (OECD), an international organization dedicated to promoting policies that will improve the economic and social well-being of people around the world, has also developed a Substitution and Alternatives Assessment Toolbox (SAATOOLBOX), which includes information and resources on chemical substitution and assessment practices and practical guidance on how to conduct assessments, including case studies. Case studies are descriptions of alternative or chemical hazard assessments that have been conducted by manufacturers, academic institutions, NGOs or government bodies, and include evaluation of alternatives for: (Ref 4.24 OECD)

- Flame retardants
- Plasticizers (phthalate-free)
- Solvents
- Nonylphenol ethoxylates for detergents

The REACH (Registration, Evaluation, and Authorisation of Chemicals) regulation in Europe, adopted in 2007, requires manufacturers and importers of certain chemicals above a one metric ton threshold to gather information on the properties of their chemical substances, and to register the information in a central database. It also calls for the progressive substitution of the most dangerous chemicals (referred to as "substances of very high concern") when suitable alternatives have been identified. REACH is discussed further under Regulatory Initiatives in Chapter 12.

4.8 REFERENCES

4.1 Canadian Center for Occupational Health and Safety (CCOHS), *Substitution of Chemicals – Considerations for Selection,* www.ccohs.ca/oshanswers/chemicals/substitution.html.

4.2 Catanach, J.S., and Hampton, S.W., *Solvent and surfactant influence on flash points of pesticide formulations.* ASTM Spec. Tech. Publ. 11, 149-57, 1992.

4.3 Dale, S.E., *Cost-effective design considerations for safer chemical plants.* In J.L. Woodward (ed.). Proceedings of the International Symposium on Preventing Major Chemical Accidents, February 3-5, 1987, Washington, D. C. (pp. 3.79-3.99). New York: American Institute of Chemical Engineers, 1987.

4.4 Davis, G.A., Kincaid, L., Menke, D. , Griffith, B., Jones, S., Brown, K., and Goergen, M., *The Product Side of Pollution Prevention: Evaluating the Potential for Safe Substitutes*. Cincinnati, Ohio: Risk Reduction Engineering Laboratory, Office of Research and Development, U. S. Environmental Protection Agency. 1994.

4.5 DeSimone, J.M., Maury, E.E., Guan, Z., Combes, J.R., Menceloglu, Y.Z., Clark, M.R., et al., *Homogeneous and heterogeneous polymerizations in environmentally-responsible carbon dioxide.* Preprints of Papers Presented at the 208th ACS National Meeting, August 21-25, 1994, Washington, DC (pp. 212-214). Center for Great Lakes Studies, University of Wisconsin-Milwaukee, Milwaukee, WI: Division of Environmental Chemistry, American Chemical Society, 1994.

4.6 Edwards, V. and Chosnek, J., *Making Your Existing Plant Inherently Safer,* Chemical Engineering Progress, January 2012.

4.7 Flam, F., *Laser chemistry: The light choice*. Science 266, 215-217, 14 October 1994.

4.8 Govardhan, C.P., and Margolin, A.L. *Extremozymes for industry: From nature and by design.* Chemistry & Industry, 689-93, 14 October 1994.

4.9 Hall, N., *Chemists clean up synthesis with one-pot reactions*. Science, 266, 32-34, 7 October 1994.

4.10 Hendershot, D.C., *Safety considerations in the design of batch processing plants.* In J. L. Woodward (Ed.). Proceedings of the International Symposium on Preventing Major Chemical Accidents, February 3-5, 1987, Washington, D.C. (pp. 3.2-3.16). New York: American Institute of Chemical Engineers, 1987.

4.11 Hendershot, D.C., *An overview of inherently safer design.* Process Safety Progress, 25 (2), 103-107, 2006.

4.12 Hochheiser, S., *Rohm and Haas History of a Chemical Company*. Philadelphia, PA: University of Pennsylvania Press, 1986.

4.13 Johnston, K.P., *Safer solutions for chemists.* Nature 368, 187-88., 17 March 1994.

4.14 Kharbanda, O.P., and Stallworthy, E.A., *Safety in the Chemical Industry*. London: Heinemann Professional Publishing, Ltd., 1980 and 1988.

4.15 Kletz, T.A., *Plant Design for Safety*. Rugby, Warwickshire, England: The Institution of Chemical Engineers, 1991.

4.16 Kletz, T.A., *Process Plants: A Handbook for Inherently Safer Design.* Philadelphia, PA: Taylor & Francis, 1998.

4.17 Kletz, T., Amyotte, P., Process Plants – A Handbook for Inherently Safer Design, 2nd Ed., CRC Press, 2010.

4.18 Lin, D., Mittelman, A., Halpin, V. and Cannon, D., *Inherently Safer Chemistry: A Guide to Current Industrial Processes to Address High Risk Chemicals*. Washington, DC: Office of Pollution Prevention and Toxics, U. S. Environmental Protection Agency, 1994.

4.19 Mandich, N.V. and Krulik, G.A., Substitution of nonhazardous for hazardous process chemicals in the printed circuit industry. Me. Finish. 90 (11), 49-51, 1992.

4.20 Manzer, L.E., *Chemistry and catalysis.* In P.T. Anastas and C.A. Farris (Eds.). Benign by Design: Alternative Synthetic Design for

Pollution Prevention (pp. 144-154). Washington, D. C.: American Chemical Society, 1994.

4.21 McCarthy, A.J., Ditz, J.M. and Geren, P.M., Inherently safer design review of a new alkylation process. 5th World Congress of Chemical Engineering, July 14-18, 1996, San Diego, CA (Paper 52b). New York: American Institute of Chemical Engineers, 1996.

4.22 Norton Engineering, *Alkylation Technology Study: Final Report*, South Coast Air Quality Management District, 9/9/16

4.23 Ontario Toxics Reduction Program, Reference Tool for Assessing Safer Chemical Alternatives, 2012.

4.24 Organization for Economic Cooperation and Development (OECD), *Substitution and Alternatives Assessment Toolbox,* www.oecdsaatoolbox.org/

4.25 Paint Removers: New products eliminate old hazards, Consumer Reports, 340-343, May 1991.

4.26 Puranik, S.A., Hathi, K.K., and Sengupta, R., *Prevention of hazards through technological alternatives.* In Safety and Loss Prevention in the Chemical and Oil Processing Industries, October 23-27, 1989, Singapore (pp. 581-587). IChemE Symposium Series, no. 120. Rugby, Warwickshire, U. K.: The Institution of Chemical Engineers, 1990.

4.27 Rogers, R.L., and Hallam, S., *A chemical approach to inherent safety.* In IChemE Symposium Series, No. 124, 235-41, 1991.

4.28 Savage, P.E., Gopalan, S., Mizan, T.I., Martino, C.J., and Brock, E.E., *Reactions at supercritical conditions: Applications and fundamentals. AIChE Journal* 41 (7), 1723-1778, 1995.

4.29 Sherringron, D.C., *Polymer supported systems: Towards clean chemistry?* Chemistry and Industry, 15-19, 7 January 1991.

4.30 Slater and Salveski, *A method to characterize the greenness of solvents used in pharmaceutical manufacture,* Journal of Environmental Science and Health Part A, July 26, 2007.

4.31 Starks, C.M., and Liotta, C., *Phase Transfer Catalysis Principles and Techniques*. New York: Academic Press, 1978.

4.32 Starks, C.M., *Phase transfer catalysis: An overview.* In C.M. Starks (Ed.). *Phase Transfer Catalysis: New Chemistry, Catalysts and Applications, September 8, 1985* (ACS Symposium Series No. 326). Washington, D.C.: American Chemical Society, 1987.

4.33 Sundell, M.J., and Nasman, J.H., *Anchoring catalytic functionality on a polymer.* Chemtech, 23 (12), 16-23, 1993.

4.34 Tietze, L.F., *Domino reactions in organic synthesis.* Chemistry & Industry, 453-457, 19 June 1995.

4.35 U.S. Environmental Protection Agency, *Design for the Environment Alternatives Assessments,* www.epa.gov/saferchoice/design-environment-alternatives-assessments.

4.36 U.S. Environmental Protection Agency and Occupational Safety and Health Administration, *Chemical Safety Alert: Safer Technology and Alternatives,* June, 2015.

4.37 U.S. Environmental Protection Agency, Office of Pollution Prevention & Toxics, *Design for the Environment Program Alternatives: Assessment Criteria for Hazard Evaluation, v2.0,* August, 2011.

4.38 U.S. Environmental Protection Agency, *Presidential Green Chemistry Challenge : 1996 Greener Reaction Conditions Award – The Dow Chemical Company,* www.epa.gov/greenchemistry/presidential-green-chemistry-challenge-1996-greener-reaction-conditions-award

4.39 U.S. Environmental Protection Agency, *Presidential Green Chemistry Challenge: 1996 Greener Synthetic Pathways Award (Monsanto),* www.epa.gov/greenchemistry/presidential-green-chemistry-challenge-1996-greener-synthetic-pathways-award

4.40 U.S. Occupational Safety and Health Administration, *Transitioning to Safer Chemicals: A Toolkit for Employers and Workers* (www.osha.gov/dsg/safer_chemicals/index.html).

4.41 Walsh, F., and Mills, G., *Electrochemical techniques for a cleaner environment.* Chemistry and Industry, 576-579, 2 August 1993.

5. Moderate - An Inherently Safer Strategy

In the context of inherent safety, *moderate*, also called attenuation, means using materials under less hazardous or energetic conditions, or designing the plant to mitigate the impact of an incident arising from the hazard (Ref 5.12 Lees). Moderation of conditions can be accomplished either physically (i.e., lower temperatures, dilution) or chemically (i.e., development of a reaction chemistry which operates at less severe conditions). Effects of incidents may be tempered (or diminished) by physical containment (containment buildings or separate equipment rooms). They may also be moderated by segregation. Where possible, providing greater distances between equipment can prevent domino or knock-on effects. Additionally, locating an entire process sufficiently far away from personnel (i.e. occupied spaces) can potentially minimize the human consequences of process safety incident. Providing buffer zones between the processes and public or environmental receptors can reduce or eliminate the effects of process safety incidents.

5.1 DILUTION

Dilution reduces the hazards associated with the transport, storage and use of a low boiling hazardous material in two ways:

1. By reducing the storage pressure.
2. By reducing the initial atmospheric concentration if a release occurs.

Some materials can be handled in a dilute form to reduce the risk of handling and storage:

- Aqueous ammonia or methylamine in place of the anhydrous material.
- Hydrochloric or hydrofluoric acid in place of anhydrous HCl or HF.
- Dilute nitric acid or sulfuric acid in place of concentrated fuming nitric acid or oleum (SO3 solution in sulfuric acid).

If a chemical process requires the concentrated form of the material, it may be feasible to store a more dilute form, and then concentrate the material by distillation or some other technique in the plant prior to introduction to the process. The inventory of material with greater intrinsic hazard (i.e., undiluted) is reduced to the minimum amount required to operate the process, however the tradeoff of including a distillation step may add a new hazardous process.

Materials that boil below normal ambient temperature are often stored in pressurized systems under their vapor pressure at the ambient temperature. The pressure in such a storage system can be lowered by diluting the material with a higher boiling solvent. This reduces the static pressure imposed on the storage container, as well as the pressure difference between the storage system and the outside environment, thereby reducing the rate of release in case of a leak in the system. If there is a loss of containment incident, the atmospheric concentration of the hazardous material at the spill location and the downwind atmospheric concentration and hazard zone are thereby reduced. For example, the use of modified HF which has a lower vapor pressure, and commensurate reduced airborne exposure hazard, has also come into more common use in recent years.

Chemical reactions are sometimes conducted in a dilute solution to moderate reaction rates, to provide a heat sink for an exothermic reaction, or to limit the maximum reaction temperature by tempering the reaction. In this example, there are conflicting inherent safety goals - the solvent moderates the chemical reaction, but the dilute system will be dimensionally significantly larger for a given production volume. Careful evaluation of all process risks is required to select the best overall approach.

5.2 REFRIGERATION

Many hazardous materials, such as ammonia and chlorine, can be stored at or below their atmospheric boiling points with refrigeration. Refrigerated storage reduces the magnitude of the consequences of a release from a hazardous material storage facility in three ways:

1. By reducing the storage pressure

2. By reducing the immediate vaporization of leaking material and the subsequent evolution of vapors from the spilled pool of liquid
3. By reducing or eliminating liquid aerosol formation from a leak

Refrigeration, like dilution, reduces the vapor pressure of the material being stored, which, in turn, reduces the driving force (pressure differential) of a release / leak from primary containment to the outside environment. If possible, the hazardous material should be cooled to or below its atmospheric pressure boiling point. At this temperature, the rate of flow of a liquid leak will depend only on liquid head or pressure, with no contribution from the vapor pressure of the material. The flow through any hole in the vapor space will be small and limited to breathing and diffusion.

Material stored at or below its atmospheric pressure boiling point has no superheat. Therefore, there will be no initial flash of liquid to vapor in case of a leak. Vaporization will be controlled by the evaporation rate from the pool formed by the leak. This rate can be minimized by the design of secondary containment, for example, by minimizing the surface area of the liquid spilled into the contained (diked) area, or by using insulating concrete containment sides and floors. Because the spilled material is cold, vaporization from the pool will be further reduced. However, materials that are lighter than air (e.g., ammonia) may, when released at cold temperatures, exhibit the dispersion characteristics of denser-than-air gases until they absorb enough heat and become neutrally or positively buoyant. This will inhibit dispersion of the material and increase its ground level concentrations for a certain distance from the release point, and hence increase any human exposure. If the material is flammable, the released vapor cloud will retain an ignitable concentration above the lower flammability limit for a longer period and greater distance.

Many materials, when released from storage in a liquefied state under pressure, form a jet containing an extremely fine liquid aerosol of small droplets, which are possibly micron and sub-micron in size. The fine aerosol droplets may not rain out onto the ground, but instead may be carried downwind as a dense cloud. The amount of material contained in the cloud may be significantly higher (typically double) than would be predicted based on an equilibrium flash calculation assuming all of the liquid phase rains out. This phenomenon has been observed experimentally for many materials, including propane, ammonia, hydrogen fluoride, and monomethylamine. Refrigeration of a liquefied gas to a temperature near its atmospheric pressure boiling point eliminates the two phase flashing jet, and the liquid released will rain out onto the ground. Containment and remediation measures such as spill collection, secondary containment, neutralization, and absorption may then be effective in reducing vaporization of the spilled liquid(Ref 5.3 CCPS).

Figure 5.1 shows an example of a refrigerated storage facility for chlorine. This facility includes a spill collection sump that is covered to reduce evaporation to the atmosphere, both by containing the evaporating vapors and by reducing heat transfer from the surrounding atmosphere. The spill collection sump is vented to a scrubber that collects the evaporated chlorine.

Marshall et al. (Ref 5.15 Marshall) provide a series of case studies that evaluate the benefits of refrigerated storage for six materials – ammonia, butadiene, chlorine, ethylene oxide, propylene oxide, and vinyl chloride. They conclude that, "refrigerated storage is generally safer than pressurized storage" for all the chemicals studied, except ammonia. Ammonia was reported to be an exception "due to a density shift with temperature making it heavier than the surrounding air." Other materials may exhibit similar results, and it is essential that the designer fully understand the consequences of potential incidents.

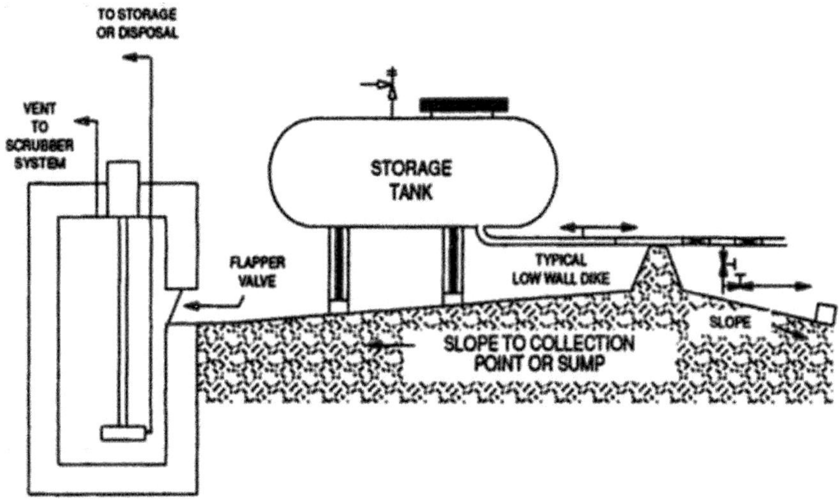

Figure 5.1: A refrigerated chlorine storage system with collection sump with vapor containment (Ref 5.16 Puglionesi))

5.3 LESS ENERGETIC PROCESS CONDITIONS

Operating under less severe conditions, as close to ambient temperature and pressure as possible, increases the inherent safety of a chemical process. Loss of containment events occurring at lower process temperatures and pressures are less energetic and result in releases that are slower and involve less total material. Related to temperature is the phase of the material. Although the temperature of a fluid must be increased (or pressure reduced) to vaporize a material, a loss of containment event in a portion of a process that contains a material in its vapor state will result in less overall material being released. For example, a piping rupture downstream of a chlorine vaporizer will result in a release of chlorine that is less than a corresponding rupture of liquid chlorine piping over the same period.

Some examples of process improvements that have resulted in the moderation of process conditions include:

- Improvements in ammonia manufacturing processes have reduced operating pressures. In the 1930s, ammonia plants operated at pressures as high as 600 bar. In the 1950s, process improvements had reduced operating pressures to 300-350 bar, and by the 1980s, processes operating in the 100-150 bar range were being built. Besides being safer, the lower pressure plants are also cheaper and more efficient (Ref 5.10 Kharbanda).

- Improvements in the production of phenol from cumene have reduced operating temperatures so that the process is not within 10°C of the temperature where a runaway reaction takes place. This eliminates the need for an emergency dump system and its associated active control measures.
- Continuous loop reactors for the nitration of hydrocarbons. Nitration processes have historically been among the most hazardous in the chemical/processing industry. The use of a loop reactor with sulfuric acid as a diluent lowers the total inventory in the process, accomplishing minimization at the same time. The reaction is limited to a few seconds in a specific part of the reactor, thus limiting the contact time between the nitric acid and the hydrocarbon. The dilution ratio of approximately 30:1 precludes concentration of reactants that can support a runaway reaction, and the maximum temperature of the process is 15 °C. Sulfuric acid is a hazardous material, but it has a low vapor pressure so, if released it does not represent a toxic inhalation hazard (Ref 5.11 Kletz). The hazard to humans is limited to direct contact with the material.
- Catalyst improvements allow methanol plants and plants using the Oxo process (a hydroformulation reaction with carbon monoxide, hydrogen and olefinic substrates) for aldehyde production to operate at lower pressures. The process also has a higher yield and produces a better-quality product (Ref 5.4 Dale).
- Improvements in polyolefin manufacturing technology have resulted in lower operating pressures (Ref 5.2, Althaus; Ref 5.4 Dale).
- Use of a higher boiling solvent may reduce the normal operating pressure of a process and will also reduce the maximum pressure resulting from an uncontrolled or runaway reaction (Ref 5.18 Wilday). For example, glymes or high boiling

isoparaffins can be substituted for diethyl ether in pharmaceutical intermediate manufacture.

- Semi-batch or gradual addition batch processes limit the supply of one or more reactants and increase safety when compared to batch processes in which all reactants are included in the initial batch charges. Therefore, for an exothermic reaction, the total energy of reaction available in the reactor at any time is minimized. However, the inherent safety benefits of semi-batch operation are realized only if the limiting reactant is consumed as it is fed and there is no buildup of unreacted material. A number of process upsets, such as contamination with a reaction inhibitor, operating at too low a temperature or forgetting to charge a catalyst to the reactor or to start the agitator, could result in a buildup of unreacted material. If any upset causing loss of reaction can occur, it is important to ensure and know that the reactants are indeed being consumed as they are fed in order to realize the inherent safety benefits of a semi-batch process. The reactor could be monitored to provide confirmation that the limiting reactant is being consumed, by monitoring some physical property of the batch that is reliably correlated to reaction progress or by on-line analysis (Ref 5.3 CCPS).
- Advances in catalysis have resulted in the development of high yield, low waste manufacturing processes. Catalysts frequently allow the use of less reactive raw materials and intermediates, and less severe processing conditions. High yields and improved selectivity reduce the size of the reactor for a specified production volume (i.e., minimization). High selectivity for the desired product also reduces the size and complexity of the product purification equipment (simplification). It may be possible to develop a catalyst sufficiently selective to make additional purifying of the product unnecessary, as in a process for HCFC-141b (CH3CFCl2) described by Manzer (Ref 5.13 Manzer 1993). Allen (Ref 5.1, Allen), Manzer (Ref 5.13 Manzer; Ref 5.14 Manzer 1994), and Dartt and Davis (Ref 5.5 Dartt) describe several catalytic processes, which are potentially environmentally-friendlier and safer.

- Atorvastatin calcium is a drug that lowers cholesterol by blocking its synthesis in the liver. The key chiral building block in the synthesis of atorvastatin is ethyl (*R*)-4-cyano-3-hydroxybutyrate, known as hydroxynitrile (HN). Traditional commercial processes for HN require a resolution step (where the racemic mixture is separated into its two enantiomers) with 50% maximum yield or syntheses from chiral pool precursors. They also require hydrogen bromide to generate a bromohydrin for cyanation. All previous commercial HN processes ultimately substitute cyanide for the halide under heated alkaline conditions, forming extensive byproducts. These processes also require a difficult high-vacuum fractional distillation to purify the final product, which decreases the yield even further. Codexis has designed an alternative HN process around the exquisite selectivity of enzymes and their ability to catalyze reactions under mild, neutral conditions. The evolved enzymes are so active and stable that Codexis can recover high-quality product by extracting the reaction mixture. The process involves fewer unit operations than earlier processes, most notably averting the need for fractional distillation of the product. The formation of byproducts and the generation of waste is reduced, avoiding hydrogen gas, and reducing the need for solvents and purification equipment. This process utilizes fewer hazardous materials and allows for more moderate operating conditions (aqueous, pH ~7, 25–40 °C, atmospheric pressure) (Ref 5.9 EPA).

5.4 SECONDARY CONTAINMENT - DIKES AND CONTAINMENT BUILDINGS

Most secondary containment systems are considered passive protective systems. They do not eliminate or prevent a spill or leak, but they can significantly moderate the impact without the need for any active countermeasures. Containment systems can be defeated by manual or active design features. For example, a dike may have a drain valve to remove accumulated rainwater, and the valve could leak or be left open. Another example is a door in a containment building that could be left open.

Harris (Ref 5.8 Harris) provides an excellent set of guidelines for the design of storage facilities for liquefied gases that can minimize the potential for vapor clouds. Figure 5.2 shows a liquefied gas storage facility that incorporates many of these principles.

- Minimize the wetted area of the substrate surface
- Minimize pool surface open to atmosphere.
- Reduce heat capacity and/or thermal conductivity of substrate.
- Prevent "slosh over" of containment walls and dikes.
- Avoid rainwater accumulation.
- Keep liquid spills out of sewers.
- Shield the pool surface from the wind.
- Provide vapor removal system to a scrubber or another emission control device.
- Provide a liquid recovery system for the contained volume to storage where possible.
- Avoid direct sunshine on containment surfaces in hot climates.
- Direct spills of flammable materials away from pressurized storage vessels to reduce the risk of a Boiling Liquid Expanding Vapor Explosion (BLEVE).
- Provide sealed below-grade collection sumps directly below or adjacent to tanks or vessels having volatile toxic materials that will rapidly collect and contain liquids and vapors that are released. The aforementioned Figure 5.1 shows an example of a collection sump system for chlorine releases (Ref 5.3 CCPS).
- Provide below-grade collection sumps for flammable or combustible materials released from tanks or vessels that will allow the released materials to burn harmlessly without any effects on other equipment, containment systems, or people. Figure 5.3 shows an example of a collection sump system, with a fire pit, for flammable/combustible liquid releases (Ref 5.3 CCPS).

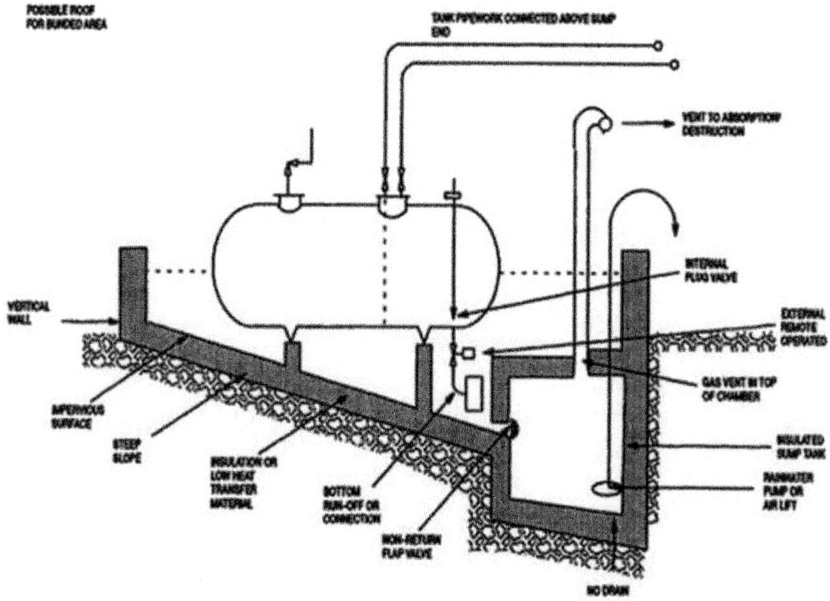

Figure 5.2: A liquefied gas storage facility (Ref 5.8 Harris)

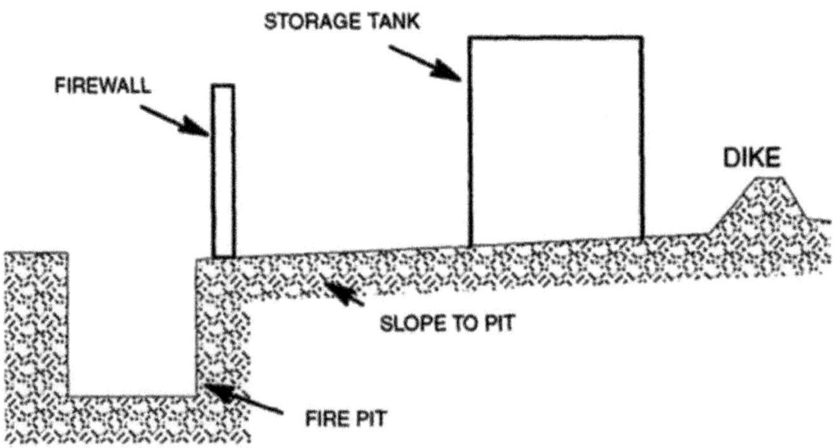

Figure 5.3: A diking design for a flammable liquid (Ref 5.6 Englund).

Containment buildings have been used to limit the impact of loss of containment incidents for many toxic materials, including chlorine and phosgene (Ref 5.3 CCPS) phosphorus oxychloride, and phosphorous trichloride. These buildings can range from a simple, light structure to reduce evaporation of a relatively non-volatile toxic material spill, to a very strong pressure vessel designed to withstand an internal explosion. Englund (Ref 5.6 Englund) describes the evolution in the design of a phosgene handling facility from an open-air plant through various stages of increasing containment, culminating in the design shown in Figure 5.4. The process is totally enclosed in a large pressure vessel capable of withstanding the overpressure of a flammable vapor deflagration.

Containment buildings offer a good example of inherent safety conflicts and tradeoffs. A containment building provides protection outside the building but can also trap and concentrate material from small leaks inside the building, increasing the risk to personnel. Provisions must be made to ensure worker protection for a process located in a containment building, such as monitoring the atmosphere in the containment structure for hazardous vapors, remotely controlling operations from outside the containment structures, restricting access, and requiring proper personal protective equipment when entry into the containment structure becomes necessary.

Frank (Ref 5.7 Frank) and Purdy and Wasilewski (Ref 5.17 Purdy) have published quantitative risk studies that evaluate the benefits of containment buildings for chlorine handling facilities. According to Frank, depending on the relative tightness of the containment building, the relative risk of fatality to a downstream receptor can be reduced by up to 60%.

Great care must be taken when evaluating tradeoffs for a containment building for a flammable and toxic material, such as hydrogen cyanide. A leak or fire inside the building could cause a confined vapor cloud explosion, destroying the building, and possibly increasing the total risk.

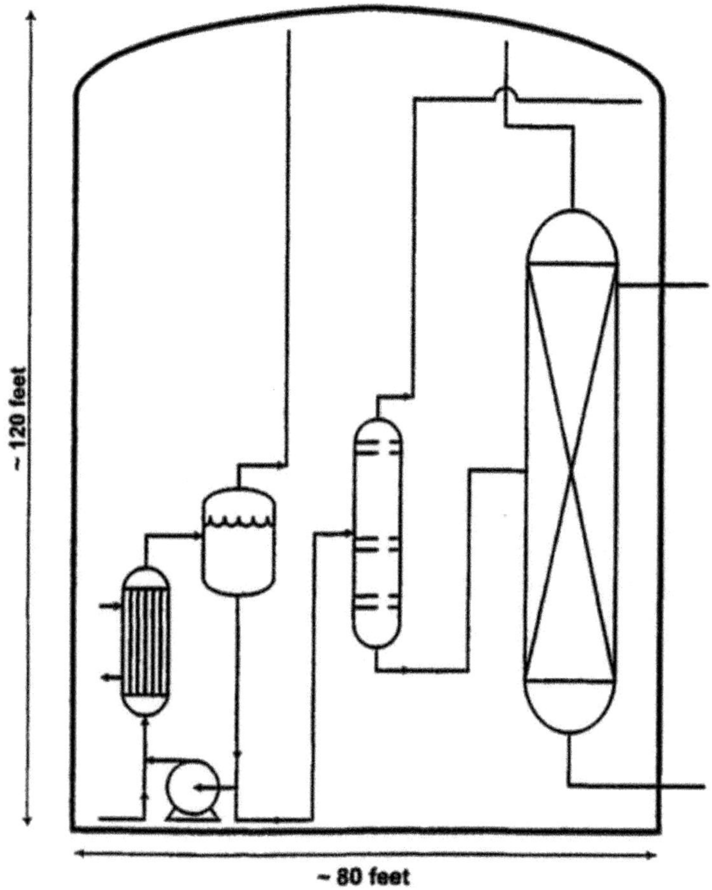

Figure 5.4: A chemical process totally contained in a large pressure vessel (based on Ref 5.6 Englund)

5.5 SEGREGATION

Like secondary containment systems, segregating the process from endpoint receptors (people, sensitive environmental areas, etc.) by the simple use of available space does not eliminate or prevent a spill or leak, but it can significantly moderate the impact without the need for any active device. This is an important design consideration for new facilities, as today's technology more easily allows for remote monitoring and operation of equipment and processes. Nonetheless, it can also be considered at other points in the life cycle of a facility. For example, refineries and chemical plants have begun to relocate their control rooms and other occupied buildings outside of the blast radius and/or toxic footprint of the process release points, to reduce or eliminate the effects of a loss of containment or explosion on personnel.

The 1999 Control of Major Accident Hazards (COMAH) regulation in the UK requires the consideration and documentation of inherently safer design alternatives during the initial design stage. In addition to the standard IS strategies, it includes plant layout as part of the overall approach, particularly to prevent the potential for domino effects by fire, explosion (pressure wave and missiles) or toxic gas cloud that could cause loss of operational control in another location. Plant layout, which is established during the design stage, is often a compromise between a number of factors such as:

- The need to keep distances for materials transfer between plant/storage units to a minimum to reduce costs and risks;
- The geographical limitations of the site;
- Interaction with existing or planned facilities on-site, such as existing roadways, drainage, and utilities routings;
- Interaction with other plants on-site;
- The need for plant operability and maintainability;
- The need to locate hazardous materials facilities as far as possible from site boundaries and from people living in the local neighborhood;
- The need to prevent confinement where release of flammable substances may occur;
- The need to provide access for emergency services;
- The need to provide emergency escape routes for on-site personnel; and,
- The need to provide acceptable working conditions for operators.

For example, these issues are partially dealt with in the US OSHA PSM and US EPA RMP regulations under facility siting, which involves the spatial relationship between process hazards and people but can also be applied to the spatial relationship between hazards and other buildings, equipment, or processes. In its most basic form, inherently safer siting calls for locating people as remotely from hazardous operations as possible, or alternatively, locating hazardous operations in remote areas, away from inhabited ("occupied") areas. For this reason, the manufacture and storage of explosives is usually conducted in unpopulated, remote areas.

5.6 REFERENCES

5.1 Allen, D., *The role of catalysis in industrial waste reduction.* In D.T. Sawyer and A.E. Martell (Eds.). Industrial Environmental Chemistry (pp. 89-98). New York: Plenum Press, 1992.

5.2 Althaus, V. E., and Mahalingam, S., *Inherently safer process designs.* In W.F. Early, V.H. Edwards, and E.A. Waltz (Eds.). South Texas Section AIChE Process Plant Safety Symposium, February 18-19, 1992 (pp. 546-555). Houston, TX: American Institute of Chemical Engineers South Texas Section, 1992.

5.3 Center for Chemical Process Safety (CCPS. *Guidelines for Engineering Design for Process Safety.* New York: American Institute of Chemical Engineers, 1993.

5.4 Dale, S.E. *Cost-effective design considerations for safer chemical plants.* In J.L. Woodward (ed.). Proceedings of the International Symposium on Preventing Major Chemical Accidents, February 3-5, 1987, Washington, D. C. (pp. 3.79-3.99). New York: American Institute of Chemical Engineers (1987).

5.5 Dartt, C.B., and Davis, M.E., *Catalysis for environmentally benign processing.* Ind. Eng. Hem. Res. 33, 2887-299, 1994

5.6 Englund, S. M., *Design and operate plants for inherent safety - Part 1.* Chemical Engineering Progress, 87 (3), 85-91 (1991).

5.7 Frank, W. L, *Evaluation of a containment building for a liquid chlorine unloading facility.* In Proceedings of the 29th Annual Loss Prevention Symposium, July 30 - August 2, 1995, Boston, MA (Paper 5b). E.D. Wixom and R.P. Benedetti (eds.). New York: American Institute of Chemical Engineers, 1995.

5.8 Harris, N.C., *Mitigation of accidental toxic gas releases*. In J. L. Woodward (Ed.). Proceedings of the International Symposium on Preventing Major Chemical Accidents, February 3-5, 1987, Washington, D.C. (pp. 3.139-3.177). New York: American Institute of Chemical Engineers, 1987.

5.9 www.epa.gov/greenchemistry/presidential-green-chemistry-challenge-2006-greener-reaction-conditions-award

5.10 Kharbanda, O.P., and Stallworthy, E.A., *Safety in the Chemical Industry*. London: Heinemann Professional Publishing, Ltd., 1988.

5.11 Kletz, T.A., *Process Plants: A Handbook for Inherently Safer Design*. Philadelphia, PA: Taylor & Francis, 1998.

5.12 Lees, F.P, Loss Prevention in the Process Industries, 4th Edition. Oxford, UK: Butterworth-Heinemann, 2012.

5.13 Manzer, L.E, *Toward catalysis in the 21st century chemical industry*. Catalysis Today, 18, 199-207, 1993.

5.14 Manzer, L.E., *Chemistry and catalysis*. In P.T. Anastas and C.A. Farris (Eds.). Benign by Design: Alternative Synthetic Design for Pollution Prevention (pp. 144-154). Washington, D. C.: American Chemical Society, 1994.

5.15 Marshall, J., Mundt, A., Hult, M., McKealvy, T.C., Myers, P. and Sawyer, J., *The relative risk of pressurized and refrigerated storage for six chemicals.* Process Safety Progress, 14 (3), 200-211, 1995.

5.16 Puglionesi, P.S. and Craig, R.A, *State-of-the-art techniques for chlorine supply release prevention*. In Environmental Analysis, Audits and Assessments: Papers from the 84th Annual Meeting and Exhibition of the Air and Waste Management Association, July 16-21, 1991, Vancouver, B.C. (91-145.5). Pittsburgh, PA: Air and Waste Management Association, 1991.

5.17 Purdy, G., and Wasilewski, M., *Focused risk management for chlorine installations.* Mod. Chlor-Alkali Technol. 6, 32-47, 1995.

5.18 Wilday, A.J., The safe design of chemical plants with no need for pressure relief systems. In IChemE Symposium Series No. 124, 243-53, 1991.

6. Simplify - An Inherently Safer Strategy

In the context of inherent safety, *simplify* means designing and/or operating the process to reduce or eliminate unnecessary complexity in order to reduce or eliminate the chemical hazard. Reducing complexity helps to accomplish a number of goals, for example: 1) minimization or elimination of extra equipment that can fail and contribute to a process safety incident, and 2) reducing or eliminating extra processing steps with hazardous conditions or which may result in a release of the hazardous chemicals or energy, or (3) reducing the use or the number of chemicals necessary in a process. A simpler process is generally safer and more cost-effective than a complex one. Kletz (Ref 6.9 Kletz 1998) offered several reasons why process designs are unnecessarily complex:

- *The need to control hazards.* Instead of avoiding hazards using inherently safer design principles, most designers choose to control them actively using controls, alarms, and safety instrumented systems.
- *The desire for technical elegance.* To some designers, simple equates to crude or primitive, whereas, if carefully designed, a simple process can achieve what it needs to do without excess equipment. A simple process design including only the essential elements to safely carry out its intended task(s) is more elegant than a complicated process that does the same thing.
- *The failure to carry out hazard analyses until late in the design.* PHAs and similar studies performed late in the design usually result in more active controls and equipment rather than more inherently safer solutions. If a preliminary PHA or similar hazard analysis technique is performed at the conceptual stage of the project, it may be easier to incorporate IS techniques, especially substitution or elimination.
- *Following standards and specifications that are no longer appropriate or not completely applicable.* Active solutions to potential hazards that are sometimes contained in design / engineering standards and specifications can accumulate in a

design and create an over-complicated process (or one which relies on control of hazards).

- *Flexibility and redundancy.* While some level of redundancy may be necessary and desirable with basic process equipment, particularly where the failure of the component will have serious effects, this should be limited to what carefully performed PHAs and other studies reveal as the correct level. For every extra pump, heat exchanger, or other basic component, additional controls, utility requirements, piping / valves and other mechanical equipment will follow, thereby greatly expanding the complexity of the process. Additionally, not every risk can or should be solved by specifying some piece of equipment to deal with it.

Kletz (Ref 6.9 Kletz 1998; Ref 6.10 Kletz 2010) also offers some suggestions on use of simple technologies in lieu of high or more recent technologies to solve certain types of problems. One such suggestion is that flare systems should be kept as simple as possible, and not be equipped with other appurtenances, such as flame arrestors, water seals, filters, etc. These components are prone to breakdowns as the result of e.g.,plugging and reducing flare capacity.

These suggestions describe a design philosophy where simple—and sometimes old technologies work just as well as newer, more sophisticated ones. Such a philosophy should be employed wherever possible before resorting to complex solutions.

Examples of simplification are discussed in the following sections. Additional examples can be found in Kletz (Ref 6.8 Kletz 1991; Ref 6.9 Kletz 1998; Ref 6.10 Kletz 2010), and in Chapter 8 of this book.

6.1 LEAVING THINGS OUT

In the spirit of Trevor Kletz's quote "What you don't have can't leak", an effective simplification technique is to combine the functionality of two or more vessels or pieces of equipment into one and leave out the redundant equipment. For example, rather than a separate knockback condenser installed on a reactor vapor line, the vapor line can in some cases be left uninsulated, and the condenser eliminated. In another example, a refrigeration compressor can have a suction-side catch pot eliminated if the vaporizer which precedes it is equipped with a disengaging space, which duplicates the functionality of the catch pot (Ref 6.10 Kletz 2010). By incorporating the technique of reactive distillation, a process which uses a reaction step with subsequent distillation can be simplified by combining the reaction and primary separation steps into one vessel, simplifying the design and eliminating several separate distillation columns. Not only is this design simpler, it also results in greatly reduced in-process inventory (Minimization!), and involves less vessels, piping, and flanges which can leak. An important consideration with reactive distillation is that it does make the process itself slightly more difficult to operate, however the tradeoff in reduction in risk more than offsets this. See Figures 6.1 and 6.2.

6.2 ELIMINATING UNNECESSARY SPARES

In many plants, spare pumps are provided as a standard design, without a proper analysis as to whether each installed spare is truly justified, based on the duty requirements. Non-installed spares are estimated to cost 5 to 6 times less than installed spares, and eliminating unnecessary installed spares simplifies piping installation, and also may eliminate unnecessary contained volume in the piping (an application of the "Minimize" technique. (Ref 6.10 Kletz 2010)

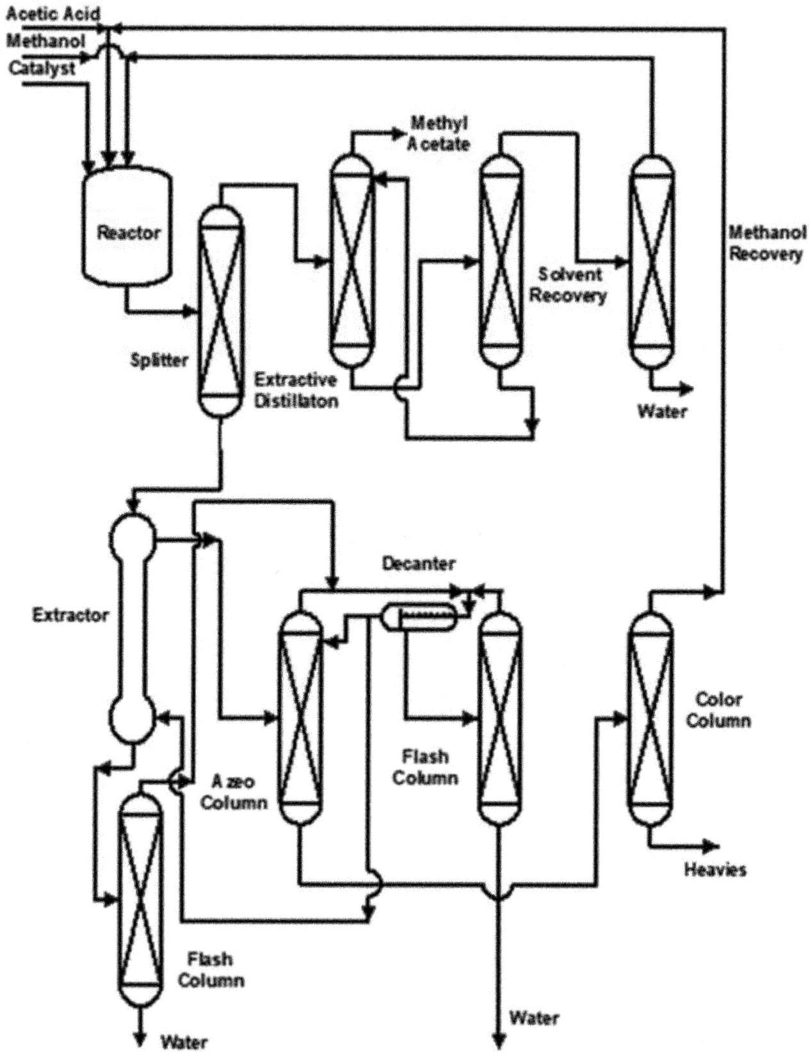

Figure 6.1 – A traditional methyl acetate process using separate reaction and distillation steps (Ref 6.19 Siirola)

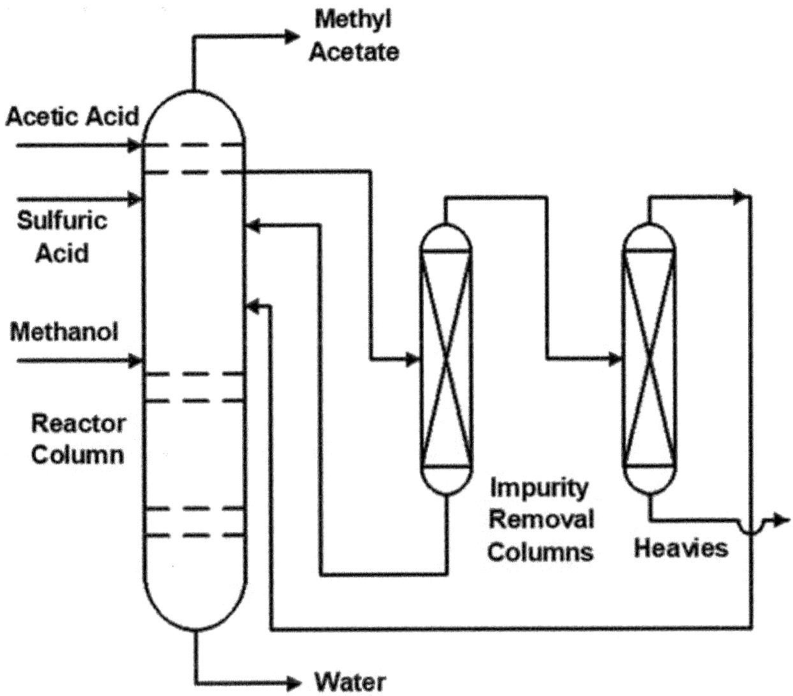

Figure 6.2 – The Eastman Chemical reactive distillation process for methyl acetate US Patent No. US4435595 A (Ref 6.1 Agreda)

6.3 INHERENTLY ROBUST PROCESS EQUIPMENT

In many cases, it is possible to design process equipment that is strong enough to contain the maximum positive or negative pressure (i.e., maximum overpressure or maximum vacuum) resulting from the worst-case process incident(s) (Ref 6.2 CCPS 1993). If such a design, under all feasible circumstances, eliminates the possibility of a loss of containment due to overpressure/underpressure, then it can be considered an inherently safer design. If not, it then only reduces the likelihood of a release, and then it becomes a form of passive safeguard design (although this may still be desirable). Containment of potential overpressure within the process vessel, or elimination of the possibility of vacuum collapse, simplifies the design by eliminating elaborate active safeguards related to pressure detection, control, and relief. Emergency relief devices, such as rupture disks or relief valves, may still be required by regulations and Recognized and Generally Accepted Good Engineering Practices (RAGAGEPs), but their size and relief capacity, as well as hazards associated with the opening of relief devices, may be reduced or eliminated. It may also be possible to eliminate catch tanks, quench systems, scrubbers, flare stacks, or other devices designed to safely discard the effluent from emergency relief systems.

If external fire is a consideration, an inherently safer design would require that the design temperature of equipment be high enough to withstand the temperature and resultant pressure generated by the fire. In general, the temperatures generated by fires will exceed the design temperatures of most materials used in the fabrication of process equipment, and therefore it is very difficult to design process equipment to be inherently robust with respect to a fire. In this instance, fire-proof insulation could be used as passive safeguard protection against external fire.. However, the potential for corrosion-under-insulation (CUI) isis an issue to consider and address, as well as understanding that the integrity of the insulation system is vital to the protection it can provide against fire impingement.

This same concept also applies to internal fires in equipment. For example, if it is possible to structurally design a column containing reactive or pyrophoric packing materials to withstand the temperatures and/or pressures from an internal fire, then the active safeguards required to maintain an oxygen-free environment would not be as critical. In this case, it would still not be advisable to allow an internal fire in a column to develop because of the other personnel and equipment hazards associated with fires. However, if the high temperatures of a fire would not affect the basic integrity of the column itself, the severity of an internal fire would be less, and the criticality of the safeguards would be lower.

Process equipment containing liquid levels should be designed to withstand the maximum hydrostatic load that could be imposed on the equipment if it were completely filled. This is particularly true of tall equipment, such as columns, towers, large reactors, etc. If such a design, under all feasible circumstances, eliminates the possibility of a loss of containment due to overfilling, then it can be considered an inherently safer design. For example, a prominent chemical company designed their columns to withstand complete filling without consequences, save for processing delays. Designing the structure of a column at a high-level hydrostatic load without completely filling it requires that active safeguards be relied upon to protect the column against structural failure.

The concept of inherent robustness also applies to designing equipment to be impervious to the corrosion mechanisms that are present given the materials of construction and within the process, and the operating conditions (i.e., temperature, pH, concentration, viscosity, etc.). The use of specialized alloys will eliminate certain types of corrosive attack. As with the pressure example above, if use of the specialized alloy or material of construction virtually eliminates the possibility of loss of containment due to corrosion/erosion, then it can be considered an inherently safer design. Fiberglass reinforced plastic (FRP), high density polyethylene (HDPE), and steel piping lined with Teflon, PFA, Kynar (PVDF), or other elastomers have increasingly been used in applications where corrosion is particularly severe. However, the use of FRP and HDPE always results in a loss of robustness with respect to temperature and pressure resistance. Lined piping systems often contain many flanges, increasing the likelihood of flange leaks. Therefore, these possibly competing inherent robustness goals (strength and corrosion resistance) must be carefully analyzed and balanced. In these instances, the design becomes a passive safeguard/layer of protection rather than an inherently safer one.

The same concept of inherent robustness can be applied for equipment handling combustible dusts, as the risk of explosion may be high. The maximum overpressure resulting from a deflagration of a combustible dust or flammable vapor in air initially at atmospheric pressure can range from 6 to 9 bar gauge. It may be feasible to build process equipment and structures that are strong enough to contain this type of event. When designing a system for combustion containment, factors such as highly reactive materials, oxygen or other oxidant enriched atmospheres, and congested geometry inside vessels or pipelines that could result in transition to detonation must be considered. These factors can significantly increase the maximum pressure of a combustion reaction.

In summary, when robust equipment design virtually eliminates equipment failure, it can be considered an inherently safer design. Therefore, it fits within the definition of simplification. It is highly effective in eliminating the possibility of an uncontrolled loss of containment. In a general sense, the removal of this possibility from a process design must be inherently safer.

6.4 PREVENTING RUNAWAY REACTIONS

Choosing the addition order of reactants to mitigate or eliminate a potential runaway reaction can reduce the hazard associated with a process and may allow for a simplified emergency relief system. It is essential that the reaction mechanisms, thermodynamics, and kinetics under runaway conditions be thoroughly understood to be confident that the design pressure is sufficiently high for all credible reaction scenarios. All causes of a runaway reaction must be understood, and any side reactions, decompositions, and shifts in reaction paths at the elevated temperatures and pressures experienced under runaway conditions must be evaluated. Many laboratory test devices and procedures are available for evaluating the consequences of runaway reactions (Ref 6.3 CCPS 1995a; Ref 6.4 CCPS 1995b). Several of these reaction hazard testing methods are summarized in Table 6.1 (Ref 6.5, CCPS 2003).

Table 6.1 Summary of Reaction Testing Methods (Ref. 6.5 CCPS 2003)

Hazards Test Stage	Method	Typical Information	Comments
Hazard Screening	Desk Calculation	Reaction enthalpy ΔH_{RXN}	Need formation energy data or must derive it. Must know precise stoichiometry

			Known reactions only, no rate information
Hazard Screening	Mixing Calorimetry	Instantaneous heat of mixing, ΔHMIXING Gas generation rates	Isothermal, from ambient to 150 °C Cannot test multi-phase systems
Hazard Screening	DSC/DTA	Reaction enthalpy, ΔH Reaction 'onset' temp, T_{ONSET}	Very quick (~2 hours), needs little sample No mixing, no pressure data, no multi-phase, although some systems mix the samples by rotating the sample container. Difficult to get representative mixture
Hazard Screening	Adiabatic Screening (includes Accelerating Rate Calorimetry)	$\Delta H_{UNDESIRED}$, T_{ONSET}, ΔT_{ADIAB} P, T, t, dP/dT, dT/dt, Simple kinetics, EA, A	Sample ~ a few grams Reasonably quick to test (~ 1/2 day) Poor/moderate sample agitation Not reliable for scale-up (high Δ-factor)

Develop Desired Reaction	Reaction Calorimetry	$H_{DESIRED}$ Power output, Q_{RXN} Heat transfer rate Accumulation, X_{AC}	Normally 0.1 to 2-liter scale Mimics normal operation Essential information for safe scale-up Very useful for process development
Detailed Hazard Assessment	Low thermal inertia (Δfactor) Adiabatic Calorimeter	$\Delta H_{UNDESIRED}$, T_{ONSET}, ΔT_{ADIAB} dP/dT; dT/dt; T_{SADT}, T_{NR}, t_{MR} estimates Vent sizing data	Sample size ~ 100ml to 1 liter Safe for general laboratory work Good mimic of large-scale runaway. Ideal for 'what-if' scenario study.
Special Studies	High Sensitivity Calorimetry	$\Delta H_{DESIRED}$ $\Delta H_{UNDESIRED}$, dT/dt; ΔT_{ADIAB}, Kinetics, E_A	Sample size 1-50 ml, ΔW/g sensitivity. Shelf life studies by accelerated aging. Combine with low adiabatic to confirm solids low self-heating rate studies.

Runaway reactions in batch reactors can sometimes be avoided by using separate reactor vessels for different stages of the process. In Figure 6.3, four reactants are added to a reactor to make a product. If materials C or D are added to the first stage (when A and B are added), or A or B are added to the second stage (when C and D are added), then a runaway reaction may occur. This situation can be rendered moot by adding materials A and B in one reactor, then piping the resulting materials to a separate reactor vessel where materials C and D are added. A runaway reaction is not possible in the two-reactor design (Ref 6.9 Kletz 1998).

6.5 SIMPLIFYING HEAT TRANSFER

Where possible, cooling systems should be designed so that they can provide adequate heat removal via natural convective cooling. This requires a thorough understanding of the hydraulic conditions created by elevation and temperature differences, and what types of mass and heat transfer mechanisms are established by these differences. Naval and commercial reactors utilize elevation and temperature differences to drive emergency cooling systems, which will keep a shutdown reactor from overheating due to decay heat. The same principles can also be applied in chemical/process system designs. For example, cooling fins can be added to vessels, and "fin-fan" type air coolers can be used which use natural air currents versus forced-fan draft for airflow.

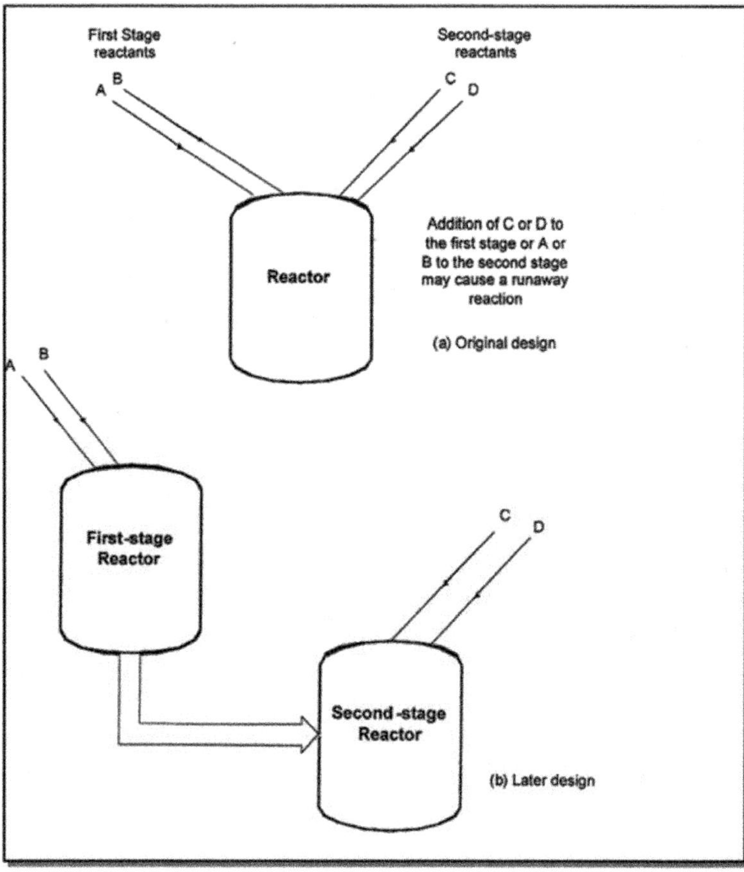

Figure 6.3: Old (a) and new (b) designs for a two-batch reaction system (Ref 6.9 Kletz 1998)

6.6 SIMPLIFYING LIQUID TRANSFER

Liquid transfer systems can be simplified to reduce the hazard. For example, transfer systems that use gravity or pressure / vacuum differentials that occur in the process, are created using inert gases, or use power fluidics, require no moving parts (i.e. pumps / compressors) or seals. Examples include:

- Use of elevated feed tanks for processes. This is an effective method of simplification; however, only if the feed materials are not hazardous, otherwise, the hazard presented by the risk posed by the transfer pumps required to fill the elevated feed tanks could well duplicate that of the feed pumps which were eliminated.

- The use of nitrogen as a motive force to unload transportation containers. This is effective only if the maximum nitrogen pressure cannot overpressure the transportation container, not counting relief valves or active pressure-reducing equipment. Keep in mind, however, with this approach, the consequences of an unloading hose failure may be higher since residual nitrogen pressure (motive force) may remain in the container or piping / hoses.

Two fluidic check valves (also known as diodes) can be combined to produce a fluidic pump. The gas supply is cyclic: when it is high, it pushes liquid into the discharge pipe; when it is low, liquid flows out of the supply tank. See Figure 6.4 below. (Ref 6.10 Kletz 2010)

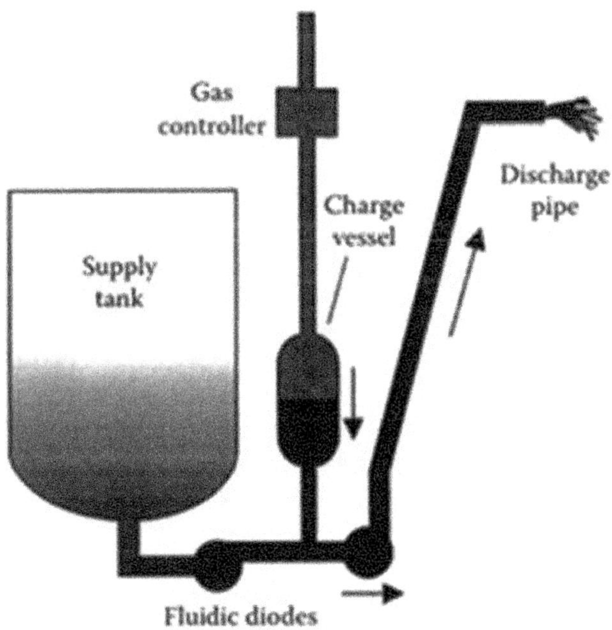

Figure 6.4: Fluidic pump system (Ref 6.10 Kletz 2010)

6.7 REACTOR GEOMETRY

Continuing research in the nuclear power industry has identified systems which utilize improved reactor and core geometry, and novel materials of construction. These research efforts are to more reliably prevent core overheating (Ref 6.6 Forsberg). Similar approaches may be applicable in the chemical/processing industry. For example, maleic anhydride is manufactured by partial oxidation of benzene in a fixed catalyst bed tubular reactor. There is a potential for extremely high temperatures to develop due to thermal runaway if feed ratios are not maintained within safe limits. Catalyst geometry, heat capacity, and partial catalyst deactivation have been used to create a self-regulating mechanism to prevent excessive temperatures (Ref 6.13 Raghaven).

6.8 OPTIMIZING CATALYST SELECTIVITY

Optimization of catalyst selectivity to maximize yield of the desired product is not a new concept; it has been practiced for years for economic optimization of processes. It can also make a process inherently safer if the change to the more selective catalyst also results in a reduction in the chemical hazard or allows for the reduction or elimination of downstream separation equipment.

6.9 SEPARATION OF PROCESS STEPS

A multi-step batch process can be carried out in a single vessel, or in several vessels, each optimized for a single processing step. The complexity of the batch reactor in Figure 6.5, with many potential process fluid and utility interactions, can be greatly reduced by dividing the same process into three vessels as shown in Figure 6.6, similar to the concept introduced in Figure 6.3 above. Again, this is an example of an inherent safety conflict. The system in Figure 6.5 requires only one reactor, although it is extremely complex, and process intermediates never leave the reaction vessel. The system in Figure 6.6 uses three vessels, each of which can be optimally designed for a single task. Although each vessel is considerably simpler, it is necessary to transfer intermediate products from one vessel to another.

If one of those intermediate products is extremely toxic, it may preferable to use the single reactor – a "one pot" process – to avoid transfer of the toxic intermediate. Additionally, this type of minimization and simplification can sometimes adversely affect the process dynamics, resulting in reduced controllability and larger disturbances from normal process operation due to changes in process feed rates or conditions, utility disturbances, or other external environmental factors. Because a "robust" process, one that can tolerate significant changes from its external environment, is also a strategy for inherently safer design, this represents an interesting conflict in designing inherently safer processes. These disturbances in process operation can result in increased safety risks, as well as poorer product quality (Ref 6.11 Luyben). As always, the inherent safety advantages and disadvantages of each system must be evaluated with careful consideration of all the hazards of a particular chemical process, as discussed in Section 2.6

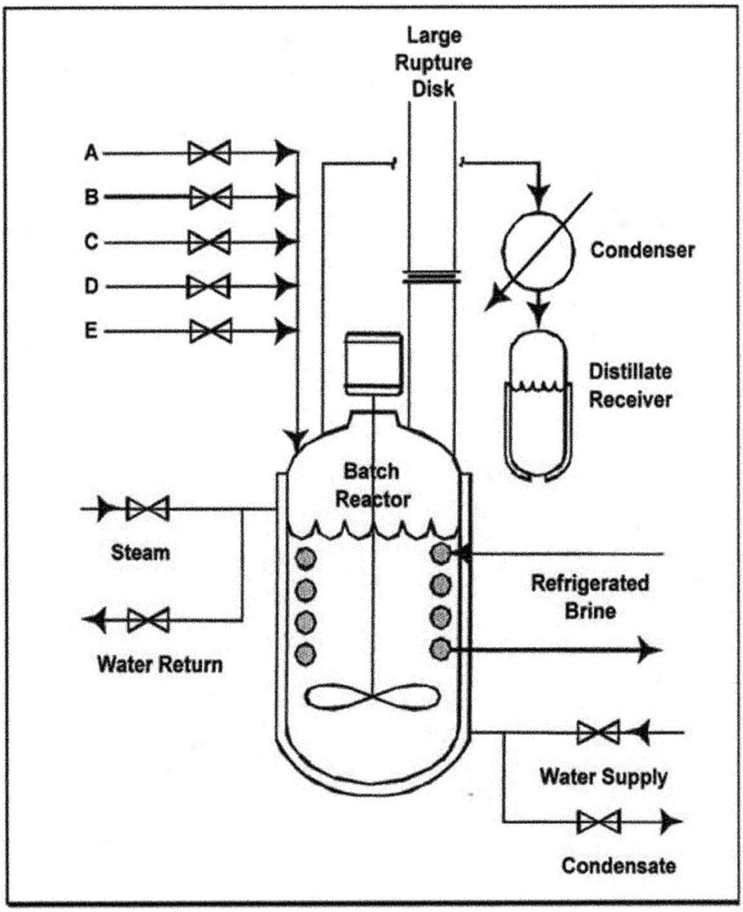

Figure 6.5: A complex batch reactor for a multistep process (Ref. 6.7, Hendershot)

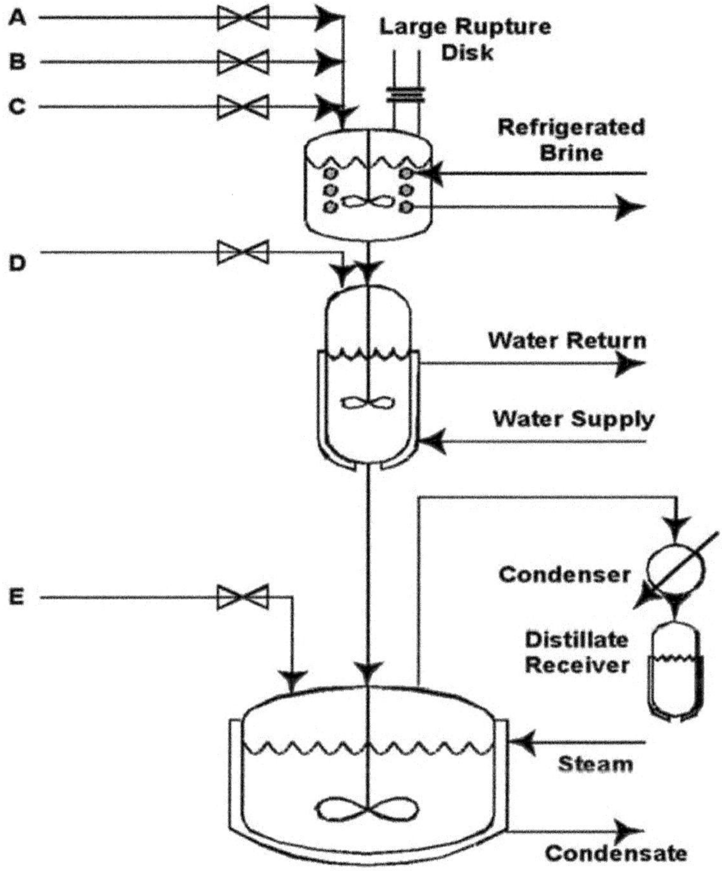

Figure 6.6: The same process as Figure 6.5 in a series of simpler reactors (Ref 6.7 Hendershot)

6.10 LIMITATION OF AVAILABLE ENERGY

Adding energy to a chemical process is often a necessary component of product manufacture. However, in some cases, the energy addition method is not carefully matched to the amount of energy needed. The method of energy addition used can result in excess energy being added because its design does not incorporate inherently safer design principles. Examples of proper matching of required energy are:

- Using immersion heaters that cannot add enough energy to cause a fire in the material being heated or damage the container.
- Using a heating medium for distillation reboiler at a temperature such that it cannot overpressure the tower in case of loss of cooling flow to the condensers.
- Limiting process heating using steam, when possible, to the saturation temperature, which adds the needed amount of heat and no more. If the heating medium maximum heat flux cannot be reduced, the heat transfer area should be adjusted to limit the energy transfer.
- Limiting pump or compressor discharge pressures to less than the downstream relief valve setpoints or the maximum allowable working pressure of any downstream components.
- Ensuring that residual heat cannot be transferred inadvertently to a material via conduction or radiation, such as a hot vessel wall that transfers heat to a material that is sufficient to cause a runaway reaction (Ref 6.9 Kletz 1998).

Some of the examples above also relate to designing equipment that is robust enough to withstand the maximum achievable temperature or pressure.

6.11 SIMPLIFICATION OF THE HUMAN-MACHINE INTERFACE

6.11.1 Overview

In the previous sections, the focus has been on simplifying designs to eliminate or reduce the chemical/stored energy hazard. This section will address the simplification of the human-machine interface; i.e., how humans interact with the process, in order to reduce the likelihood of errors. The human-machine interface includes all aspects of the process (equipment layout, accessibility, operability, maintainability, functionality of controls, etc.), not simply the computer screen or control panel from which the process is operated. This can be considered a subset of Human Factors. For a more comprehensive discussion of Human Factors, see the CCPS book, *Human Factors Methods for Improving Performance In the Process Industries*, 2006.

6.11.2 Equipment Layout, Accessibility, and Operability

Process equipment should be designed in such a way that it is simple and intuitive as possible and be installed such that all equipment is easily accessible. Ground-level installation of instrumentation and other equipment is preferable where possible. Ergonomics, which is one aspect of human factors, should be applied in the layout of equipment, valves, controls, and anything else that operating, and maintenance personnel need to access. Designs that provide good easy access to equipment, while ideally avoiding undo bending, climbing, and stretching, make it more likely that process equipment will be operated properly, and safety equipment will be properly operated in an emergency, particularly if additional highlighting aids are provided such as through signage, color-coding, clear labeling, and/or lighting. Designs should also avoid requiring an operator to change locations when monitoring and controlling a process. For example, an operator should be able to see the temperature indicator for a process from the same location where the operation of the cooling water control valve is performed. Designs and systems should minimize potential harmful exposures in both normal and emergency operations. This consideration affects the location of normal and emergency drains and vents.

The chemistry of a process can be made inherently safer by selecting materials that can better tolerate human error in handling, mixing, and charging. For example, if a concentrated reagent is used in a titration, precision in reading the burette is important. However, if a dilute reagent is used, less precision is needed.

6.11.3 Maintainability

A space station design that requires less "spacewalk" time is inherently safer than a design that requires more. If the astronauts do not need to go outside their space vehicle, they bear less risk. For chemical plants, designs or operating regimes that reduce or eliminate the need for vessel or other process equipment opening, entries, or hazardous levels of maintenance are, in general, inherently safer than those that do not. For example, in one chemical plant, rail cars, tank trucks, and some reactors and storage tanks were cleaned manually by personnel who entered the vessel; fatalities occurred from unexpected or undetected low oxygen content or toxicity. An inherently safer system is a rotating pressurized water spray head that does the cleaning without vessel entry. Another example is to eliminate filters that require changing, to reduce the potential for exposure. This may require a redesigned filter (such as a self-cleaning design), or a process change that eliminates the need for a filter at all.

Equipment that can be reached for inspection, repair, or monitoring from permanent platforms is more likely to be safely inspected, calibrated, repaired, and replaced than equipment that requires climbing with a safety harness or scaffold.

Calibrating equipment usually requires disconnecting it from the process. Equipment that requires less calibration is inherently safer.

- For example, a furnace oxygen analyzer is not protecting the furnace while it is being calibrated. Equipment that can function in abnormal operating conditions is inherently safer than equipment that fails in those conditions.
- An oxygen analyzer that was designed to stop displaying the oxygen content in a furnace when the oxygen content went below 4%. While the oxygen analyzer shutdown tripped the furnace, it left the operators without any indication of the furnace oxygen level blind during the shutdown and delayed the re-start. An analyzer that continued to display the actual oxygen concentration during the upset would be inherently safer.

Equipment should be designed so that there is only one right way to reassemble it.

- If it is important for a pipe sleeve to be right side up, then it could be notched or pinned so it will only go in right side up. One plant experienced a vessel leak caused by a siphon-break hole in a dip tube which was not oriented away from the vessel sidewall. The constant impingement of corrosive liquid into the sidewall from the siphon-break hole resulted in accelerated corrosion/erosion.
- It is common practice to use unique fittings for nitrogen utility service, different from the fittings used for compressed air or water, in order to ensure against cross contamination of the nitrogen system.

- One plant found key relief valves installed backwards after testing / maintenance because the inlet and outlet flanges were identical. They revised the valve and piping flanges, so the relief valves could only be installed in the correct orientation.

6.11.4 Error Prevention

To prevent errors, it is important to make it easier to do the right thing and more difficult to do the wrong thing (Ref 6.12 Norman). If the design and layout of procedures do not clearly indicate what should be done, the resulting confusion can increase the potential for error. Likewise, the design of training programs and materials, including verification of knowledge and skills, can increase or decrease the potential for error.

Systems in which it is easy to make an error should be avoided. For example, to reduce the risk of contaminated product and reworked batches, it is generally better to avoid bringing several chemicals together in a manifold. However, manifolding can be done safely, and may be the best design when all factors are considered, particularly when clear labeling and / or color coding is employed. The alternatives to a manifold should be considered systematically and a decision made on the most inherently safe design.

6.11.5 Design of Equipment and Controls – Making Status Clear

Within the context of inherent safety, "making status clear" means designing controls, indications, alarms and other man-machine interfaces so that there is no confusion, especially during accident or upset conditions, of the status of plant equipment. This concept is a subset of simplification and is a staple of human factors engineering analysis of process equipment. Some examples of making status clear include:

- Arranging indications and control devices so that they are logical and avoid human error. There may be differences in convention with respect to what is "logical", depending on the geographic or industry-specific frame of reference. For example, light switches in the UK are typically switched "down" to turn on the lights, whereas in the US, it is the opposite.
- Designing distributed control system and other digital control system output terminals to be clear and unambiguous through the judicious choice of colors, fonts, and other display

characteristics. Again, there may be differences in convention with respect to these choices. In addition, color-blindness needs to be addressed in the selection of colors.

- Avoiding information overload and difficult navigation in the design of digital control system output terminals.
- Designing indications (gauges and meters) so that they match the ranges of the process parameters being monitored.
- Designing control systems that avoid alarm overload and nuisance alarms. There are formal alarm management programs available to review the control systems and their visual and audible outputs to help rationalize alarm systems.
- Designing field conditions and equipment that allow operators to easily determine the equipment condition. This goal covers environmental conditions, such as lighting and ease of access, as well as intelligent equipment design conventions, i.e., valve handles and stems and operating mechanisms that quickly and easily indicate the position of valves at a visual glance.

6.12 SUMMARY

This chapter describes the last of the four main design strategies for development of inherently safer processes: Simplify. This strategy can be applied at any phase of the process life cycle. Examples of the Simplify strategy, are drawn from all phases in the process life cycle. In the next chapter, we will discuss layers of protection strategies which should be implemented to address the residual risk following the application of the inherent safety strategies discussed here.

6.13 REFERENCES

6.1 Agreda, V. H., Partin, L.R. and Heise, W.H., *High-purity methyl acetate via reactive distillation*. Chemical Engineering Progress, 86 (2), 40-46, 1990.

6.2 Center for Chemical Process Safety (CCPS) *Guidelines for Engineering Design for Process Safety*. New York: American Institute of Chemical Engineers, 1993.

6.3 Center for Chemical Process Safety (CCPS), *Guidelines for Technical Planning for On-Site Emergencies*. New York: American Institute of Chemical Engineers, 1995

6.4 Center for Chemical Process Safety (CCPS), *Guidelines for Chemical Reactivity Evaluation and Application to Process Design*. New York: American Institute of Chemical Engineers, 1995.

6.5 Center for Chemical Process Safety (CCPS), *Essential Practices for Managing Chemical Reactivity Hazards.* New York: American Institute of American Institute of Chemical Engineers, 2003.

6.6 Forsberg, C.W., Moses, D.L., Lewis, E.B., Gibson, R., Pearson, R., Reich, W.J., et al., Proposed and Existing Passive and Inherent Safety-Related Structures, Systems, and Components (Building Blocks) for Advanced Light Water Reactors. Oak Ridge, TN: Oak Ridge National Laboratory, 1989.

6.7 Hendershot, D.C., Safety considerations in the design of batch processing plants. In J. L. Woodward (Ed.). *Proceedings of the International Symposium on Preventing Major Chemical Accidents, February 3-5, 1987, Washington, D.C.* (pp. 3.2-3.16). New York: American Institute of Chemical Engineers, 1987.

6.8 Kletz, T.A., *Plant Design for Safety*. Rugby, Warwickshire, England: The Institution of Chemical Engineers, 1991.

6.9 Kletz, T.A., *Process Plants: A Handbook for Inherently Safer Design*. Philadelphia, PA: Taylor & Francis, 1998.

6.10 Kletz, T.A. and Amyotte, P., Process Plants: A Handbook for Inherently Safer Design, Second Edition. CRC Press, 2010.

6.11 Luyben, W.L. and Hendershot, D.C., *Dynamic disadvantages of intensification in inherently safer process design*. Ind. Eng. Chem. Res., 43 (2), 2004.

6.12 Norman, D.A., *The Psychology of Everyday Things*. New York: Basic Books, 1988.

6.13 Raghaven, K.V., *Temperature runaway in fixed bed reactors: Online and offline checks for intrinsic safety.* Journal of Loss Prevention in the Process Industries, 5 (3), 153-159, 1992.

6.14 Siirola, J.J., *An industrial perspective on process synthesis.* In AIChE Symposium Series, 91, 222-233, 1995.

7. Applying Inherent Safety Strategies to Protection Layers

In the previous four chapters, the IS strategies of Minimization, Substitution, Moderation, and Simplification were applied to the design of a process, with the focus being to reduce or eliminate the chemical or energy hazard. These IS strategies can also be applied to traditional protection layers. In Chapter 2, Section 2.7, Layers of Protection are defined as "a concept whereby a device, system, or human action is provided to reduce the likelihood and/or severity of a specific loss event." Layers of Protection themselves are not an inherent safety design concept; they are "bolt-ons" to an existing design. Unlike inherently safer designs, they do not "virtually eliminate" either the inherent chemical hazard itself, or the possibility of a release of chemicals or energy from the particular part of the process in which they are employed. They provide an additional safety margin and are either active or passive. Active layers of protection require external input from other systems (e.g., power) or humans to function, while passive layers do not require any external input to function. They reduce overall risk by (typically) reducing the frequency of a potentially hazardous loss of containment and/or release of chemicals or energy, but do not in and among themselves reduce the inherent chemical hazard. However, they can be made more effective by the application of IS strategies, as shown in the examples below:

- *Substitution*: A carbon steel piping system typically requires painting to protect it from the elements and stave off exterior corrosion, which if allowed to progress unchecked could lead to a loss of containment. Substituting a higher alloy (e.g., stainless steel) for the carbon steel can eliminate the need for painting and greatly reduce or eliminate the exterior corrosion. Alloy selection is critical depending on the atmospheric conditions. It is quite possible that the higher-alloy system will cost less over the service life due to lower maintenance costs, even with a higher installed cost.

- *Moderation*: A sampling procedure for a hot process stream requires that the operator don heavy and cumbersome protective gloves and a face shield prior to opening the sample valve. A sample cooler (with local temperature indicator) can be installed, reducing the risk of thermal burns to the operator
- *Minimization*: A safe startup procedure that requires the operator to walk up and down the stairs three times to manipulate valves in the correct sequence—and where a hazard could occur if taken in the incorrect sequence—can be made safer by locating the valves so that operator must walk up the stairs only once during the startup, reducing the frequency of errors per operation.
- *Simplification*: (Ref 7.5 Kletz 1998; Ref 7.6 Kletz 2010) Fire protection systems should be as simple as possible. For example, technology for detecting fires by using sophisticated detectors to measure heat, ultraviolet radiation for fires, light absorption by smoke, detection of smoke using ionizing radiation, etc. is often less reliable than a simple detector that allows the fire to burn through a filament and break an electrical circuit, or to melt a plastic tube that contains air under pressure and the resultant low pressure then activates an alarm or trips a sprinkler or deluge system. Fire suppression systems that involve complex piping systems with many sprinkler nozzles can be prone to plugging and difficult to maintain. A simpler, single-flow, manually-operated monitor device that can be directed to drench an area is a simpler solution, though it usually requires a larger supply of water. Fire insulation (fireproofing) is even simpler, is completely passive if installed and maintained properly, and not prone to water supply failures or human error, although the potential for corrosion-under-fireproofing (CUF), as well as the need for periodic inspection and repair, needs to be considered and addressed.

7.1 OPERATING PROCEDURES

Human intervention, consistent with documented operating procedures (including emergency procedures), is a key layer of protection for any process. Procedures that are not followed due to obsolescence, inaccuracy, unavailability, or difficulty in implementing often present process safety risks. The *Simplification* and *Minimization* strategies can be applied when developing or revising operating procedures. Applying inherently safer techniques to the design of procedures requires consideration of the following (Ref 7.3 CCPS 2006):

- *Completeness and accuracy*: Does the procedure have enough information for the user to perform the task safely and correctly?
- *Appropriate level of detail*: Has the level of detail considered the experience and capabilities of the users, their training, and their responsibilities? Leaving detail out of a procedure in lieu of relying on user training is a critical aspect of writing procedures.
- *Conciseness*: Conciseness means eliminating detail and language that does not contribute to work performance, safety, or quality. Conciseness also means segregating “need-to-know" from the "nice-to-know" information. There is a balance between conciseness and appropriate level of detail.
- *Consistent presentation*: This element ensures that the procedure is readily comprehensible. It demands the use of:
    - A consistent terminology for naming components and operations, with corresponding labels in the field.
    - A standard, effective format and page layout.
    - A vocabulary and sentence structure suitable for the intended user. Writing documents at a targeted level of reading comprehension is possible.
- *Administrative control*: All procedures should be reviewed thoroughly before use and periodically thereafter. A “Job Cycle Check” is an effective means followed in the industry to ensure personnel are periodically practicing the procedures, and also helps get feedback on ease of operating with the procedures.

CCPS (Ref 7.2 CCPS 2006) includes guidelines for when a procedure is required, as well as an example procedure checklist.

The use of Information Mapping (Ref 7.4 Horn), a technique for organizing and labeling information for easy comprehension, use, and recall can be a useful tool in the development of procedures that incorporate IS concepts. Information Mapping is a structured approach for the creation of clear, concise and highly usable information focused on the target audience. It does this by analyzing, organizing and presenting the information based on the needs of the user and the purpose of the information, and is both subject matter and media independent. It is a way of thinking and communicating in which information developers approach the content with a set of systematic principles and techniques to ensure that it can be readily used. For example, operating procedures which have been information mapped will have a set template, format, structure, and contents. Personnel from differing units across a plant site (or in differing plant sites across an organization) are able to quickly find important information (such as emergency shutdown or safety and health effects), since it should always be in the same location/chapter/section of each procedure.

7.2 MAINTENANCE PROCEDURES

Maintenance procedures also require the same attention to human-machine interface considerations as operating procedures, as maintenance errors can be just as costly as operating errors. Designing or purchasing equipment that is easily maintained, and that is difficult or impossible to assemble or install incorrectly, improves the inherent safety of the equipment, and thereby the process. For example, seal-less pumps eliminate the potential for seal leaks and their associated hazards. This is the application of the *Substitution* IS strategy. However, it should be recognized that there is frequently a tradeoff between hazards; the seal-less pump that eliminates seal leaks may bring its own set of hazards, such as rapid temperature build-up if it runs dry or is operated in a dead-head condition. Additional protection layers may be needed to avoid the new hazardous consequences.

7.3 RELOCATION

Relocating process equipment to a less-hazardous location can lower the design requirements and simplify the installation. This is the application of the *Moderation* strategy. For example, electrical control equipment or switchgear can be relocated outside of the classified electrical area, rather than be designed for it. This makes the installation inherently safer (completely removes the risk of ignition rather than reducing it), and less expensive (standard electrical enclosure vs. one designed for classified locations). (Ref 7.6 Kletz 2010).

Relocating personnel whom could be potentially impacted from a fire, explosion, or toxic release is another moderation strategy which can be employed. One refinery moved their control building and plant personnel offices to a remote location (across the street, well away from the toxic and blast/fire zones) and purchased property around the site to create a buffer zone. This approach is a common facility siting technique which does not remove the chemical hazard but separates people from the hazard.

7.4 CONTAINMENT

If it is not feasible to contain a runaway reaction within the reactor, it may be possible to *moderate* the consequences, by piping the emergency device effluent to a separate pressure vessel for containment and subsequent treatment. Quench drums, vapor-liquid separation vessels, vapor-liquid separators, and other similar devices can be used to contain the effluent from exothermic / runaway reactions (Ref 7.1 CCPS 1993).

Adequate secondary containment for tanks, vessels, or for entire process units is also an application of the *moderation* IS strategy, as it prevents the spread of liquid releases and minimizes the surface area for evaporation. Common containment structures that surround multiple tanks should be avoided or minimized if possible. The ongoing integrity of containment structures is also an important issue, particularly for earthen berms that can settle, erode, or otherwise weaken or lose their design capacity. Penetrations through secondary containment walls should be avoided or sealed properly.

Another example of applying the *moderation* IS strategy to a layer of protection is the use of blast walls, heat shields, and other barriers to absorb the energy from explosions and limit their radius of effect, or to absorb other potentially hazardous energy sources such as sound and thermal energy, as illustrated by the following:

- At one plant, operators were required to monitor a bulk solids railcar unloading operation. The pneumatic blower and hydraulic vibrator used for the task created a very high noise area around the railcar, requiring the operator to wear both earplugs and earmuffs as they monitored the unloading operation. It became common practice to "monitor" from farther and farther away from the railcar (to escape the noise), at a distance well in excess of the regulatory requirements. An operator's shed was installed with very effective sound insulation, which (along with the necessary air conditioner!) allowed the operators to closely monitor the process in comfort and safety.
- A facility which manufactures rocket propellant designed their processing building (in which the propellant was formulated and mixed) with large earthen berms surrounding the building, to absorb the force of any explosions.

7.5 MORE ROBUST PROCESS EQUIPMENT AND DESIGN

In parallel with the concept of inherently robust equipment design introduced in Chapter 6, in which losses of containment were made virtually impossible, applying the *substitution* strategy, by using more robust materials of construction or more robust designs can reduce the likelihood of system or component failure, consequently reducing the likelihood of losses of containment, toxic releases, fires, and explosions. In previous examples, higher alloy materials of construction have been shown to provide better corrosion resistance but can also provide increased toughness and fatigue resistance. Improved casing designs for rotating equipment can reduce the likelihood of losses of primary containment in the event of rotating element failure. Pump designs with double mechanical seals are more robust than single seal pumps. Seal-less pumps greatly reduce the risk of a process fluid leak, but they also introduce new hazards and concerns, such as overheating and internal leakage, which may be very rapid. Newer seal-less pump designs incorporate advanced thermal protection shutdown devices to prevent overheating. Other more robust pump designs, depending on the application, include diaphragm pumps, jet pumps, and eductors. Using smart transmitters instead of older analog transmitters in a given application make the control function more reliable.

The shell and tube sides of liquid-to-liquid heat exchangers, and the tubes of air-cooled heat exchangers can be designed to contain the maximum attainable pressure on either side. Such a design eliminates reliance on pressure relief devices or other active safeguards for overpressure protection, although these may still be necessary per regulatory requirements for heat exchanger shells which are designed as pressure vessels. (Ref 7.5 Kletz 1998; Ref 7.6 Kletz 2010)

External sight glasses provide simple level detection and can outperform other level devices when multi-layer process streams are encountered (such as a hydrocarbon-intermediate layer-water mixture in a decanting vessel). However, they can be prone to external damage with the possibility of loss of containment. Newer, armored sight glass designs (whether external or built into the face of the vessel) greatly reduce the hazard of breakage while still allowing for reliable level detection. In addition, there are new options available (surface coatings and/or different glass formulations) to reduce or eliminate chemical attack of the glass itself.

7.6 SIMPLIFIED PROCESS EQUIPMENT AND DESIGN

In addition to making process equipment and designs more robust, they can be simplifiedsimplified, eliminating needless complication, and reducing the likelihood of failure (less is more!). Consider the following examples:

- Buildings containing compressors that process flammable gases should be open to natural ventilation. This usually means that the walls are eliminated and replaced with partially open panels, and high points are vented properly. Completely enclosing a flammable gas compressor and relying on a series of active controls to detect and trip the machine when leaks occur is not as reliable. (Ref 7.5 Kletz 1998; Ref 7.6 Kletz 2010)
- Oxidation processes usually require measurement of process flow stream oxygen levels. For mixed gas streams, where other vapors or liquids are present, this measurement can be difficult. Complicated systems that scrub these other materials out of the stream to be measured generally do not work well, so oxygen monitors provide inaccurate measurements. Sometimes a

simple solution, like an uninsulated vertical length of pipe that allows liquids to condense and flow back into a vessel or pipe, solves the problem without the complexity of a scrubbing or separation system. (Ref 7.5 Kletz 1998; Ref 7.6 Kletz 2010)

- Flare systems should be kept as simple as possible, and equipped with as few additional appurtenances as possible, to avoid unplanned breakdowns, plugging and reduced flare capacity. While some additional equipment may be necessary for protection, such as flame arrestors, water seals, filters, etc., these components should be carefully evaluated for their necessity in risk reduction as well as necessary protections and maintenance."

7.7 DISTRIBUTED CONTROL SYSTEMS

The *moderation* strategy can be applied to distributed control systems (DCS). These systems normally use input and output modules that contain multiple inputs or outputs. Failure of the module therefore has the potential to simultaneously disable many control loops. Attention to the assignment of input / output points to the modules makes the plant more tolerant of a failure in one of them (Ref 7.1 CCPS 1993; Ref 7.3 CCPS 2012). Distributed control systems also represent a source of common cause failure because so many of the controls-related safeguards (non-SIS) pass through, or are part of, the DCS. Therefore, judicious choice of control signal groupings for I/O module assignment and the physical separation of I/O modules and other components using barriers so that short circuits cannot disable multiple related control functions simultaneously will reduce the possibility of common cause failures and result in inherently safer control systems. Kletz (Ref 7.5 Kletz 1998; Ref 7.6 Kletz 2010) offers some suggestions for improving process computer systems:

- Software should be scrutable and well documented so that process engineers and others who need to know how the process is controlled can understand what the control system is doing.
- Old software should not be re-used on new applications without very careful review. The European space rocket Arianne failed due to re-used software. One chemical plant re-used a batch

process control program for an upgraded reactor system. The system failed during water-batching (prior to commissioning) due to control valves having different fail-safe positions.

- Safety instrumented systems (SIS) must be independent of basic process controls.
- The process must be designed to be tolerant of the failure of control system components or faulty software.
- The software should be tolerant of hardware failures and correct for them automatically, via self-diagnostics.

Alarm functions are typically built into distributed control systems. The *minimization* strategy can be applied by reducing the quantity of process alarms to the smallest amount which will provide sufficient alerting to operations personnel, then prioritizing these alarms, and ensuring that a program is in place to manage nuisance/spurious alarms. It is easier to train well on the response to a smaller number of (actual) alarms, ensuring that the proper, timely response will be made during process upsets or emergency conditions.

7.8 SUMMARY

This chapter describes the application of inherently safer strategies to traditional protection layers, in order to improve their effectiveness and robustness. In this approach, the likelihood of the incident occurring is reduced, perhaps by several orders of magnitude, but remains "non-zero", in contrast with the direct application of IS strategies to the process, in which the likelihood is virtually reduced to zero.

7.9 REFERENCES

7.1 Center for Chemical Process Safety (CCPS), Guidelines *for Engineering Design for Process Safety*. New York: American Institute of Chemical Engineers, 1993.

7.2 Center for Chemical Process Safety (CCPS), *Human Factors Methods for Improving Performance in the Process Industries.* New York: American Institute of Chemical Engineers, 2006.

7.3 Center for Chemical Process Safety (CCPS), *Guidelines for Engineering Design for Process Safety, Second Edition* New York: American Institute of Chemical Engineers, 2012.

7.4 Horn, R.E., *Developing Procedures, Policies & Documentation*, Info-Map, Waltham, 1992, page 3-A-2

7.5 Kletz, T.A. *Process Plants: A Handbook for Inherently Safer Design.* Philadelphia, PA: Taylor & Francis, 1998).

7.6 Kletz, T.A. and Amyotte, P., Process Plants: A Handbook for Inherently Safer Design, Second Edition. CRC Press, 2010.

8. Life Cycle Stages

8.1 GENERAL PRINCIPLES ACROSS ALL LIFE CYCLE STAGES

As discussed previously, a process goes through various stages of evolution, including:

- Concept
- Research
- Design development
- Detailed engineering design
- Procurement, construction, and commissioning
- Operations and maintenance
- Change management
- Decommissioning

This progression is typically referred to as the process life cycle. In this chapter, each stage will be described and how IS concepts and strategies can be employed in each stage will be detailed. Throughout the life cycle of a process, opportunities will arise to apply the concepts and practices of inherently safer strategies. These opportunities should be evaluated to determine if the strategies can be applied, and whether the risks and costs are commensurate with the possible reductions in risk. In addition to the eight stages listed above and in Figure 8.1, a discussion of IS aspects of Transportation has been added to this chapter. Transportation is associated with the life cycle of a process, as it is a part of the operation's value chain activities. IS strategies can be applied onsite when planning and executing the transportation of hazardous materials within the facility as well as to and from the facility.

This chapter demonstrates that applying inherently safer strategies can enhance process safety, while also improving economic and other objectives, such as quality, productivity, security, energy conservation, and pollution prevention. This application, utilizing formal review methods by trained individuals as identified in Chapter 10, will link the general principles of inherently safer concepts to all life cycle stages.

It should be noted here that there is a tendency for some inherent risks to be volume or quantity dependent. A material that is considered "safe" at lab scale may carry high risk when managed at bulk manufacturing quantities, or even at pilot plant scale. Eventual scale must be considered when evaluating risks even at the research and development phase. For example, doing a process hazards review of a pilot plant run is essential for safe pilot plant operations, but may not reveal inherent hazards of bulk scale production. Virtual "scaling-up" must be part of the early life cycle inherent safety review. A similar issue may arise as a result of lab equipment being different from manufacturing equipment. A reaction that is safe in lab equipment may not be in the plant. These issues need to be taken into consideration in moving from lab scale to pilot plant to full-scale manufacturing.

As described in Chapter 2, the life cycle of chemical processes consists of eight stages. Figure 8.1 repeated from Chapter 2 depicts these life cycle stages.

8.2 CONCEPT

Processes are conceived in a combination of technical activities "on paper" and in the laboratory. It is during the Concept stage that a product is conceived based on a perceived or real industrial or business need. This translates to the need for particular physical and chemical properties to achieve the desired compound. This, in turn, begins to narrow the choice of feed materials and possible intermediates to arrive at a desired final product as the basic chemistry is worked out.

Final, or even initial equipment designs, are not chosen during this stage of the process life cycle, so the IS strategies of *Minimization*, *Simplification*, and to a large degree, *Moderation* may not be applicable during this stage. For example, an objective to "intensify" a process and reduce the scale poses a challenge to the designers to meet that goal and results in reduced quantities of chemicals and potential exposures. However, this is the stage when *Substitution* can have its largest impact. Selecting an intrinsically safer, less hazardous chemical to provide the needed reaction(s), if possible, at this stage is far easier than during subsequent life cycle stages.

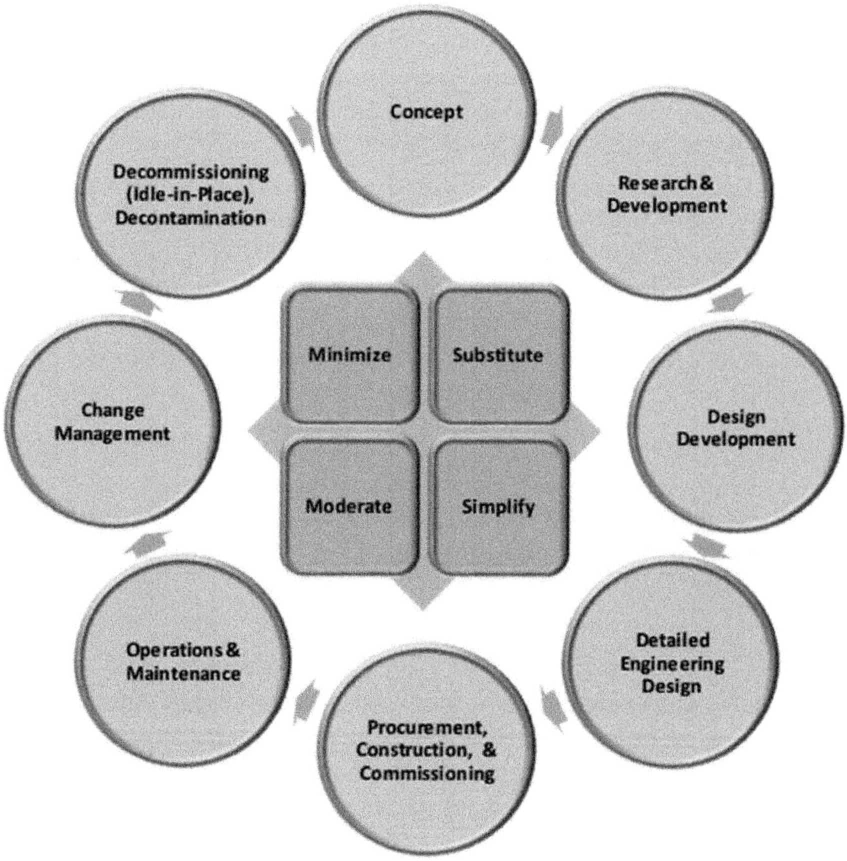

Figure 8.1: Stages in the Life Cycle of a Chemical Process

The notion that there is always a safer chemical to use is not correct; it is not always possible to substitute a less hazardous chemical to achieve the reaction needed to produce the desired molecule(s). Sometimes there is only one known way to produce a desired product. However, if there is an alternative chemistry that involves a different way to produce a given chemical this is the optimum stage to employ it. Sometimes this will require first-principles basic research to be conducted. Such activities have a fair amount of uncertainty because they may not result in the desired product. Therefore, process engineers and scientists who undertake research at the conceptual stage typically choose basic chemistry that has already been proven.

8.3 RESEARCH

Research chemists, working with other research and development (R&D) engineers and scientists, can play a fundamental role in applying inherent safety to process development. They are responsible for decreasing, to a practical minimum, the risks imposed by a future product, as well as the process used to make and distribute it. They have this important responsibility early in the development of a process and can significantly affect the outcome. Changes to the process design to incorporate inherently safer strategies during the research stage of the life cycle are typically easier to accomplish and relatively inexpensive. As the life cycle proceeds to successive stages, the difficulty of implementation and the costs grow dramatically. There is usually still opportunity to use the *Substitution* strategy during the Research stage because the equipment and its characteristics have not been chosen yet. It is during this stage that the *Minimization* and *Moderation* strategies can be employed.

To apply inherent safety appropriately, research chemists must make an in-depth investigation to finalize the process chemistry, and into the entire process that will be required to use that chemistry. To choose the inherently safest chemistry, the team needs to consider:

- The process safety and occupational safety hazards of the chemistry
- The effects of the chemistry on the environment
- Security considerations, i.e., whether the chemical could be used for making improvised chemical weapons
- The upstream, downstream, and ancillary unit operations (including waste disposal) required to use the chemistry
- The operating and reserve inventory needs for the chemistry which includes storage and transfer needs for feedstocks, intermediates, and final products; inventory requirements are often affected by transportation dispatching schedules as much as the projected process throughput

- The transportation considerations for the raw materials and final products, including minimizing hazardous materials transportation

On the following page is a very simplified description of how IS can be included in the research function of the company.

Implementation of Inherent Safety in Chemical Research

First, inherent safety is included in the procedures that researchers must follow when developing a process chemistry and/or a process design.

The researchers need to review hazards and document a process hazards analysis (PHA) for each experimental set-up and significant change in that set-up. A checklist is a required part of that PHA effort and IS questions have been included in the instructions for completing the checklist.

The format (template) for technical reports on product, chemistry, and process development includes a section on IS, as does the report format for applying for permission to file for a patent. The principal instruction on the template for each type of report is similar to the following:

If the process or chemistry being addressed is aimed at the implementation of a new manufacturing process or to improvements/changes to an existing manufacturing process, a discussion of the anticipated hazard/risk level *at the commercial scale* must be included. This discussion should be from an IS perspective and include consideration of the quantities of hazardous materials involved and the severity of process conditions.

A standard "index" sheet (Ref 8.44 Heikkila), a form that dictates IS be considered, is also required. The index sheet is a chart that gives 5 levels of definition, from low to high, for toxicity, flammability, and reactivity (the material factor), quantity (the quantity factor), and reaction severity, pressure, temperature, corrosivity/erosivity, dust content, operability, and experience (the process factor). The researcher is asked to assign a "level of severity" to each factor, and then total all the figures. A simplified example is shown below:

Index Sheet for the Proposed Di-lithium Crystal Process:

- Toxicity/Flammability/Reactivity: 5
- Quantity: 4
- Reaction Conditions (T, P, corr/ero, dust, operability): 2
- TOTAL: 11
- Comments: Need additional proposals

From an IS standpoint, the higher the resulting number, the higher the hazard of the chemistry or process. Several *alternative chemistries* and/or processes *must be proposed*, and an index sheet completed for them as well. If competing chemistries and/or processes exhibit a lower total "level of severity," then the researcher is obligated to defend his choice. No chemistry that exhibits "severe" factors in all categories is accepted.

8.3.1 Inherently Safer Synthesis

Personnel engaged in research activities, e.g., chemists, engineers, and other scientists have many possible opportunities to incorporate inherent safety in the development of process technology, including:

- synthesis routes
- catalysis leading to less severe operating conditions
- the use of a less reactive reagent, or enzyme-based chemistry and bio-synthesis
- reduction or elimination of hazardous solvents
- reduction of solvent hazards by tempering a reaction by using a more volatile solvent that will boil and more reliably remove the heat of reaction
- immobilization of hazardous reagents and catalysts by attaching active groups to polymeric or immobile backbones
- dilution of reagents
- reactions in water as opposed to those that proceed in a hazardous solvent
- elimination of hazardous unit operations

- near-critical and super-critical processing
- replacement of batch reaction processes with semi-batch or continuous processes reducing the quantity of reactants present
- use of processes that are less sensitive to critical operating parameter variations

These possible choices of process technology and materials are not always available for each final product desired; some of these choices may not be viable in certain applications and situations. However, research personnel should thoroughly vet all possibilities using the technology knowledge in their respective companies, other companies if this can be arranged, and search the literature diligently for as much information as possible to reduce the hazards and select the range of inherently safest chemistry possible while achieving the desired product. Where available and accessible, information describing incidents or near misses that involve the candidate process technologies should be reviewed. Those technologies that have contributed to the root cause(s) of process safety incidents or near misses should be withdrawn from consideration.

8.3.2 Types of Hazards Associated with Research

Table 8.1 contains a representative list of the types of process safety hazards and hazardous events that researchers attempt to find and to minimize in searching for the best chemistry.

Table 8.1. Types of Process Safety Hazards

Fires	Flash fires
	Pool fires
	Jet fires
	Solid fires (pyrophoric materials or flammable metals)

Explosions	Vapor clouds
	Confined deflagrations
	Detonations
	Exothermic runaway reactions
	Physical overpressure of pressure vessels
	Brittle fracture
	Polymerizations
	Decompositions
	Boiling liquid, expanding vapor explosions (BLEVE) Undesired reactions catalyzed by materials of construction or by ancillary materials, such as pipe dope and lubricants
Toxicity-related hazards	Toxic to humans (acute) causing reversible or irreversible injury or fatality Environmentally toxic to plant, animal or fish life (large scale events)

Important flammability characteristics are the lower and upper flammability limits, the flash point, the minimum ignition energy, the minimum oxygen concentration, and the autoignition temperature. Values of these properties are available in many different publications and sources, including:

- *Safety Data Sheets (SDS).* The SDS is a part of the Globally Harmonized System (GHS) of Classification and Labelling of Chemicals managed by the United Nations (Ref 8.76 GHS) that was set up to replace the assortment of hazardous material classification and labelling schemes previously used around the world. Incorporation of GHS into national regulations and standards was due by 2008, and by 2017, the GHS requirements had largely been adopted by the major nations of the world. In the U.S., SDSs are required by US OSHA's Hazard Communication Standard (Ref 8.70 HAZCOM). The European Union REACH regulations (Ref 8.38 REACH) provide the current requirements for SDSs in the EU. SDSs are the primary reference for information on the properties of chemicals because manufacturers are required by regulation to create these documents and provide them to all users upon request. In recent years, a number of online resources have been developed providing easy access to SDS information. A note of caution: Despite improvement in SDS quality and completeness in recent years, they still sometimes provide inconsistent information from manufacturer to manufacturer and are sometimes still incomplete. However, for basic information such as flammability limits and exposure limits, they are the primary resource.

- Several **CCPS publications** on fires and explosions provide flammability properties information for liquids and vapors (Ref 8.27 CCPS 2003b); (Ref 8.17 CCPS 2010) and for solid materials and dusts (Ref 8.11 CCPS 1999), (Ref 8.27 CCPS 2003b), (Ref 8.15 CCPS 2017).
- The former **U.S. Bureau of Mines (USBM)** has published a large amount of information regarding the flammability properties of liquids, gases, and dusts (Ref 8.54 Kuchta); (Ref 8.29 Coward), (Ref 8.85 Zabetakis). Most of the USBM's safety, health, and materials programs were taken over by the Department of Energy.
- The **U.S. Coast Guard** has published a document, *Chemical Data Guide for Bulk Shipment by Water* (Ref 8.78 USCG) that provides the flammable, toxic, and reactivity properties of chemicals that are typically shipped aboard vessels.
- The **National Institute of Occupational Safety and Health (NIOSH)**, which is part of the Centers for Disease Control and Prevention (CDCP), has published the *Pocket Guide to Chemical*

Hazards (Ref 8.64 NIOSH) that provides the flammable, toxic, and reactivity properties of chemicals.

- Other open literature sources of information: (Ref 8.83 Yaws) (Ref 8.81 Patty's), (Ref 8.80 ATSDR), (Ref 8.66 RTECS), (Ref 8.56 Sax's). In addition to these well-known references, in recent years many resources that describe the properties of hazardous materials have become available via the Internet. They are too numerous to list here but can be found using common Internet search engines.

In addition to the process safety hazards and events described in Table 8.1, there may be other hazards and events of interest during the choice of process technology, e.g., occupational hazards, chronic health effects (carcinogenic, mutagenic effects, etc.), long term environmental effects (e.g., greenhouse gas effects, ozone depletion, etc.), product hazards (customer injury or illness, waste disposal issues, etc.). The PHA and risk assessment procedures or the company or facility in question, as well as the defined scope of individual PHA and other risk analyses will determine which types of consequences are included in a given study.

The properties included in the references described above have typically been developed under standardized test conditions; therefore, process developers should be cautioned to evaluate the properties published in the open literature at the projected process conditions, as these conditions may differ from the published values.

Vapor Clouds. The temperature and pressure caused by the ignition of a flammable vapor cloud are usually represented by static pressure, impulse (pressure integrated over time), and radiant heat flux result from the deflagration or detonation of a flammable vapor cloud. An ignited vapor cloud whose flame speed meets or exceeds the speed of sound (i.e., a detonation) will cause destructive overpressures, while ignited clouds whose flame speed is less than the speed of sound (i.e., a deflagration) can also cause damaging overpressures. The actual response of any structure to blast overpressures is a function of the pressure over time (i.e., the impulse). A vapor cloud that ignites but whose flame speed is not rapid enough to cause an overpressure (i.e., a fireball) will be destructive to both people and property within its radius and will also generate significant radiant heat which will represent a hazard beyond the boundaries of the fireball itself (Ref 8.17 CCPS 2010).

The ignition of a flammable material in congested spaces, or the interior of process equipment (vessels, as well as piping) can result in overpressures that are significantly greater than unconfined vapor cloud explosions. Vapor clouds do not need to be completely confined to result in these amplified effects. Outdoor releases which are partially confined by buildings, equipment, or have obstacles to the flame front such as obstructions or topography have been shown to create conditions to enhance both the possibility of ignition and the effects of resulting explosions (Ref 8.61 NFPA 2014); (Ref 8.57 Lewis).

Pool Fires. Pool fires are of much longer duration than vapor cloud ignition events and the thermal radiation intensity near the pool is usually high. In general, pool fires result in property damage, but do not usually result in significant numbers of casualties. However, one by-product of a pool fire is its possible effects on adjacent process equipment. The most severe effect from this external heat source is a possible Boiling Liquid Expanding Vapor Explosion (BLEVE). These events, which result in very large fireballs, occur when process vessels containing flammable materials with high vapor pressures (generally light hydrocarbons or chemicals with similar flammable properties, such as vinyl chloride monomer) are exposed to significant amounts of external heat and fail rapidly and catastrophically (Ref 8.17 CCPS 2010). In general, the exposed vessel must be very close so that the flames from the pool fire impinge on the vessel or are exposed to the heat flux from the vessel for this hazard to be realized. Another by-product of pool fires is the potential for toxic combustion products to be released into the atmosphere.

Jet Fires. Jet fires represent a special type of flammable hazard. A release of flammable liquid or gas under pressure creates a roughly conical tongue of flame that, like a pool fire or fireball, creates flame impingement and thermal radiation hazards inside and outside of the cone boundaries (Ref 8.17 CCPS 2010).

Runaway/Exothermic & Decomposition Reactions. A runaway reaction in a vessel or a physical overpressurization of a vessel can cause it to lose its structural integrity. Reaction stability is a complex function of temperature, concentration, impurities, and degree of confinement. Knowledge of the reaction onset temperature, the rate of reaction as a function of temperature, and heat of reaction is necessary for analysis of a runaway reaction. Process conditions that result in the rapid decomposition of reactants can also physically overpressurize a vessel (Ref 8.13 CCPS 1995).

Toxic Hazards. The dispersion and consequences resulting from the release of toxic materials require complex analyses that attempt to simulate many physical and chemical phenomena in nature and model the movement and change in concentration of released materials in the atmosphere versus time (Ref 8.25 CCPS 1996) (Ref 8.12 CCPS 2000). Toxicological effects for humans are often expressed as a concentration (i.e., parts per million), and a number of resources are available to understand acute and chronic toxic effects. As with information regarding flammability properties, the SDS is the primary reference for toxicological data. However, caution must be exercised in using toxicological data. Some data are intended to describe chronic exposures to workers, while others are intended to measure the short-term (acute) exposures associated with accident situations. Most of these data have been extrapolated from laboratory animal experiments and need to be corrected for the size and physiology of humans (Ref 8.81 Patty's); (Ref 8.72 Rand) (Ref 8.79 USDOE); (Ref 8.1 ACGIH) (Ref 8.64 NIOSH); (Ref 8.12 CCPS 2000).

Physical Hazards. Other hazards can present a process safety risk. For example, the simple overpressurization of a vessel or tank above its maximum allowable working pressure (MAWP) (or design pressure) can result in a rupture and release of its contents. The codes for pressure vessel design, e.g., the ASME Boiler & Pressure Vessel Code, NR-13 in Brazil, and the European Pressure Equipment Directive all contain safety margins such that vessel failure should occur at some pressure above the MAWP, however, the operating history, corrosion and damage mechanism environment, inspection practices, and maintenance will determine how effective these safety margins are over the life of the vessel. Another example of severe physical hazards is the hazard represented by high speed rotating equipment, e.g., the failure of a turbine blade at operating speed.

8.3.3 *Hazards Identification Methods*

Process hazard analysis during the research stage includes the identification of hazards that can be reduced or eliminated by inherently safer design, and those which are controlled by safety systems and administrative procedures. Several research methods can be used to identify and evaluate hazards and are described below.

Molecular Structure and Compounds. Certain molecular groupings are likely to introduce hazards into a process. The research chemist should identify these groupings. A search of the open literature will assist in identifying which types of compounds are likely to create potential hazards. Table 8.2 presents molecular structures and compound groupings associated with known hazards. The groupings in the table were developed from CCPS (Ref 8.13 CCPS 1995) and Medard (Ref 8.58 Medard). The table is not all-inclusive.

The hazards of new compounds may not be known but the hazards of analogous compounds, or of those with the same or similar molecular groupings may be known. Testing may be necessary to determine the exact hazardous characteristics of any given compound.

Table 8.2: Representative Potentially Hazardous Molecular Groupings.

The primary associated hazard is indicated to the right of the molecular grouping

Ammonia	Toxicity and fire
Chlorinated hydrocarbons	Toxicity
Cyano compounds	Toxicity
Double and triple bonded hydrocarbons:	Fire and explosion
Epoxides:	Explosion

Hydrides and hydrogen	Explosion
Metal acetylides	Explosion
Nitrogen compounds Amides and imides and nitrides Azides Azo- and diazo- and diazeno-compounds Difluoro amino compounds Halogen-nitrogen bond containing compounds Hydrazine-derived nitrogen compounds Hydroxy ammonium salts Metal exolates Nitrates (including ammonium nitrate) Nitrites Nitroso compounds N-metal derivatives Polynitro alkyl and aryl compounds Sulfur-nitrogen bond containing compounds	All explosion
Oxygenated compounds of halogens	Explosion
Oxygenated manganese compounds	Explosion
Peroxides (and peroxidizable compounds)	Fire and explosion

Polychlorinated biphenyls (PCBs)	Environmental
Poly cyclic aromatic hydrocarbons	Environmental

Reactivity of Types of Compounds. Many additional hazards derive from the hazardous reactivity of combinations of chemicals. The open literature contains numerous lists of the reactivity of different types of chemical combinations. Table 8.3 presents examples of combinations of compounds that are known to be reactive. More complete discussions and lists of highly energetic chemical interactions are published by CCPS (Ref 8.13 CCPS 1995), Yoshida (Ref 8.84 Yoshida), Medard (Ref 8.58 Medard), FEMA (Ref 8.40 FEMA), and Bretherick (Ref 8.9 Bretherick). A complete review of the topic of chemical reactivity is also available from (Ref 8.13 CCPS 1995) (Ref 8.20 CCPS 2003a).

The Interaction Matrix. The chemical interaction or reaction matrix is a recognized useful hazard identification tool. These matrices simply list the materials onsite or in a given operation on both axes of a table/matrix with the intersection noted in the body of the matrix to indicate where the two materials can react adversely with each other. Notes are used to clarify the conditions of the reactions and hazards presented by them. Since a matrix is a two-dimensional tool, it cannot directly show combinations of more than two materials. Table 8.3 is an example of chemical reactivity matrix. Matrices are shown in CCPS *Essential Practices for Managing Chemical Reactivity Hazards* (Ref 8.20 CCPS 2003a).

A general method for producing a plant/site-specific chemical compatibility matrix is given by the American Society for Testing and Materials (ASTM) International (Ref 8.7 ASTM), and an example matrix for a process is presented by Gay and Leggett (Ref 8.41 Gay). Necessary utilities, such as inerting nitrogen, and the materials of construction should be listed as components in a matrix, as should the operator and other populations that are impacted by the process. Tests or calculations may be appropriate if the effect of an interaction identified in a matrix is unknown as described below. The U.S. Coast Guard (Ref 8.78 USCG) has published a comprehensive general chemical compatibility matrix for use in determining the possible reactions between different types of marine cargoes. This matrix stems from work performed by the USCG and the chemical industry.

Table 8.3 An Example Chemical Reactivity Matrix

	Chemical A	Chemical B	Chemical C	Materials of Construction	Personnel
Chemical A	-				
Chemical B		-			
Chemical C	x	x	-		
Materials of Construction			x	-	
Personnel		y	y		-

x or y = Interaction or incompatibility
(e.g, incompatible due to exothermic reaction, corrosive, health hazard, etc.)

Software and online resources are available to assist in making and maintaining reactivity/interaction matrices. Examples include:

- American Institute of Chemical Engineers (AIChE), Center for Chemical Process Safety (CCPS), Chemical Reactivity Worksheet (CRW) (Ref 8.67 CRW4)

- American Society for Testing and Materials (ASTM) International, CHETAH: Chemical Thermodynamic & Energy Release Evaluation (Ref 8.6 ASTM CHETAH).
- US Seal (www.usseal.com/jmchem.html).
- iProcessamart.com (www.iprocessmart.com/techsmart/compatibility.htm)
- IDEX Health and Science (www.idex-hs.com/education-and-tools/educational-materials/chemical-compatibility)

Reactivity Testing. There are a number of testing methods available to determine the thermal stability and the onset temperature of exothermic reactions, as well as the rate of reaction and heat generated per unit mass of the material(s) involved. These are summarized below, and described in full by (Ref 8.20 CCPS 2003a), Englund (Ref 8.35 Englund 1990), and Fauske (Ref 8.39 Fauske):

- Differential scanning calorimetry
- Differential thermal analysis
- Insulated exotherm test
- Decomposition pressure test
- Carius sealed tube test
- Mixing cell calorimetry
- Vent sizing package
- Accelerating Rate Calorimeter®
- Reactive System Screening Tool/Advanced Reactive System Screening Tool

Table 8.4: Reactive Combinations of Chemicals

Substances		Type of Hazard
A +	B =	Hazardous Event
Acids	Chlorates	Spontaneous ignition
	Chlorites and Hypochlorite	Spontaneous ignition

	Cyanides	Toxic and flammable gas generation
	Fluorides	Toxic gas generation
	Epoxides	Heat generation, polymerization
Combustibles	Oxidizing agents	Explosion
	Anhydrous Chromic Acid	Spontaneous ignition
	Potassium permanganate	Spontaneous ignition
	Sodium peroxide	Spontaneous ignition
Alkali	Nitro compounds	Easy to ignite
	Nitroso compounds	Easy to ignite
Ammonium Salts	Chlorates	Explosive ammonium salts formed
	Nitrites	Explosive ammonium salts formed
Alkali Metals	Alcohols, Glycols	Flammable gas and heat generation
	Amides, Amines	Flammable gas and heat generation
	Azo- and diazo-compounds	Flammable gas and heat generation

Inorganic Sulfide Metals	Water	Toxic and flammable gas generation
	Explosives	Heat generation and explosion
	Polymerizable compounds	Polymerization and heat generation

Additional testing methods exist to determine the shock sensitivity of materials, flash point, autoignition temperature, and flammability limits of liquids and vapors, as well as the explosive properties of dusts, such as the minimum ignition energy (MIE), minimum explosive concentration (MEC), limiting oxygen concentration (LOC), and rate of pressure rise (Kst).

Other Hazard Identification Tools. The What if Checklist, and Hazard and Operability (HAZOP) methods are well-publicized and widely-used hazard identification tools. These methods are typically used to analyze the potential hazards associated with a preliminary or detailed process design. However, their basic brainstorming concept and format are also applicable to the analysis of inherent safety opportunities. In addition, the Preliminary Hazard Analysis (PrHA) technique was specifically designed to evaluate the potential hazards associated with the very early stages of a process design, while the basic processing technology and chemistry are being chosen. CCPS presents guidance on the use of these tools (Ref 8.14 CCPS 2008) as well as Chapter 12 of this guideline. Note that the application of these analytical methods will not only offer the opportunity to examine the inherently safer design associated with the basic chemistry of a process design, but also help identify active and passive safeguards that would be appropriate for any given process chemistry. Chapters 10 and 12 provide additional information and direction on the choice of process hazard analysis techniques when evaluating inherently safer strategies.

Alternative Chemistries and IS Measurement Methods. Alternative process chemistries will likely pose different hazards, as well as different degrees of the same hazards. The comparison of alternatives will inevitably lead to tradeoffs. The chosen candidate, though not devoid of hazards altogether, will likely be the one with the least risk. There is, at present, no universally accepted analytical method to conclusively judge among alternatives. Edwards (Ref 8.34 Edwards) proposed such a tool consisting of seventeen potential parameters, including pressure, temperature, yield, toxicity, flammability, corrosivity, etc. These parameters were split into ranges and a relative scoring index, consisting of an ordinal number (i.e., 1, 2, 3, etc.), was assigned to each range. Each step in the alternative process is analyzed, with the result being a summed relative score. The alternative process with the lowest score is the most inherently safe. Lawrence (Ref 8.55 Lawrence) also proposed a similar relative IS scoring index consisting of a process index and a chemical index that were then combined. These original relative scoring methods have been modified and extended to add additional parameters, such as type of equipment, equipment layout, safety of process structures, and chemical interaction to the analysis (Ref 8.44 Heikkilä). A further refinement to the basic relative scoring system was added by Gentile (Ref 8.42 Gentile) to account for the discontinuities associated with using fixed ranges for the important parameters (i.e., the relative score for temperature changes from 2 to 3 for a 1□ change from 199 □C to 200 □C). This refinement is accomplished using fuzzy logic and moves the indexing scores in a continuous manner rather in discrete, sudden jumps.

Gupta (Ref 8.43 Gupta) modified the relative scoring approach, which is biased by the number of steps in the alternative being analyzed, and by the assumed equal weight of hazard assigned to each dimensionless value given to each range of the parameters of interest. For example, does a "3" assigned to a temperature range of 200-299 □C have the same level of hazard as a "3" assigned to a pressure range of 150-200 psig? To avoid these issues, the safety-related process parameters of interest (i.e., pressure, temperature, and a non-dimensional combined index that characterized flammability, explosiveness, and toxicity) for each step in the process were simply plotted and compared to each other. When this graphical method was used to analyze the alternatives for producing methyl methacrylate (MMA), compared to those generated by the Lawrence relative scoring method for the same alternatives, a different conclusion was reached. Of six possible routes for producing MMA, the acetone cyanohydrin alternative was the least inherently safe alternative using the relative scoring method, but the best alternative using the parameter graphical method. The differences involve the importance of the process parameters selected to characterize the risk. The graphical method displays the importance of these values in each step, whereas the relative scoring method does not highlight them as clearly. In choosing which type of alternative analysis is appropriate, the key is to determine which parameters are most important and how they are measured with respect to each other. In the example described herein the number of steps, and hence the number of hazardous chemicals, dominate the relative scoring method results, thereby emphasizing substitution and simplification to eliminate or replace the materials used and reduce the number of steps. The operating pressure dominates the graphical method results, thereby emphasizing moderation to operate at lower values of pressure.

Khan (Ref 8.50 Khan) compared several available IS index methods - the Lawrence IS Index; the *Dow Fire & Explosion Index* (Ref 8.32 Dow FEI) and *Dow Chemical Exposure Index* (Ref 8.31 Dow CEI); the Mond Fire, Explosion, and Toxicity Index (Ref 8.47 ICI); and the Safety Weighted Hazard Index (Ref 8.49 Khan 2001) - to an ethylene oxide production process. Several of these indices predate the formalization of the IS strategies described above. The conclusion was that none of the four indices evaluated were comprehensive methods that were easy, required little interpretation, and could be applied at all times during the life cycle of a process. The Lawrence IS Index allowed all of the IS strategies to be analyzed, but only at the process development stage. The other methods had varying success in capturing the IS strategies adequately in their respective analytical methodologies. In using these types of methods to analyze IS alternatives, it is important that the results be analyzed with a full understanding of the strengths and weaknesses of the method employed. When the overall risks are high using more than one method may be appropriate.

The INSIDE (INherent Safety In DEsign) project was a European government/industry project established by the Commission of the European Community in 1994 to encourage and promote inherently safer chemical processes and plants. The goal of the *INSIDE* Project was to develop practical ways to encourage the use of inherent safety in process development and plant design. The result of this work has been a collection of tools and methods known as the *INSET* Toolkit (Ref 8.48 INSET) that is specifically designed for this purpose. The *INSET* Toolkit provides chemists and engineers with the tools and methods to systematically identify, evaluate, optimize and select inherently Safety, Health and Environmental protection (SHE) processes and designs. Whether a project is completely new, or an existing process in a new plant, or whether a company is considering modifications to an existing plant and process, safety, health and environmental hazards are treated in an integrated way to ensure the conflicts and synergies between them are recognized and effectively managed. The Toolkit pays special attention to the key early stages of a project when almost all the main decisions that determine the SHE performance of the plant are taken. It should be noted that the tools focus on only the inherent safety aspects in the decision-making, not on the total safety, health, or environmental picture. Also, the INSET tools do not replace the need to apply conventional safety studies, hazard analyses, and risk assessment.

The INSET Toolkit is arranged to be used in four stages, as described in Table 8.4. Specific tools are used in each stage to achieve the analysis described in Table 8.5. The INSET guidance provides detailed guidance on applying each tool.

Table 8.5 INSET Toolkit Stages (Ref 8.48 INSET)

Stage I: Chemistry route selection
This stage is where potential chemistry routes to manufacture the product are sought; for some products there could be hundreds of routes available. Some simple screening is carried out on these to see which (say five) should be evaluated further.
Stage II: Chemistry route detailed evaluation
This stage involves taking a few potential chemistry routes, gathering the relevant chemical data, and assessing the routes in detail. A final selection of the best route, or perhaps two routes, to be further optimized/developed or to be used directly as the basis of the plant process should result. It is particularly applicable where the available options have many conflicting aspects and there is no immediately obvious route alternative.
Stage III: Process design optimization
The selected route(s) from *INSET* Stage II are assessed to optimize the conditions and take account of the practicalities of industrial scale processing and the implications of using particular processing equipment.
Stage IV: Process plant design
The initial process design is developed and "challenged" to identify further changes in sequencing, feed profiles, conditions, unit operations and equipment selection in order to improve the performance. The detailed aspects of equipment sizing and pipework fittings are subsequently evaluated to try to identify means of reducing the process inventories and eliminating complexity and hence the possible leak points.

Additionally, the *Dow Fire and Explosion Index* (FEI) (Ref 8.32 Dow FEI), the Dow Chemical Exposure Index (Ref 8.31 Dow CEI), the Mond Index (Ref 8.47 ICI), and the Mond-like acute toxicity Hazard Index (Ref 8.75 Tyler) are analytical methods that can assist in choosing the most appropriate process chemistry among alternatives. Guidance on alternative selection is recognized as a future need in Chapter 16.

Life Cycle Costs. The life cycle cost of a process is the net total of all expenses incurred over its entire lifetime. The choice of process chemistry can dramatically affect this life cycle cost. A quantitative estimate of life cycle cost cannot be estimated at the research stage with sufficient accuracy to be of practical value. However, a qualitative attempt to estimate the life cycle costs of competing chemistries can be beneficial. Implicit in any estimate of life cycle cost is the estimate of risk. One alternative may seem more attractive than another until the risks associated with product liability issues, environmental concerns, and process hazards are given due consideration. Cost benefit analyses (Ref 8.28 CCPS, 1995a) are useful in predicting and comparing the life cycle costs of alternatives.

Examples of Inherently Safer Synthesis Routes. As summarized by Bodor (Ref 8.8 Bodor), the ethyl ester of DDT is highly effective as a pesticide and is not as toxic. The ester is hydrolytically sensitive and metabolizes to non-toxic products. The deliberate introduction of a structure into the molecule to facilitate its hydrolytic deactivation to a safer form can be a key in creating a chemical product with the desired pesticide effects, but without the undesired environmental effects. This technique is also being used extensively in the pharmaceutical industry and is applicable to other chemical industry sectors as well.

Synthetic rubber latex is produced using a process with a large and hazardous inventory of butadiene and styrene. In a modified process, the reactor has an initial charge of water and emulsifier. Monomers are added to the reactor as one pre-mixed stream, and emulsified aqueous sodium persulfate is added as the other stream. The improved scheme, discussed by Englund (Ref 8.36 Englund 1991a), contains less hazardous material and operates at a lower, more controllable temperature. It illustrates that large and established processes may be made safer by applying inherently safer technologies.

8.4 DESIGN DEVELOPMENT

Chemists and process engineers have essential roles to play in applying inherently safer strategies during the process design development stage. At this point, the chemistry has already been chosen, thus defining the basic hazards of the materials. Process development personnel need to focus primarily on process synthesis, unit operations, and the type of equipment required for an inherently safer process. A thorough understanding of the necessary operational steps, as well as the alternate operational steps, is essential to develop an efficient and safe process. Many of the methods discussed in Section 8.3 for the research stage of the process life cycle are useful in the process development stage as well. The Dow FEI (Ref 8.32 Dow FEI), Dow CEI (Ref 8.31 Dow CEI), and Mond Index (Ref 8.47 ICI), Tyler, (Ref 8.75 Tyler), SWeHI (Ref 8.49 Khan 2001), IS Index (Ref 8.55 Lawrence), graphical IS comparisons (Ref 8.43 Gupta), and INSET Toolkit (Ref 8.48 INSET) are examples of these methods, as well as the variety of available PHA techniques (Ref 8.14 CCPS 2008). It is also appropriate to revisit the basic chemistry to study alternate options that may be available, given the unit operations that are anticipated to be used in the process design.

8.4.1 Unit Operations - General

There are a variety of ways of accomplishing a particular unit operation. Alternative types of process equipment have different characteristics that can be used to achieve a higher level of inherent safety, such as inventory, operating conditions, operating techniques, mechanical complexity, and self-regulation. Self-regulation means that the process/unit operation naturally moves itself toward a safe condition, rather than an unsafe condition when certain process parameters or conditions reach a certain point, e.g., as the temperature of a reaction increases the reaction rate decreases, thereby lowering the temperature naturally. For the unit operation of reaction, the designer could select a continuous stirred tank reactor (CSTR), a small tubular reactor, or a distillation tower to process the reaction, each of which has a different impact on the instantaneous inventory of materials in the process.

Before studying alternative types of equipment, the process requirements must be understood. For example:

- Is a solvent necessary?
- Must the products or by-products be removed to complete the reaction?
- What mixing and/or time requirements are necessary?
- What sequencing is necessary for material additions?
- Is the reaction exothermic, endothermic, or adiabatic?

These and other relevant questions must be answered before alternate reaction schemes can be evaluated. Similarly, different unit operations are available to accomplish the same processing objective. For example,

- Should a filter, a centrifuge, or a decanter be used to separate a solid from a liquid?
- Should crystallization or distillation be used for a purification step?

It is inherently safer to develop processes with wide safe operating limits that are less sensitive to variations in the operating parameters, as shown in Figure 8.3. Sometimes this type of process is referred to as a "forgiving" or "robust" process. If a process must be controlled within a very small temperature band in order to avoid hazardous conditions, that process would have narrow safe operating limits. For some reactions, using an excess of one reactant can enlarge the safe operating limits.

8.4.2 Unit Operations - Specific

Some examples and considerations for specific common unit operations are described as follows.

Reaction. Reactor design is particularly critical because reactors involve chemical transformations, and often potentially significant energy releases. Evaluation of the safety characteristics for a given reactor design requires an understanding of what physical or chemical processes control the rate of reaction (catalysis, mass transfer, heat transfer, etc.), as well as the total potential energy consumption or generation involved in the reaction. Energy generating pressure and/or undesired side reactions should also be evaluated. This information is usually necessary for evaluating the suitability of various reactor types (CSTR, batch, tubular, various novel designs such as eductors in loop reactors, static mixers, extruders) for the desired reaction. Mixing and mass transfer are often the critical elements in reactor design as chemicals often react quickly once the molecules are brought together under the correct conditions.

Not all reactions take place in a designated reactor. Some occur in piping systems, a heat exchanger, a distillation column, or a tank. It is vitally important to understand the reaction mechanisms and know where the reactions will occur before selecting the final design.

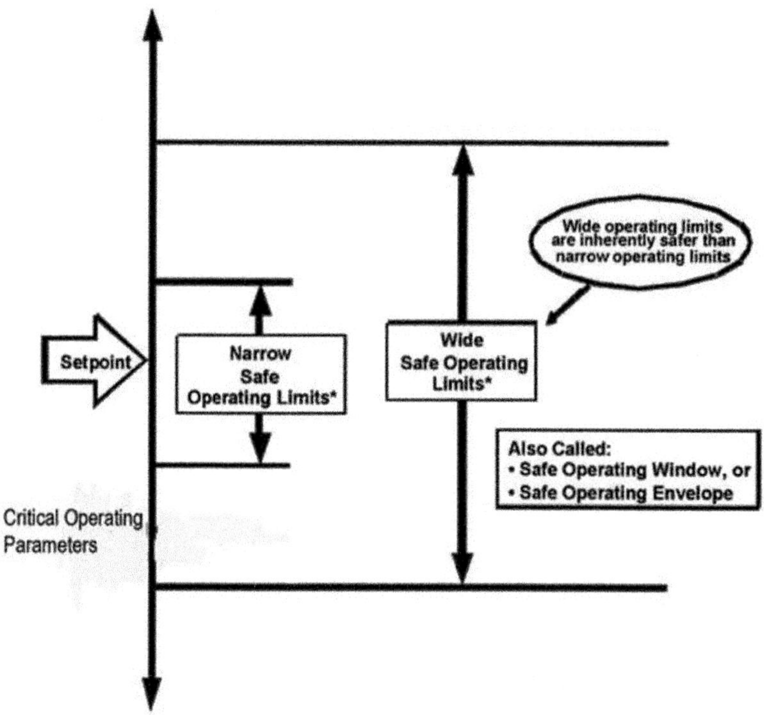

Figure 8.2: Process Design Safe Operating Limits

Some batch reactions have the potential for generating very high energy levels. If all of the reactants (and catalysts, if applicable) are charged into a reactor before the reaction is initiated, and two or more of the materials in the reactor react exothermically, a runaway reaction may result. The use of continuous or "semi-batch" reactors to limit the energy present and to reduce the risk of a runaway reaction should be considered. The term "semi-batch" refers to a system where one reactant and, if necessary, a catalyst is initially charged to a batch reactor. A second reactant is subsequently fed to the reactor under conditions such that an upset in the reacting conditions can be detected and the flow of the second reactant stopped, thus limiting the total amount of potential energy generated in the reactor.

Additional discussion regarding reactor design strategies is covered in Section 3.2 on *Minimization* (as an inherently safer design strategy), and in Section 8.4 on the process design.

Distillation. There are options to minimize hazards when distilling materials that may be thermally unstable or have a tendency to react with other chemicals present. These options include:

- Trays without outlet weirs
- Sieve trays
- Wiped film evaporators
- An internal baffle in the base section to minimize hold-up
- Reduced base diameter (Ref 8.52 Kletz 1991)
- Vacuum distillation to lower temperatures
- Smaller reflux accumulators and reboilers (Ref 8.30 Dale)
- Internal reflux condensers and reboilers (Ref 8.30 Dale)
- Column internals that minimize holdup without sacrificing operational efficiency (Ref 8.30 Dale)

Another option is to remove toxic, corrosive or otherwise hazardous materials early in a distillation sequence, reducing the spread of such materials throughout a process (Ref 8.82 Wells).

Low-inventory distillation equipment, such as the thin film evaporator, is also available and should be considered for the distillation of hazardous materials. This equipment offers the additional advantage of short residence time and is particularly useful for reactive or unstable materials.

Solids Handling. The handling of solid materials frequently has the potential for producing dust clouds that can lead to explosion hazards (if the combustible dusts have the needed properties to support dust explosions). Dusts also present potential industrial hygiene hazards. Handling solids in the form of larger particle size granules or pellets, rather than a fine powder, reduces the potential for worker exposure and also increases the minimum explosive concentration and decreases the residence time that the dust will be suspended in the air, thus reducing the hazard of dust explosions.

If the solid is unoxidized and hence combustible, the dust explosion hazard can be greatly reduced or even eliminated by using a larger particle size material. However, it is important to remember that particle attrition can occur during handling and processing, resulting in the generation of small particles that can become suspended, increasing dust explosion hazards. The sequence of size reduction steps or even the required particle size must be studied to minimize the number of processing steps that involve very small particles. Another option would be to change the form to a shape less prone to dust generation, such as pellets, beads, prills, etc. Handling of solids as a wet paste or slurry rather than dry solids or powders can also reduce hazards, reducing worker exposure and dust explosion hazards, such as with dyes (Ref 8.10 Burch). For example, using wet benzoyl peroxide instead of the dry form reduces the hazards of this extremely reactive material (Ref 8.84 Yoshida).

It may even be possible to eliminate solids handling by processing in a solution. However, this may require an assessment of the hazards of a toxic or flammable solvent in a process compared to the hazards of the solvent-free process. Additionally, hybrid mixtures of solid material dusts with flammable gases and liquids are generally more hazardous than the dust by itself (Ref 8.22 CCPS 2004), (Ref 8.15 CCPS 2017).

Inherently safer approaches to dust explosion hazard control also include building equipment and structures strong enough to contain an explosion. A more thorough discussion of dust explosion hazards and their control have been published by CCPS (Ref 8.11 CCPS 1999) (Ref 8.27

CCPS 2003b), Eckhoff (Ref 8.33 Eckhoff), and NFPA (Ref 8.63 NFPA 2004), (Ref 8.64 NFPA 2006).

Heat Transfer. Some processes have large heat transfer requirements. This may result in large inventories of material within the heat transfer equipment. If the material is thermally unstable, it would be inherently safer to reduce the residence time in the heat exchanger. Options to minimize heat exchanger inventory include the use of different types of heat exchanger designs. As noted in Chapter 3, inventories in shell and tube heat exchangers can be reduced by the use of 'turbulators" in the tubes to enhance heat transfer coefficients, and by placing the more hazardous material on the tube side.

Heat transfer equipment has a great variation in heat transfer area per unit of material volume. Process inventory can be minimized by using heat exchangers with the minimum volume of hazardous process fluid for the heat transfer area required. Table 8.5 compares the surface compactness of a variety of heat exchanger types. Falling film evaporators and wiped film heat exchangers also reduce the inventory of material on the tube side.

In the past several decades, a number of advances in compact heat exchanger technology have occurred. However, due to lack of awareness, use only in non-process plant applications, absence of reliable design methods, and lack of testing and investigation in actual process operating situations, these fairly advanced heat exchange technologies have not seen widespread use in the process Industries. Compact heat exchangers are generally defined as having a compactness ratio of 700 m2/m3 or greater, and a hydraulic diameter of 1 mm to 10 mm. By contrast, shell and tube heat exchangers have much lower compactness ratios and larger hydraulic diameters. Therefore, their efficiencies are much lower, and their footprint is much larger. The following is a summary of the current types of compact heat exchangers and their applications (Ref 8.74 Stankiewicz):

- *Plate heat exchangers (PHE)*: Originally developed for use in the food industry (milk pasteurization), PHEs are made of pressed corrugated plates with heat transfer surface areas of 0.02-3.0 m2 per plate and hydraulic diameters of 2-10 mm. Compactness ratios can be as high as 1000 (see Table 8.5). PHEs have seen use in chemical, district heating, and power applications, but mainly for single-phase duty. However, dual phase use of PHEs has been applied in the refrigeration industry. In automotive applications as heaters and evaporators, PHEs have been used to achieve a reduction factor of volume-to-heat capacity of 2.

- *Spiral heat exchangers*: Spiral heat exchangers consist of plates that are welded together and rolled to create a spiral passage. Heat transfer surface areas vary widely (0.05-500 m2). Spiral heat exchangers have been used in phase change applications.
- *Plate and shell heat exchangers*: This type of heat exchanger consists of a bundle of plates inserted into a shell. In the process industry, these plate and shell heat exchangers have been used as boilers because the shell side can be designed to easily accommodate a higher pressure. The variety of geometries that can be used make it possible to achieve very high compactness ratios (see Table 8.5). Alternative materials of construction other than aluminum have been used to accommodate high operating temperatures and pressures and/or high corrosivity environments (copper, stainless steel, titanium).
- *Plate-fin heat exchangers (PFHE)*: Aluminum PFHEs were first developed for the aircraft industry because of the need for a high compactness and light weight. Today, PFHE technology is used for air separation, hydrocarbon separation, and the liquefaction of natural gas (LNG).
- *Flat tube-and-fin heat exchangers*: This type of heat exchanger has been developed primarily in the automotive industry for use in air conditioners and engine cooling. These are primarily gas-liquid, with a few liquid-liquid heat exchange situations. The gas side is the finned side, and the liquid side is the small diameter channel side. Mechanically, these heat exchangers can accommodate high pressures (20-140 bar).
- *Microchannel heat exchangers*: This type of heat exchanger is characterized by hydraulic diameters of less than 1 mm. To manufacture these devices chemical etching, or micromachining, technology is required. Microchannel heat exchangers have been used in severe applications such as high temperature nuclear reactors, and offshore platforms. The most

common type is the printed circuit heat exchanger in which a channel is chemically etched into a plate. Hydraulic diameters in the 50-200□m range are possible. The plates are then stacked and bonded together. Very high temperature (900 □C) and pressure (500-1000 bar) applications are possible with this type of heat exchangers. The main limitation of a microchannel heat exchanger is high pressure drops across the channel.

One caution associated with compact heat exchangers, particularly plate and spiral exchangers, is that they often contain a significant amount of gasketed surfaces, which may increase the likelihood of leakage.

Additional work has been performed in recent years to develop multi-function heat exchangers that combine heat exchange and reaction unit operations in one device (Ref 8.74 Stankiewicz). Several such devices already exist, including catalytic plate reactors, where a plate-type heat exchanger is coated with a reaction catalyst. A heat exchange reactor has to meet several design objectives:

- The residence time in the device must be sufficient to complete the desired reaction.
- The fluid temperature must be controlled, implying high heat transfer coefficients.
- If the feed and reactant are not pre-mixed well, the channel geometry must create turbulence that is sufficient to accomplish adequate mixing.
- The pressure drop across the device must be acceptable.

Some compact heat exchanger designs meet these characteristics and have high heat and mass transfer coefficients even at low flows and have flows that are turbulent enough (Re > 300) to ensure adequate mixing. However, additional design work needs to be done before such devices are ready for widespread application in the process industries.

Piping. Inventory in piping systems can represent a major risk. For example, a quantitative risk analysis of a chlorine storage and supply system identified the pipeline from the storage area to the manufacturing area as the most important contributor to total risk (Ref 8.45 Hendershot). To minimize the risk associated with transfer lines, their lengths should be minimized by careful attention to unit location and pipe routing. Pipe size should be sufficient to convey the required amount of material and no larger. However, small bore piping is less robust and less tolerant of physical abuse when compared to large diameter piping, and additional attention to proper support and installation is required. In some cases, such as chlorine for water treatment applications, it may be possible to transfer material as a gas rather than a liquid, greatly reducing inventory in the transfer line.

Chemical inventory can also be minimized by removing dead leg piping that is no longer needed in the process. Dead legs may experience enhanced galvanic cell corrosion. Other damage mechanisms, e.g., corrosion under insulation are usually more pronounced in piping dead legs. Additionally, dead leg piping can trap heavier substances (by means of density or solid settling) that can subsequently create unknown hazards, such as freezing and expansion of water in dead leg piping that can cause pipe failure upon thawing (Ref. 8.77 CSB).

Piping systems should be designed to minimize the use of components which are likely to leak or fail. Sight glasses and flexible connectors such as hoses and bellows should be eliminated wherever possible. Where these devices must be used, they must be specified in detail, so they are structurally robust, have the same temperature and pressure ratings as the fixed piping (or as close as possible), are compatible with process fluids, and are installed to minimize the risk of external damage or impact. In addition, these more fragile elements of the piping system should be inspected on a more frequent basis than the rest of the piping system. Caps and plugs should be utilized for vents and drains that are used infrequently.

Where flanges are necessary, spiral wound gaskets and flexible graphite type gaskets are preferred. The construction of these gaskets makes them less likely to fail catastrophically resulting in a large leak. Proper installation of spiral wound gaskets, particularly the torquing of the flange bolts, is important in preventing leaks.

Table 8.6: Surface Compactness of Heat Exchangers. Adapted from Kletz (Ref 8.53 Kletz 2010)

Type of Exchanger	Surface Compactness (m^2/m^3)
Shell and tube	70-500
Plate	120-225 up to 1,000
Spiral plate	Up to 185
Shell and finned tube	65-270 up to 3,300
Plate fin	150-450 up to 5,900
Printed circuit	1,000-5,000
Regenerative - rotary	Up to 6,600
Regenerative - fixed	Up to 15,000 *
Human lung	20,000

* *Some types have a compactness as low as 25m^2/m^3.*

8.5 DETAILED ENGINEERING DESIGN

As the process moves from the process development stage to the detailed design stage, the chemistry, unit operations, and type of equipment have been set. The design stage focuses primarily on detailed equipment specifications, piping and instrumentation design, and layout details. Although the opportunity to incorporate inherently safer strategies still exists during detailed design, many of the *Moderation* and *Substitution* opportunities are no longer possible because the process chemistry and the safe operating envelope will already have been specified. However, if the equipment/unit layout was not established in earlier stages, an opportunity to apply the minimization strategy may exist at this stage. And, many opportunities exist to apply the simplification strategy during detailed design.

Plant designs should be based on a risk assessment that considers the process and the site in detail, as well as all of the principles of inherently safer operation. Earlier decisions may limit the options during the detailed design stage, but inherently safer principles can still be applied. The detailed design step is the last step at which changes can be made at moderate cost because most of the equipment is purchased after this detailed design is approved. Once the equipment has been purchased and fabricated, and the facility is constructed, the cost of modification increases substantially.

8.5.1 Process Design Basis

To reduce the potential for large releases of hazardous materials:

- If not already accomplished, minimize or eliminate the in-process inventory of hazardous materials to the lowest amount necessary consistent with the minimum operational needs. This includes inventory in the process equipment, piping, as well as in vessels and tanks. Elimination of intermediate storage tanks will likely require improvements in the reliability of the upstream and downstream equipment in order to preclude unit or plant shutdowns caused by running out of material.
- Review secondary containments, impoundments, and spacing for tanks storing flammable materials. A sump inside a dike facilitates the collection of small spills. Sump drains, or pumps can direct material to a safe and environmentally acceptable place. See NFPA 30, the flammable liquids code (Ref 8.60 NFPA 2018).
- Review the layout to minimize the length of piping containing hazardous materials and to eliminate unneeded piping dead legs.
- In batch operations, minimize pre-charging the most energetic chemical. Consider adding energetic material in a "semi-batch" mode. That is, add most of the ingredients initially, followed by the addition of the energetic material under flow control. A Safety Instrumented System (SIS) is used to isolate the feed to

the reactor when high temperature or high pressure indicates abnormal operation outside the safe operating limits. A passive safeguard (pipe size, orifice, limited pump capacity) could be considered to limit available energy. Low temperature can be dangerous if the energetic material "pools" unreacted in the reactor and then the reaction initiates. The pooled material could have enough potential energy to result in catastrophic releases.

When dealing with flammable materials, selection from inherently safer design options may vary according to the site and process. For example:

- Use non-flammable materials, if this is still an option at this stage of the life cycle.
- Design the vessel to withstand the maximum pressure generated (i.e., inherent robustness).

Additionally, a number of active and passive layers of protection should also be provided as appropriate for the storage and handling of flammable materials, including using inerting vessels, installing explosion suppression and directing relief vents to a safer location (Ref 8.61 NFPA 2014), (Ref 8.60 NFPA 2018), (Ref 8.17 CCPS 2010), (Ref 8.20 CCPS 2003a).

8.5.2 Equipment

Some engineers specify the maximum allowable working pressure (MAWP) for pressure vessels, considering the intended operating temperatures and pressures only. While this saves on vessel thickness, it can result in a vessel design that is not tolerant of process upsets resulting from control system, communications, or utilities failure. If these process upsets potentially result in process safety incidents, protection layers are required to reduce the risk. An inherently safer choice is to minimize the use of layers of protection by designing robust equipment. For example, if the vessel MAWP is higher than the maximum achievable pressure, a SIS to shut down the pressure sources on high pressure are not necessary. The proper specification of process equipment should consider potential overpressure scenarios, such as those listed in API Standard 521: Pressure-relieving and Depressuring Systems.

Vessels with passive design can fully withstand any achievable overpressure without exceeding the yield stress of the materials. If the overpressure in a vessel remains in the elastic range, the metal returns to its normal crystalline state after stretching. Systems designed to "bend but not break" slightly exceed the plastic region of the metal and are deformed (hardened). The vessel is then actually made stronger by this process; however, the new hazard that is introduced is that the vessel will not stretch and will usually burst if the scenario is repeated. Thus, vessels subjected to plastic range stresses require more frequent inspections for deformation and integrity. A truly passive design is not only safer, it is more cost effective when the lifetime test and inspection requirements are considered.

A robust design must be extended to include all system hardware elements. Little is gained if containment is lost when connected piping, joints, or instruments are designed for lower pressure than the vessel and then fail due to overpressure. In designing the process and equipment, the same engineering principles described in Section 8.3 should be used to minimize the accumulation of, and to contain, energy or materials:

- Specify design pressures high enough to contain pressures generated during exothermic reactions and avoid opening the relief valve and/or exceeding the maximum allowable working pressure of the vessel (the safe upper pressure of the safe operating envelope).
- Use physical limits of pipe size, restrictive orifices, and pump sizing to limit excessive flow rates.
- Use gravity flow in the equipment layout where feasible to minimize the need for pumps or solids handling equipment for hazardous materials.
- Review injection points and piping runs for erosion concerns. Design injection points, elbows, turns, and other erosion-prone areas for lower velocities.
- Use materials of construction with low corrosion rates.
- Use materials of construction that are applicable over the full range of operating conditions, such as normal startup, normal shutdown, emergency shutdown, and system draining. For

example, carbon steel may be acceptable for normal operating conditions, but may be subject to brittle fracture at low temperatures under abnormal conditions, as in the case of a liquefied gas. Cold water, less than 60 oF, which is routinely used during hydrotesting may cause brittle fracture of some carbon steels.

- Avoid materials of construction that crack or pit; uniform corrosion is safer than non-uniform, localized corrosion patterns.
- Avoid incompatible chemicals that could come into contact under abnormal conditions.
- Do not use copper fittings in acetylene or ammonia service, or titanium in dry chlorine service. These principles also apply to gaskets, lubricants and instruments.
- When reviewing the materials of construction also consider external corrosion concerns. Insulation containing chlorides can initiate rapid stress cracking of austenitic stainless steel, as can installations in marine environments.
- If possible, eliminate low-strength equipment such as sight glasses, hoses, rotameters, bellows, expansion joints, and most plastic equipment.
- Minimize possible contamination via fewer cross-connections and fewer hose stations. Minimize the number of hoses required in loading/unloading facilities. Cross-contamination, sometimes even from small amounts of material acting as a catalyst, can result in undesired hazardous reactions. To prevent contamination due to rainwater and spills, consider storing a material that can react vigorously with water indoors. Also, do not store water-reactive materials below grade.
- Flexible connections should never be used as a cure for improper piping alignment and piping support concerns. Figure 8.4 illustrates both good and poor piping alignment. Where expansion joints are required in piping systems containing toxic materials, consider using double-walled expansion joints with pressure indication between the two walls for leak detection.

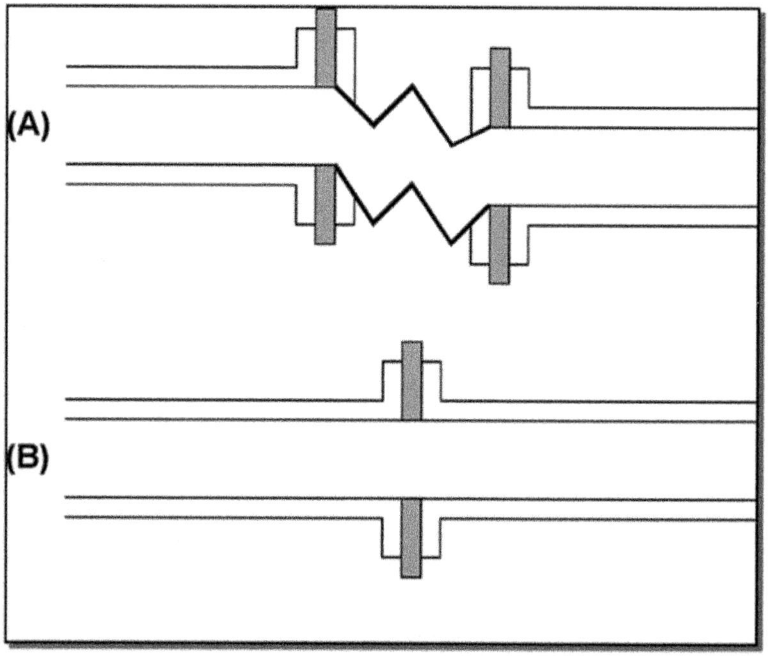

Figure 8.3: Proper and improper piping design. (A) shows improper use of a flexible connection to compensate for poor piping alignment, while (B) shows proper design and alignment

- All welded pipe is preferable to flanged piping, and threaded piping should be avoided for flammable and toxic materials.
- Pressure testing should be conducted via hydrotesting rather than pneumatically.
- Commissioning should be conducted using water or another non-toxic fluid rather than a hazardous process fluid.
- Pinch valves that have no packing can leak if the tube breaks.
- There are trade-offs for magnetic drive and canned pumps versus centrifugal pumps with double-mechanical seals. The former have no seals to leak but need active SISs to prevent high temperature when handling temperature-sensitive materials. Similarly, diaphragm pumps, which have no shaft seals to leak, have potential for process leaks out of the exhaust line and air

leaks into the process on diaphragm failure. Double diaphragm pumps need two failures before material leaks out the exhaust line.

- Consider a weak roof circumferential seam for low pressure storage tanks, if the tank could split under internal pressure. The roof seam should fail first, not the bottom seam. (Note: Users are cautioned that the details of the construction of the weak seam are extremely critical. The design may specify a steel of a certain grade, using the minimum strength of that grade's specification. The steel actually supplied may have a strength significantly higher than the minimum grade specification. For all the joints of the tank except the weak seam, higher strength is better and makes the tank inherently safer. However, for the weak seam, higher strength increases the burst pressure and negates its benefit. Likewise, the thickness of the steel plate may be greater than that specified, increasing the strength of the weak seam).

8.5.3 Process Controls

Many of the inherently safer design aspects discussed here appear in CCPS's *Guidelines for Safe Automation of Chemical Processes* (Ref 8.21 CCPS 2016a) and *Guidelines for Safe and Reliable Instrumented Protective Systems* (Ref 8.23 CCPS 2016b).

The ultimate goal of inherently safer design is elimination of all hazards, with no need for active forms of risk control. However, some control systems are always required in a layers of safety strategy. Using inherently safer strategies in the design logic of a specific control system can make it safer than other alternatives.

Response time considerations lead to the establishment of "never exceed" limits and "never deviate" actions. The never exceed limit is that point at which unsafe consequences will occur. The mandatory action point is a value set low or high enough to allow time for the instrumentation-based or human controls to prevent the hazard. An example of response lag is a system's ability to maintain a heat balance. In reacting systems, control is lost when the system generates heat faster than it can be transferred to other systems or to the surroundings. This excess heat produces an immediate pressure rise in systems containing volatile materials. In other systems, the reactants decompose

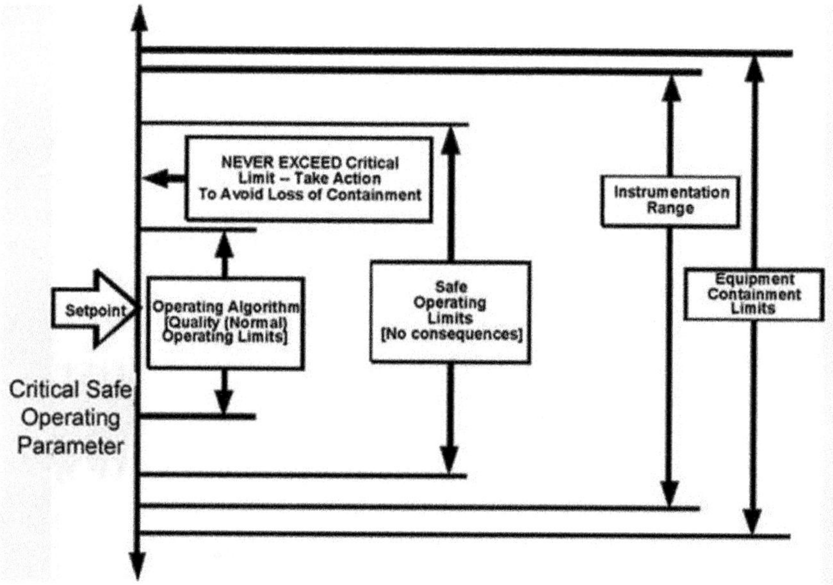

Figure 8.4: Operating Ranges and Limits

to gases resulting in a pressure rise. The mandatory action temperature limit is set low enough to allow time for the response to prevent loss of containment from high pressure. Another example, an upset in feed to a tank could lead to an overflow on high level. The control system design provides enough time for the tank level control to sense the upset and to take corrective action on the flow into or out of the tank before it overflows. For such a tank, the maximum setpoint for the level must be reduced to allow adequate response time.

Basic Process Control Systems (BPCS) and Safety Instrumented Systems (SIS). There are few chemical plants that are so robust that an active control system is not required. Using both active and passive controls can assure product yield and quality and maintain safe operating conditions. This type of control system is known as a basic process control system (BPCS). The BPCS acts to alarm and moderate a high or low operating condition within the never exceed limits. However, when a high risk that is considered to be intolerable cannot be lowered with existing control systems or other layers of protection, a SIS is provided to rapidly shutdown or otherwise automatically place the process in a safe state if the BPCS fails to maintain safe operating conditions. A BPCS should not be used as the sole source of a process safety shutdown. Many of the following guidance items related to the design, operation, and testing of BPCSs and SISs are not inherently safer technology in a strict sense, because they relate to active safeguards. However, much of this guidance can also be considered part of the inherently safer strategy of *Simplification*.

- Inherently safer SISs should be fully independent of the process control system logic residing in the BPCS, including input/output (I/O) cards and logic solvers. Common-mode failures can also result if the BPCS and SIS share components, including power supplies and any other utility system, such as instrument air.
- The SIS should normally be fail-safe, i.e., it is designed to achieve or maintain the safe state of the process on loss of power, or a de-energize to trip design.
- The BPCS should also be programmed to take its outputs to the safe state if input or output signals are lost. Due to the potential in most facilities for power or other utility loss, it can be very difficult to achieve adequate risk reduction with non-fail-safe design. Fail-safe designs are significantly less complex, thereby helping to implement the IS *Simplification* strategy.
- Another example of *Simplification* is that operators should always receive notification from the BPCS or the SIS as the mandatory action points are approached. An everyday example is a high temperature warning light on an automobile. It gives no indication of an abnormally high temperature until the temperature reaches the high alarm point. A temperature indicator gives advance warning before the alarm point. The change in display provides feedback of its functionality by normal fluctuations.
- When choosing SIS input variables, where possible, use direct readings of the process parameter being controlled, and not an indirect reading. If pressure is being controlled, measure pressure directly rather than inferring it indirectly from temperature. This choice eliminates the lag time in processing that occurs when an indirect variable is chosen. Direct readings also eliminate potential errors in the inferred relationship

between the variables. Another example of direct readings is in valve position signals. The position of a valve should be derived from devices that directly measure the valve stem position or stroke. The signals should not be derived from control signals that tell the valve to go to a particular position. In other words, the open or closed indication for a valve or control functions derived from them should only be given when the valve physically reaches the intended position, not when it is "told" to move by the control system.

Operating Modes of Control Systems. Process control and safety shutdowns must be provided during all modes of operation, not only in the normal, steady-state operating mode. This includes startup, shutdown (normal and emergency), inspection/testing, and temporary operations (if any)

- The design of the control system should achieve the highest level of *Simplification* for emergency shutdown and other abnormal/upset situations. If an emergency shutdown capability includes a single operator action to trip (a single mouse click or the activation of a single manual button/switch), it should be easily found, very clearly labeled, easily operated, and unambiguous as to intended function.
- Important controls for higher risk situations should not, if possible, actually cause the process parameter to approach a critical safe limit when undergoing testing. For example, using inherently safer principles, a high-level shutdown that can be tested without actually raising the level in a tank or vessel is safer than one that requires raising the level. Another example involves the overspeed testing of high-speed rotating machinery. The governors should be able to be tested without actually forcing the machine to reach the trip speed (most electronic governors incorporate this design feature). Older, mechanical and electro-mechanical governors do not have this capability.
- In order to test the full control loop/function, the final controlled element(s) must also be tested. To do so without creating a transient or frequent unit/plant shutdowns, it is often necessary to perform full SIS functional tests during shutdown periods.

Depending on the nature of the plant response, it may be possible to initiate a planned shutdown by tripping the equipment during a test. This should only be attempted if the resulting shutdown will be orderly and stable and will not cause transients in other process parameters that are outside their normal limits.

Alarm Systems. With the digitization of chemical/processing facility control systems in recent decades, every signal into the control system can be alarmed. This has raised a genuine safety concern as operators can easily become overloaded sensorially by too many alarms sounding and indicating simultaneously. Additionally, the design of the control systems with respect to alarm indication on the control screens, audible and visual characteristics, and alarm silencing is a continuing concern, although improvements in this area have been made over the years. BPCS/SIS digital and analog alarm displays should be grouped to be readily identifiable by color, physical position, and distinctive sound annunciation. An on-screen list of potential action alternatives should be displayed. The navigation of digital BPCSs and SISs should be intuitive and user-friendly, particularly with respect to alarm screens. The ergonomics of control panels and other operator interfaces are further discussed in Chapter 6.

Further analysis, design, testing, alarm system rationalization, and other life cycle guidance is available in ANSI/ISA 84.00.01 (Ref 8.3 ANSI/ISA 84.01), ANSI/ISA 18.2 (Ref 8.2 ANSI/ISA 18.2, and in CCPS (Ref 8.23 CCPS 2016b) for both BPCSs and SISs.

8.5.4 Utility & Supporting Systems

Utility systems are important with respect to process safety, both as a source of possible release and process safety scenarios themselves, and also when they fail and the effect on the process they support. Factors to consider regarding the support of utility systems include:

- Do not allow nitrogen or pressurized air supplies to overpressure tanks or vessels. These components should be designed to withstand the maximum pressure that can be generated by the pneumatic system, assuming that all pressure control equipment/regulators have failed. An alternative solution is to alter the maximum pressure that can be generated in the pneumatic system to be lower than the maximum allowable working pressure or design pressure of any equipment that can be exposed to the pneumatic pressure. If the utility system pressure cannot be lowered, then passive safeguards such as restriction orifices should be used if possible. If these inherently robust solutions are not possible due to operational requirements, then active controls, such as pressure relief devices or dual (staged) regulators will be required.

- Use water or steam as a heat transfer medium rather than high temperature flammable or combustible oils (Ref 8.51 Kharabanda), (Ref 8.52 Kletz 1991), (Ref 8.53 Kletz 2010).
- Use heat transfer oils with high flash points and boiling points and operate them below their boiling point. Alternatively, use molten salt if water or steam is not feasible as a heat transfer medium (Ref 8.30 Dale); (Ref 8.52 Kletz 1991), (Ref 8.53 Kletz 2010).
- Seek alternatives to chlorine for water treatment and disinfecting applications. For example, in recent years sodium hypochlorite has been used successfully in both industrial and municipal water treatment applications in place of elemental chlorine. This is the most prevalent successful example of use of the IS strategy of *Substitution*. Other alternatives for water treatment include calcium hypochlorite, or non-chemical treatments such as ozone, ultraviolet radiation and heat treatment (Ref 8.68 Negron), (Ref 8.59 Mizerek).
- Use magnesium hydroxide slurry to control pH, rather than concentrated sodium hydroxide (Ref 8.36 Englund 1991a).

8.5.5 Batch Processes

Batch processes present some unique risks because of their characteristics and use. Some batch process systems are dedicated to one product or one product family. However, many batch processes are multi-purpose, multi-product, and multi-stage. This provides the maximum flexibility to owners, particularly in the specialty chemicals sector, to produce a wide range of different products in the same unit/process. Pilot plants also exhibit the same multi-purpose/product/stage characteristics to allow for maximum research and experimentation flexibility. As a result of these flexibilities, there are also some increased hazards/risks:

- Inventories of chemicals in batch processes are sometimes higher than in continuous processes as the entire batch inventory is present in the process at one time. In some continuous processes in situ generation of the most hazardous substances can sometimes be accommodated and the instantaneous inventories of these substances significantly reduced.
- Operators often have a more direct role in the process operations and control than for continuous operations. Multiple connections to and from raw material storage vessels, the reactors/mixing vessels, and intermediate/final product storage vessels exist. Valve manifolds are used to select the proper routing of materials, and especially those with multiple rows or identical valve operators present an increased likelihood of human error.
- Batch processes that are multi-product, multi-stage very often introduce operational flexibility by using flexible hoses that are manually shifted from different storage tanks(s) of feedstocks to a reactor where the product is to be made.
- Transient operating modes predominate in a batch or pilot plant. The entire batch run is usually an elongated transient operation with very little operation that can be characterized as steady-state. Transient conditions increase the likelihood of an upset.
- The status of plant equipment changes with time, and the same equipment may be exposed to a wide variety of changing conditions during the batch. A non-hazardous deviation, such as a charge of an incorrect material, at one place and time, may represent a substantial risk in a different place and time.

Because of these characteristics, there are a number of risks that can be abated, or even eliminated, using inherently safer strategies. Examples include:

- The design of vessels, piping, and other piping system components should be chosen such that the pressure and temperature ratings are higher than the maximum anticipated conditions for any anticipated product. For pilot plants, this is difficult to predict when the plant is initially being designed. However, the nature of the products being tested should provide an idea of the pressure and temperature ranges that the equipment will be exposed to over its life cycle. Unanticipated process conditions that arise during the life of the batch or pilot plant will have to be evaluated and accommodated using the management of change process.

- The design of vessels, piping, and other piping system components should be chosen such that the most severe corrosion, erosion, and other similar conditions are accommodated in both the materials of construction, as well as the thickness of the equipment.
- To the extent possible, the use of flexible hoses and other "weak" components should be minimized or eliminated. This is often very difficult to do in a multi-product facility.
- Where possible, unique hose fittings should be used so that only one set of connections is possible with a given flexible hose.
- Careful and distinctive labeling, particularly using different colors and shapes, will simplify the operations of valve manifolds and reduce human errors. The same is true of other equipment where there are multiple choices and the operating features are identical. See Chapter 6 for further discussions of the human factors associated with the design of such equipment to make it simpler to operate, more error resistant, and more error tolerant.

8.5.6 Other Design Considerations

Additional guidelines on the proper design features for chemical/processing facilities include:

- *Guidelines for Safe Storage and Handling of High Toxic Hazard Materials* (Ref 8.24 CCPS 1988a).
- *Guidelines for Vapor Release Mitigation* (Ref 8.26 CCPS 1988b).
- *Guidelines for Evaluating Process Plant Buildings for External Explosions and Fires* (Ref 8.18 CCPS 1996b).
- *Guidelines for Engineering Design for Process Safety* (Ref 8.16 CCPS 2012) provides an Inherently Safer Technology Checklist. This

checklist has been adapted and is provided in Appendix A of this book.

These references do not address inherently safer strategies explicitly. However, they do provide relevant guidance that fits within the definitions of the four strategies.

8.6 PROCUREMENT, CONSTRUCTION, AND COMMISSIONING

By the time equipment is being procured and erected and installed, easier and more cost-effective opportunities for applying the IS strategies of *Substitution*, *Moderation*, and *Minimization* have passed. During the commissioning of a process lessons may be learned about the operation that offer an opportunity to adjust material inventories that were not apparent during design and these might represent an opportunity to use the *Minimization* strategy. However, these lessons will have to be reflected in operating and logistics procedures that impose administrative limits on inventories from exceeding certain amounts. The tanks sizes, and hence the maximum achievable inventories will have been established during the preliminary and detailed design phases.

A design which has deficiencies requires change during construction, which is a costly and less efficient method of implementing a project. The changes made in the field may not be well designed due to perceived time pressures, nor do they typically receive the same level of engineering or hazard review. Moreover, field changes may not be recognized as significant. In the Kansas City Hyatt Regency Skywalk incident, the initial design of a hanging support was "not buildable" and the contractor improvised an alternate design. The second design was not strong enough to withstand the forces involved when a number of guests started dancing on it. The Skywalk collapsed, resulting in injuries and death to many of the guests (Ref 8.71 Petroski 1985).

If the opportunity still exists during the construction phase of a project, the layout should be established or adjusted to minimize the effects of spacing on the severity or likelihood of a process safety related release. If the layout can be modified to eliminate or minimize the effects of domino or knock-on effects, this should be accomplished. The application of formal (and quantitative) facility siting analysis and practices following the BP Texas City explosion in 2005 is a result of the industry standards that were adopted in the follow-up to that incident. Many of the practices described in these standards are applied to the construction and commissioning phases of the life cycle when temporary and soft-sided (i.e., tents) structures are used (Ref 8.4) API 753), (Ref 8.4 API 756).

The layout considerations also include supporting maintenance activities such as avoiding crane lifts over operating equipment, especially equipment that contains hazardous materials and, in particular, hazardous materials held under elevated pressure or temperature. Also, if possible, the layout should maximize space and ample space should be allowed for maintenance access that avoids damaging other equipment. Most of these layout issues must be addressed during detailed design in order to assure that they can be accommodated.

During the construction and commissioning phase of a new process the *Simplification* strategy may be available. However, using inherently robust equipment designs, simpler reactor designs, heat and mass transfer designs, the separation of processing steps into simpler and smaller steps, improved and simpler control systems logic, and other basic IS improvements must be addressed during the design phases of a project (see Chapter 6). Most of the possible improvements that will present themselves will be modifications to layers of protection. Some of these modifications discovered during construction and commissioning will increase the robustness of these safeguards and independent protection layers (IPL) and are hence inherent in concept. However, some of them will be the addition of more layers, which is not inherent in nature.

As the equipment is erected and the process operated for the first time the opportunity or need for these simplifying modifications and additional layers of protection will become apparent. Chapters 6 and 7 provide additional guidance and examples of *Simplification* and the application of IS concepts to layers of protection. Also, as the operational practices and procedures are further developed at the end of the commissioning phase and the start of the Operations phase, the application of IS concepts to these practices and procedures will possibly present themselves. Chapter 7, and also Chapter 14 contain guidance for the use of IS in operations and in the programmatic aspects of process safety programs.

8.7 OPERATIONS & MAINTENANCE

The longest stage in the life cycle of a process is the operations and maintenance stage. This phase will likely last for decades and will span many changes in personnel, operating and maintenance philosophy, business/financial changes, and perhaps ownership. There are two issues that are important with respect to inherent safety that should be addressed during this phase:

- Preserving the inherent safety features and practices provided during the process development phase of process life.
- Seeking opportunities for continued improvement in inherent safety.

8.7.1 Preservation of Inherent Safety

A primary objective of any process safety program is to maintain or reduce the level of process safety risk in the process. The design basis of the process, especially those inherently safer features that are built into the design and installation of the process, should be clearly documented as inherent in nature and how/why they are inherent. This is particularly important for IS measures because they are not currently required by Recognized and Generally Accepted Good Engineering Practices (RAGAGEPs) or other standards that can compared to the design later. Design measures that are inherent safety strategies are "inherent" to the final design decisions made and deeply embedded in the basic features of the process. There is generally no standard, report, calculation, of other reference that shows the IS benefits that were incorporated into the final design, unless someone thinks to create such documentation. Therefore, without such specific IS documentation it will not be possible to completely evaluate any anticipated change, especially those that may occur years later, to determine whether it strengthens or weakens the inherent safety features or practices or has a neutral effect. Complete IS documentation is also important because the persons (or even the organizations) who provided inherent features or practices will likely not be available to explain the inherent safety aspects of the design.

Adequate IS related documentation is the first layer of defense against losing the technical rationale for decision-making that resulted in IS measures. The Management of Change (MOC) program is the next layer of defense. The importance of MOC is addressed in Section 8.7.

Using inherently safer principles, there are some Simplification improvements that are possible during construction and commissioning or operations stages of life that have minimal impact of the design and functionality of the process and can be accomplished at modest cost. Most of these improvements make the process more tolerant of human error, which is the basic definition of Simplification:

- Altering valve design if possible so that the valve position is obvious at a glance. See Chapter 11 for the example of a ball valve handle that was installed using the incorrect convention of the valve handle being out-of-line with the piping and flow direction when the valve is open.
- Use of different flange sizes or characteristics so that valves cannot be installed backwards, particularly check valves.
- Replacing slip blinds with spectacle flanges so that the status of the blind is always obvious at a glance.
- Replacing FRP (fiber reinforced plastic) tanks and piping with translucent materials so that flow and level are visible without the aid of flow or level instrumentation. If non-translucent FRP materials were initially used, this will likely have to wait until the components require replacement due to normal wear, or at the first available planned outage.
- Replacement of flexible hoses with fixed piping where possible. Piping with flexible joints that can accommodate equipment vibration and the use of articulated arms for loading and unloading are examples.
- Using quick-disconnect couplings only where absolutely necessary and replace unnecessary couplings with bolted joints that require a controlled work environment and the issue of a safe work permit to disassemble.

These and other examples of *Simplification* are presented by Kletz and Amyotte (Ref 8.53 Kletz 2010).

Although it is usually difficult to modify the process chemistry and operating conditions during the operations and maintenance stage of process life, there may be opportunities to make partial changes that have profound process safety impacts. It may not be possible to substitute chemicals, but it may be possible to use them in more moderate conditions. Examples: even a small adjustment in pH may have a significant effect on the rate of corrosion; small adjustments in flow may reduce erosion problems in certain components. Larger scale projects later in life, such as debottlenecking, may offer the opportunity to change heat and mass transfer equipment to simpler, more inherently robust, and more IS friendly designs (see Chapters 6 and 7).

8.7.2 Inherent Safety - Continuous Improvement

At a minimum, a facility with a process safety program will perform PHAs or similar studies at least once every five years. Many companies and facilities also perform PHAs as part of projects. During these studies, each safety device or procedure should be evaluated to see if it can be eliminated or modified by applying inherently safer principles of *Simplification*. These include the use of:

- Valve designs that offer a visual indication of actual position and not the commanded position.
- Mechanical connections (or disconnections) for blanking, draining, cleaning, and purging connections so that maintenance activities cannot be started without first disconnecting lines that might add hazardous materials to the equipment. One example is to route a nitrogen line across or through a manway, so the vessel cannot be entered without disconnecting the nitrogen line.
- Accessible valves and piping to minimize errors.
- Adequate spacing to avoid crowded vessel access.
- For operability and maintainability, minimize the complexity of any instrumented protection layers, where possible (Ref 8.23 CCPS 2016b).
- Logical numbering of a group of equipment. Figure 8.5 is an example of poor assignment of equipment numbers for pumps.
- Logical control panel arrangements. Figure 8.6 is an illogical arrangement of burner controls on a kitchen stove.

- Organized and complete technical information (i.e., the process safety information) for the process equipment that includes not only the basic design information, data, and specifications, but also additional information that will thoroughly support operations and maintenance activities, such as maintenance materials, vulnerabilities of the materials of construction to corrosion and damage mechanisms (often referred to as Corrosion Control Documents or CCDs), lubricants, packing materials, etc.

During PHAs and at other opportunities when operating procedures undergo review, they should be reviewed using inherently safer principles for the following safer provisions:

- Using checklists in operator rounds.
- Using a second person to double-check certain log entries.
- Scheduling to minimize cross-contamination and clean-ups.
- Recording a value for a process variable rather than just a check mark.
- Reformatting procedures to make them more user-friendly. For example, placing an easily recognized tab at the page containing the emergency procedures. See Chapter 6 for further discussion of human factors.

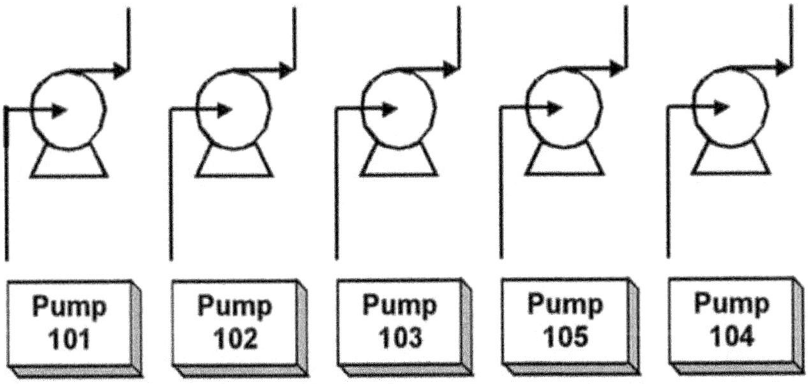

Figure 8.5: An example of poor assignment of equipment identification numbers

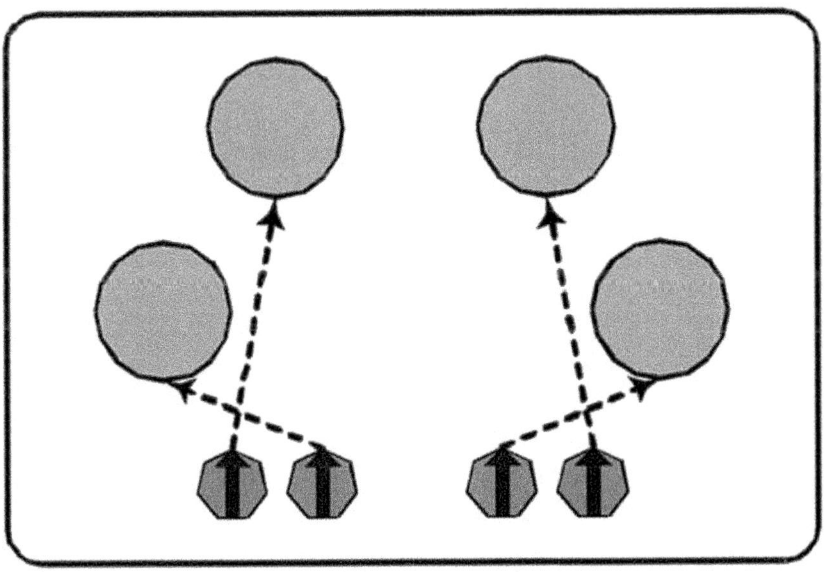

Figure 8.6: An illogical arrangement of burner controls for a kitchen stove. From Ref 8.69 Norman.

In addition to operating procedures and practices, asset/mechanical integrity program practices and procedures also offer an opportunity for the application of the inherently safer strategy of simplification:

- Maintenance procedures should be written using the same philosophy described above for operating procedures. Data sheets and other forms/media for recording inspection, testing, and preventive maintenance results should be complete and user-friendly.
- Spare parts and materials must be correctly specified and installed during maintenance. Sometimes people use what they have, not what was specified. At a minimum, repair and spare parts management procedures should be modified so that work orders for maintenance always include the part number of the replacement materials, and this same number is used to label the part and/or its storage location in the warehouse.
- Spare parts management procedures should be modified to track the expiration dates of any material in stock that has a shelf life. This can be accomplished using common computerized inventory systems, and also by simply placing distinctive labels on the warehouse storage locations. Dedicated spares should be labeled as such prior to storage, and separate storage areas for dedicated spares should be considered.
- Plant equipment should be designed (or redesigned) for ease and reduced risk in performing inspections, tests, and preventive maintenance tasks. For example, if the pipe thickness can be measured from a platform, it is more likely to be done (and done on time) than if a crane is required.
- As corrosion and damage mechanisms that affect the facility become known and studied, and industry consensus is published on the appropriate tests and inspections, appropriate alloys to prevent or resist these should be employed as the opportunity presents itself. Clearly, an entire unit is not going to being demolished and re-built using a different alloy. However, areas of the affected units that are most sensitive to the corrosion mechanism in question, as well as any equipment that

experiences excessive corrosion and must be replaced, alloy(s) that have been shown to resist or prevent the corrosion mechanism in question should be used. This decision depends on the remaining life of the equipment and the possible severity of the consequences associated with the release of the material. This is an example of making the process more inherently robust (Ref 8.46 Hurme).

These suggestions for improving operational and maintenance practices can also be considered as applications of the four IS strategies, although they do not directly relate to the chemical hazards or physical design of the process itself. See Chapter 14 for a more detailed discussion of the application of IS concepts and strategies to the programmatic aspects of other process safety program elements.

8.8 CHANGE MANAGEMENT

Management of Change (MOC) programs are used to manage process changes on process equipment (not replacement-in-kind), utility systems, operating and other procedures, as well as supporting policies, practices, and procedures. They are elements of all process safety programs required by regulation, and are often key elements of voluntary, consensus process safety programs. MOC programs impose a review and approval process that helps ensure that changes are thoroughly vetted before they are physically implemented to manage risk.

MOC programs should also preserve and protect against elimination of inherently safer features. For example, debottlenecking projects are intended to increase throughput and process efficiency. However, increasing unit or plant throughput requires that equipment size be increased at various points in the process(es). Increasing the size of a valve or installing a larger pump could result in high pressure in a vessel, thus increasing the risk of a release. Sanders (Ref 8.73 Sanders) presents a number of examples of changes affecting the safety of a plant. MOC procedures and forms should be modified to include appropriate reviews and verifications that confirm:

- that inherently safer strategies previously incorporated have not been compromised, and,

- that the design, installation, operation, and maintenance of the modified equipment or practice incorporates the four inherently safer strategies, where appropriate.

Plant modifications also provide an opportunity to incorporate inherently safer strategies that were overlooked or deferred during the initial design. For example, if a pressure vessel must be replaced because it is at or near retirement thickness, this is an opportunity to specify a replacement vessel that has a higher pressure rating that cannot be reached or exceeded by the worst-case credible transient event, or to upgrade to a different material of construction that is more resistant to known corrosion or damage mechanisms. The MOC program should trigger review of such considerations during the change review and approval process.

A final check that inherent safety issues have been properly addressed on projects can also be added to pre-startup safety review (PSSR) procedures and checklists.

8.9 DECOMMISSIONING

The application of inherently safer strategies during the decommissioning phase of the life cycle is just as important as its design or operations phases. This is because the equipment is not under the daily attention and care of personnel who are responsible for its safe continued operation. Decommissioned equipment may be "abandoned," with respect to operating or maintenance personnel, for years before some action is taken to recommission or dismantle it. Inspection, testing, and preventive maintenance activities are likely to have been discontinued because the hazardous materials have presumably been completely removed from the equipment. Sometimes, the safe work practices and other carefully applied rules for performing work on the equipment, including line breaking/equipment opening may have been abated as the equipment is no longer considered hazardous from a chemical release viewpoint. If the decommissioning phase has lasted years, there simply may be no one who can remember the state in which the equipment was left. In particular, the inherently safer strategies of *Moderation* and *Substitution* are important while in a decommissioned state. The key actions necessary to ensure that a decommissioned unit/plant or individual piece of equipment remains in a safe and stable condition during the decommissioning process are:

Mechanical Isolation. The decommissioned equipment must be completely and reliably isolated mechanically from other portions of the plant that are still active. Reliable isolation does not include closed valves (although they will likely all be closed). To reliably isolate equipment in a decommissioned state, blinds/blanks or physical disconnection must be used. Maintaining the decommissioned equipment in a completely isolated state from any other equipment is an application of the *Moderation* strategy (i.e., keeping the conditions in the decommissioned equipment at zero inflows or outflows so it does not represent a leakage path of hazardous materials to the environment).

Venting/Purging. If the equipment must be vented to the atmosphere or purged with an inert gas during the decommissioned state to keep it in a safe condition, these states must be maintained at a minimum with locked open valves. Preferably vents should be configured so that there is no isolation possible in the vent path, i.e., all valves have been removed. Maintaining the decommissioned equipment in a vented (or purged) state is also an application of the *Moderation* strategy (i.e., keeping the conditions in the decommissioned equipment at low or even ambient temperatures, pressures).

Electrical Isolation. The decommissioned equipment must be completely isolated electrically from the active portions of the plant. This includes power supplies, as well as control signals which will prevent any equipment in the decommissioned portion of the plant from operating or changing state. The electrical isolation should be accomplished by removing circuit breakers or fuses. At a minimum locked open breakers, switches, and other electrical devices should be used.

Removal of Hazardous Materials. The decommissioned equipment must be completely free of inventories of hazardous materials, including residual amounts. In this context the inherently safer strategy of *Substitution* is used to replace the inventory of hazardous materials with ambient air, or an inert purge liquid or gas (e.g., nitrogen) that will preserve a safe atmosphere inside the decommissioned equipment. However, the equipment necessary to maintain a purged condition might be considered an active safeguard. Completely removing the hazardous materials from a decommissioned process that is left in place is the elimination of a hazard and a first order IS concept as described in Chapter 2. Even if some small amount of hazardous materials cannot physically be removed because they have solidified and adhered to inside surfaces of the equipment or cannot be reached without dismantling the equipment, the removal step is a strong application of *Minimization* during decommissioning.

Documentation of Status. The exact state of the decommissioned equipment must be clearly documented, using the management of change process or an equivalent, so that at some point in the future, possibly years downstream, any actions taken to recommission, modify, or dismantle the equipment can be done safely. This is a form of the inherently safer strategy of *Simplification*.

Example 8.1

A 50-gallon stirred pot reactor was used for the production of sodium aluminum hydride. In the presence of water, sodium aluminum hydride reacts exothermally enough heat to cause the hydrogen that is released to explode. The reactor was emptied, cleaned thoroughly (by report), and then placed in an outdoor surplus equipment yard with the nozzles open to "weather" the equipment. About one year later, a maintenance man was ordered to clean up the reactor in preparation for reuse. He was told to put on full protective fire gear before opening the vessel. He did not don this PPE and proceeded to open the vessel and wash it out with a fire hose. An explosion resulted when water dislodged crusted-over sodium aluminum hydride trapped in a nozzle. The worker was burned, requiring a two-week hospital stay and several months of recuperation.

Attention must be given to the long-term protection of people or the environment from the hazards of abandoned equipment. Equipment that meets the criteria for disposal in a landfill, i.e., it has been properly cleaned, may not be suitable for other uses. Problems such as the one related in the following example can be avoided by making the abandoned equipment unusable if it cannot be cleaned adequately.

Example 8.2

Reactors equipped with heavy agitators used for tetraethyl lead manufacture during World War II were disinterred from bomb rubble and were found by the people who dug them up to be ideal for processing fish paste for human consumption. The reactors were washed, but this did not prevent poisoning a number of people.

There is a temptation for managers to delay the dismantling of decommissioned or abandoned plants as long as possible, usually due to the costs of demolition. However, experience teaches that there will never be a time for chemical plant closure activities that will be less expensive or less hazardous than immediately after the plant is closed, because it is the time when:

- the most people who know how to handle the materials and where the materials are located are available.
- units are still intact, decontamination can be most easily performed, the procedures are well-known, and the equipment is available.
- waste disposal contracts that cover the materials in the plant are still open or can be re-opened easily.
- design documents, waste manifests, maintenance records, and other files are most likely to be readily available.
- equipment has probably not corroded to the point that it can't be handled safely—valves will open, nuts are not frozen, and instruments are in working condition.
- you are least likely to encounter tanks, drums, etc., with contents nobody can identify without an expensive analytical investigation.

8.10 TRANSPORTATION

Transportation of chemicals within a facility and to/from a facility is not a distinct phase in the life cycle of a chemical process, as it is a vital support activity that occurs constantly, in most facilities every day.

Addressing transportation risk at various life cycle stages can increase the inherent safety of the overall operation. This may mean minimizing the number of railcars sitting on track within or outside the plant, the selection of suppliers close to the plant, or other options. This section provides an overview of inherently safer concepts and their application to potential transportation risks.

In choosing the location of a new plant, or in assessing risks related to an existing plant, transportation risk must be considered. The design of new chemical processing units should include at the earliest opportunity a qualitative or quantitative risk assessment of the value chain for the material, including production, use, and transportation of materials in order to minimize overall risk. Risk assessments of existing processes, including an assessment of transportation risk, may result in the conclusion that the process should be moved. Process risk assessment techniques are available in numerous references (for example Ref 8.12 CCPS 2000) and the analysis of transportation risk is documented in *Guidelines for Chemical Transportation Risk Analysis* (Ref 8.14 CCPS 2008).

The assessment of transportation risk must include consideration of the capabilities, equipment, and practices of the entire value chain associated with the manufacture and distribution/shipment of chemicals. Other than ensuring that the carriers, which are separate entities, have the correct and valid licenses, certifications, training programs, emergency response plans, and other documented practices. the fixed facilities where chemicals are manufactured cannot control the policies, practices, procedures, or equipment of the carriers other than to deny them business. Once the carrier leaves the facility with a shipment, the fixed manufacturing facility has released the materials into the care of the carrier and cannot directly control the safety or security of the shipment outside its property, even though they usually are still the owners of the shipment until it is delivered to the customer.

The chemical/processing industry, i.e., the shippers by DOT definition and the carriers (i.e., railroads, trucking companies, marine transportation firms, air cargo companies, etc.) have historically worked together to improve both transportation safety and response to emergencies. The American Association of Railroads (AAR), among others, has worked with the American Chemistry Council (ACC) to recommend improvements in equipment, routing, and procedures to enhance safety. The Chlorine Institute has for years acted to provide inherently safer transport of chlorine. The ACC Responsible Care , CHEMTREC, and TRANSCAER programs have resulted in significant improvement over the years in chemical transportation safety and emergency response.

There are regulations in many countries governing the transportation of chemicals, and any evaluation of transportation risks and options must include consideration of those regulations. In addition, some companies have policies that require going beyond legal requirements for specific materials. With the addition of security concerns associated with the shipment of chemicals in the transportation sector, it is important that the implementation of inherently safer strategies take into account, at a minimum, the onsite transportation infrastructure.

8.10.1 Location Relative to Raw Materials

It may be possible to reduce or eliminate transportation risk by locating the plant where hazardous raw materials or intermediates are produced, if the risk from transporting the raw materials or intermediates outweighs the risk of transporting the final product(s). An example of this IS practice followed the Bhopal incident in 1984. The few facilities in the U.S. that produced methyl isocyanate (MIC), the highly toxic chemical released at Bhopal, began manufacturing MIC and then using it in situ, thereby stopping the transportation of MIC to another location where it was used to manufacture pesticides. The entire process of manufacturing and using the material was accomplished in one location (note that currently in the U.S. MIC use in the production of carbamate based pesticides has been replaced with a different process chemistry, which is an application of *Substitution*). Locating the starting and ending points in the value chain of a given chemical product at the same site will probably provide additional opportunities for risk reduction by inventory reduction, i.e., *Minimization*. Of course, increasing the number of processes at a particular site may increase the overall risk at that site.

Example 8.3

A plant produced methyl methacrylate by reacting hydrogen cyanide with acetone to produce acetone cyanohydrin, followed by further processing to produce methyl methacrylate. The hydrogen cyanide was produced at another site and was transported to the methyl methacrylate plant by rail. A hydrogen cyanide plant was subsequently installed at the methyl methacrylate plant site to eliminate the need for shipping hydrogen cyanide or acetone cyanohydrin.

Example 8.4

A company produced bromine in Arkansas and brominated compounds in New Jersey. A risk assessment resulted in a recommendation to consider the transfer of the bromination processes to the bromine production site in Arkansas. Economic considerations and the decrease in risk justified such a transfer, and it was implemented. Although safety was not the only consideration, it was an important factor in this decision.

8.10.2 Shipping Conditions

The physical condition and characteristics of the material shipped should be considered in transportation risk assessments on a case-by-case basis. There may be options available to reduce the transportation risk by reducing the potential for releases or the severity of the effects of releases. A few possible ways of improving safety by modifying conditions, and hence using the IS strategy of *Moderation* or *Minimization* are:

- Refrigerate and ship the material at atmospheric pressure or at reduced pressure.
- Ship materials in a concentrated state to reduce the number of containers, then dilute the concentrate at the user site.
- Ship and use the material or a substitute in diluted form, i.e., aqueous ammonia instead of anhydrous ammonia, or bleach instead of chlorine.

- Ship and use the intermediates rather than the raw materials.

8.10.3 Transportation Mode and Route Selection

Select a transportation mode to minimize risks to the extent practicable. Drums, ISO containers, tank trucks, rail tank cars, barges, and pipelines offer tradeoffs in throughput/inventory, container integrity, size of potential incidents, distance from supplier or customer, and the frequency of incidents. Barges may have fewer accidents than tank trucks, but the environmental and economic consequences of a major release from a barge in a major waterway, particularly one that supplies potable water to surrounding populations, may be severe enough to make the tank truck shipments a more attractive choice.

The transportation mode used will affect the shipper's options with regard to the selection of the shipment routing. Using truck shipments instead of rail to ship drums, ISO containers, and tank trailers, may allow the shipper to specify highway routes rather than rail to avoid high risks. Shipment via highway has more route options than rail which are fixed and has less options from a given pairing of point of origin and final destination. The time of day and duration of travel is also easier to specify with truck shipments than with rail, as security escorting, if it deemed necessary, is usually easier with trucks than rail shipments.

Railroads choose the routing of rail tank cars and shippers have little or no control to select routings that represent lower risks. However, large amounts of hazardous chemicals can be shipped via rail through densely populated urban areas. In July 2013, a runaway train carrying flammable Bakken crude oil derailed in the center of the Quebec town of Lac-Megantic. The resulting fire and explosion killed 47 people. Alternative routings and improved tracking of rail shipments by the railroads has helped reduce hazards such as long-term storage of tank cars containing toxic or flammable materials on spurs adjacent to residential areas.

The routing of barge shipments is essentially fixed by the location of the shipper and the receiver and the waterway(s) that connect them, and there is generally no choice of routing via pipeline. Data on accident rates by mode and references are given by CCPS (Ref 8.14 CCPS 2008) and can be used to select the safest shipping mode.

Carriers and shippers have worked together to improve transportation safety. The American Association of Railroads has agreed to designate routes that handle 10,000 loads per year or more of chemicals as "Key Routes." Key Routes receive upgraded track, enhanced equipment to detect flaws in equipment or trackage, and lower speed limits.

"Just-in-time" supply of materials may affect the mode of transportation and could actually increase associated risks. For example, if drums of a chemical are stockpiled near a user, the material is not under the same level of control that could be provided by either the supplier or user if the inventory was maintained in a storage tank at one or the other facility. This type of risk should be included when contemplating "just-in-time" shipments.

8.10.4 Improved Transportation Containers

Transportation risk can be reduced by applying inherently safer design principles to transportation containers. Some examples of design improvements are as follows:

- The shipment of environmentally-sensitive materials in general purpose rail cars has been phased out, and DOT specification 105 pressure cars are used instead (*Substitution*).
- Thermal insulation can be used to maintain lower temperatures in the containers and to provide improved protection from fire (*Moderation*).
- Baffles and subdivided internal barriers reduce the amount of chemical that can be released if only one compartment is breached (*Minimization*).
- Non-brittle containers can be used to improve resistance to impact or shock damage.

In addition, transportation containers can and have been made more inherently robust and enhanced layers of protection have been added to them as follows:

- Rail car design, particularly for certain cargoes, has been enhanced so that they can withstand more serious and energetic crashes and derailments without breaching. For example, the end bells on many chemical rail cars have been reinforced to withstand higher head-on crash speeds

- Remote-controlled or spring-loaded (energize to open) shutoff valves can reduce the severity of incidents.
- Tank cars, trailers, and other containers can be specified without bottom outlets, or be provided with skid-protection for bottom outlets.
- All rail tank cars must be equipped with roller-bearing wheels.
- Using containers designed for the maximum pressure that the contents can generate given their vapor pressure at the maximum anticipated ambient conditions enroute will eliminate the need to refrigerate the container for safety.
- Rail cars for chemicals have been designed with "shelf" couplers or double shelf couplers and reinforced tank ends to reduce releases from collision accidents.
- Some barges have double hulls.
- "Low-boy" trailers are used for truck movement of ISO containers. Low center of gravity trailers are also available for tanks. The lower center of gravity reduces the risk of a turnover.
- Baffles can be used in large containers to improve stability.
- Non-brittle containers can be used to improve resistance to impact or shock damage.

8.10.5 Administrative Controls

In addition to improving safety during transportation by optimizing the mode, route, physical conditions, and container design, the way the shipment is handled should be examined to see if safety can be improved. For example, one company performed testing to determine the speed required for the tines of the forklift trucks used at its terminal to penetrate its shipping containers. Subsequently, they installed governors on the forklift trucks to limit this speed below what was required for penetration. They also specified blunt tine ends be installed on their forklifts. This is an example of making the design of a process more inherently robust.

Another way of making transportation safer using inherently safer principles, by using procedural means, is a program to train drivers and other handlers in the safe handling of the products, to refresh that training regularly, and to use only certified drivers (which is a regulatory or legal requirement in many jurisdictions).

8.10.6 Management of Transportation Containers On-site

Transportation containers are delivered to the site to offload raw or intermediate materials or arrive at the site to load intermediate materials or final products, or both. These containers are also sometimes used as temporary storage vessels when they are connected directly to feed the processes. Chlorine is typically offloaded in this manner, as well, as certain industrial gases (i.e., boron trifluoride, hydrogen) which are offloaded from tube trailers connected directly to the process. The presence onsite of fully or partially loaded rail cars, tank truck trailers (bulk or tube trailers), or barges represents an inventory management issue. Often, the number of these containers present on-site, either connected or staged/stored awaiting connection, represent the highest inventory of hazardous materials on-site. This is particularly true of rail cars, where the number present onsite is often a function of both the required production rates and the railroad's dispatching practices. This results, in many cases, in multiple fully loaded 90-ton chlorine rail cars being stored onsite awaiting use or transportation offsite.

The risk of on-site transportation containers loaded with hazardous materials can be reduced as follows:

- Reduce the number of loaded transportation containers to the absolute minimum, consistent with required operational needs. This will require negotiating with the railroad for modified switching schedules, which may be difficult to achieve. However, a facility may discover that the switching schedules in place are based on railroad dispatching schedules that are obsolete and changes can be made. If changes can be made this could represent a significant application of the IS strategy of *Minimization*.
- Store loaded railcars with raw materials, intermediate materials, or final products that are hazardous materials on sidings that are within the site security fence and controlled by site security personnel.

- Assure that railcars permitted on-site have current and valid stencils or other records that indicate that the required pressure and relief device tests have been performed successfully.
- Use gravity, or pneumatic pressure in conjunction with full-flow vents or vapor recovery systems to offload transportation containers and avoid overpressure or damaging vacuum conditions (*Moderation*).

One of the significant issues that arises in the use of IS concepts in transportation, and in particular the application of the *Minimization* strategy, is risk transfer. This occurs because facility production rates and throughputs are generally constant assuming stable economic and market conditions. Consequently, minimizing onsite inventories or limiting the number of loaded rail cars onsite will reduce the inventory of chemicals available for release at any given time if an incident occurs. However, this means more frequent chemical shipments and hence more loaded trucks, rail cars, and other containers operating on the transportation infrastructure carrying these materials. These shipments are mobile, operate at significant speeds (particularly on highways), and pass through or near populated areas and perhaps in close proximity to sensitive public and environmental receptors (e.g., schools, rivers, etc.). While the transportation infrastructure is fairly safe in developed countries, the risk of accidental release in transportation versus release in a fixed facility is higher. Overall, the risk of chemical release increases when the number the chemical shipments increases. This increase should be weighed when making decisions about *Minimization* onsite. See Chapter 13 for a more detailed discussion of risk transfer and inherent safety.

8.11 REFERENCES

8.1 American Congress of Government Industrial Hygienists (ACGIH), Threshold Limit Values for Chemical Substances and Physical Agents and Biological Exposure Indices, 2017.

8.2 American National Standards Institute/Instrument Society of America (ANSI/ISA), Management of Alarm Systems for the Process Industries, ANSI/ISA-18.2-2016, Instrument Society of America, 2016.

8.3 American National Standards Institute/Instrument Society of America (ANSI/ISA), Application of Safety Instrumented Systems for the Process Industries, ANSI/ISA-84.00.01-2004, Instrument Society of America, 2004.

8.4 American Petroleum Institute, Management of Hazards Associated with Location of Process Plant Portable Buildings, API-753, 2012.

8.5 American Petroleum Institute, Management of Hazards Associated with Location of Process Plant Tents, API-756, 2014.

8.6 American Society for Testing and Materials (ASTM International). CHETAH: Chemical Thermodynamic & Energy Release Evaluation, Ver 10.0, 2016.

8.7 American Society for Testing and Materials (ASTM International), E2012-06 Standard Guide for the Preparation of a Binary Chemical Compatibility Chart, ASTM International, 2006.

8.8 Bodor, N., Design of biologically safer chemicals, Chemtech, 25 (10), 22-32, 1995.

8.9 Bretherick, L., Handbook of Reactive Chemical Hazards, 5th Edition, London, UK: Butterworths, 1995.

8.10 Burch, W., Process modifications and new chemicals, Chemical Engineering Progress, 82 (4) 5-8, 1986.

8.11 Center for Chemical Process Safety (CCPS 1999), Avoiding Static Ignition Hazards in Chemical Operations. American Institute of Chemical Engineers, 1999.

8.12 Center for Chemical Process Safety (CCPS 2000). Guidelines for Chemical Process Quantitative Risk Analysis 2nd Ed., American Institute of Chemical Engineers, 2000.

8.13 Center for Chemical Process Safety (CCPS 1995), Guidelines for Chemical Reactivity Evaluation and Application to Process Design. American Institute of Chemical Engineers, 1995.

8.14 Center for Chemical Process Safety (CCPS 2008). Guidelines for Chemical Transportation Risk Analysis, American Institute of Chemical Engineers, 2008.

8.15 Center for Chemical Process Safety (CCPS), Guidelines for Combustible Dust Analysis, American Institute of Chemical Engineers, 2017.

8.16 Center for Chemical Process Safety, Guidelines for Engineering Design for Process Safety, American Institute of Chemical Engineers, 2012.

8.17 Center for Chemical Process Safety (CCPS 2010), Guidelines for Evaluating the Characteristics of Vapor Cloud Explosions, Flash Fires, and BLEVEs, 2nd Ed., American Institute of Chemical Engineers, 2010.

8.18 Center for Chemical Process Safety (CCPS 1996b). Guidelines for Evaluating Process Plant Buildings for External Explosions and Fires, American Institute of Chemical Engineers, 1996.

8.19 Center for Chemical Process Safety (CCPS 2008a), Guidelines for Hazard Evaluation Procedures, 3rd Ed., American Institute of Chemical Engineers, 2008.

8.20 Center for Chemical Process Safety (CCPS 2003a), Essential Practices for Managing Chemical Reactivity Hazards, American Institute of Chemical Engineers, 2003.

8.21 Center for Chemical Process Safety (CCPS 2016a). Guidelines for Safe Automation of Chemical Processes, American Institute of Chemical Engineers, 2016.

8.22 Center for Chemical Process Safety (CCPS 2004), Guidelines for Safe Handling of Powders and Bulk Solids, American Institute of Chemical Engineers, 2004.

8.23 Center for Chemical Process Safety (CCPS 2016b), Guidelines for Safe and Reliable Instrumented Protective Systems. American Institute of Chemical Engineers, 2016.

8.24 Center for Chemical Process Safety (CCPS 1988a). Guidelines for Safe Storage and Handling of High Toxic Hazard Materials, American Institute of Chemical Engineers, 1998.

8.25 Center for Chemical Process Safety (CCPS), Guidelines for Use Vapor Cloud Dispersion Models, American Institute of Chemical Engineers, 1996.

8.26 Center for Chemical Process Safety (CCPS 1988b). Guidelines for Vapor Release Mitigation. American Institute of Chemical Engineers, 1988.

8.27 Center for Chemical Process Safety (CCPS 2003b), Understanding Explosions, American Institute of Chemical Engineers, 2003.

8.28 Center for Chemical Process Safety (CCPS 1995a). Tools for Making Acute Risk Decisions with Chemical Process Safety Applications. American Institute of Chemical Engineers, 1995.

8.29 Coward, H., Jones, G. Bureau of Mines Bulletin 503, Limits of flammability of gases and vapors. National Technical Information Services, 1952.

8.30 Dale, S., Cost-effective design considerations for safer chemical plants. J.L. Woodward (ed.), Proceedings of the International Symposium on Preventing Major Chemical Accidents, February 3-5, 1987, Washington, D.C. (pp. 3.79-3.99). American Institute of Chemical Engineers, 1987.

8.31 Dow Chemical Company (Dow CEI). Dow's Chemical Exposure Index Guide, 1st Edition. American Institute of Chemical Engineers, 1994.

8.32 Dow Chemical Company (Dow FEI). Dow's Fire and Explosion Index Hazard Classification Guide, 7th Edition. American Institute of Chemical Engineers, 1994.

8.33 Eckhoff R., Dust Explosions in the Process Industries. Butterworth Heinemann, 1997.

8.34 Edwards, D., Lawrence, D., and Rushton, A., Quantifying the inherent safety of chemical process routes, 5th World Congress of

Chemical Engineering, July 14-18, 1996, San Diego, CA (Paper 52d). New York: American Institute of Chemical Engineers, 1996.

8.35 Englund, S., (1990) Opportunities in the design of inherently safer chemical plants, Advances in Chemical Engineering, 15, 69-135, 1990.

8.36 Englund, S. (1991a), Design and operate plants for inherent safety - Part 1, Chemical Engineering Progress, 87 (3), 85-91, 1991.

8.37 Englund, S. (1991b), Design and operate plants for inherent safety - Part 2, Chemical Engineering Progress, 87 (3), 85-91, 1991.

8.38 European Union, Regulation (EC) No 1907/2006, Registration, Evaluation, Authorisation and Restriction of Chemicals (REACH), 2007.

8.39 Fauske, H., Managing Chemical Reactivity–Minimum Best Practice, Process Safety Progress 25 (2), 120-129, American Institute of Chemical Engineers, 2006.

8.40 Federal Emergency Management Agency (FEMA), U. S. Department of Transportation (DOT), and U. S. Environmental Protection Agency (EPA), Handbook of Chemical Hazard Analysis Procedures. FEMA Publications Office, 1989.

8.41 Gay, D., Leggett, D., Enhancing thermal hazard analysis awareness with compatibility charts, Journal of Testing and Evaluation 21 (6), 477-80, 1993.

8.42 Gentile, M., Development of a hierarchical fuzzy logic model for the evaluation of inherent safety, PhD Thesis. Texas A&M University, College Station, TX, 2004.

8.43 Gupta, J., Edwards D., A simple graphical method for measuring inherent safety, Journal of Hazardous Materials, 104, 15-30, 2003.

8.44 Heikkilä, A., Inherent safety in process plant design. VTT Publications 384, Technical Research Centre of Finland, Espoo, D Tech Thesis for the Helsinki University of Technology, 1999.

8.45 Hendershot, D., The use of quantitative risk assessment in the continuing risk management of a chlorine handling facility, The Analysis, Communication, and Perception of Risk (pp. 555-565). Plenum Press, 1991.

8.46 Hurme, M., and Rahman, M., Implementing Inherent Safety Throughout Process Life cycle, Journal of Loss Prevention in the Process Industries, 18, No. 4-6, 2005.

8.47 Imperial Chemical Industries (ICI), The Mond Index, Second Edition. Imperial Chemical Industries PLC, 1985.

8.48 INSET, Toolkit Combined Version in Single Document, Volumes 1 and 2 – The Full Toolkit, The Inherent SHE In Design (INSIDE) Project, 2001.

8.49 Khan, F.I. et al., Safety Weight Hazard Index (SWeHI): A new user-friendly tool for swift yet comprehensive hazard identification and safety evaluation in chemical process industries, Trans IChemE, Part B, 79B (2), 65, 2001.

8.50 Khan, F., *Evaluation of Available Indices for Inherently Safer Design Options, Process Safety Progress (22) 2, American Institute of Chemical Engineers, 2003.*

8.51 Kharbanda, O., Stallworthy, E., Safety in the Chemical Industry. London: Heinemann Professional Publishing, Ltd, 1988.

8.52 Kletz, T., (1991) Plant Design for Safety, The Institution of Chemical Engineers, 1991.

8.53 Kletz, T., Amyotte, P., (2010) Process Plants – A Handbook for Inherently Safer Design, 2nd Ed., CRC Press, 2010.

8.54 Kuchta, J., Bureau of Mines Bulletin 680, Investigation of fire and explosion accidents in the chemical, mining, and fuel-related industries - a manual., National Technical Information Services, 1985.

8.55 Lawrence, D., Quantifying inherent safety for the assessment of chemical process routes, PhD Thesis, Loughborough University, Loughborough, UK, 1996.

8.56 Lewis, R., Sax's Dangerous Properties of Industrial Materials, Wiley, 2005.

8.57 Lewis, B., and von Elbe, G., Combustion Flames and Explosions of Gases, 3rd Edition, Academic Press, 1987.

8.58 Medard, L., Accidental Explosions, Volume 2: Types of Explosive Substances. Ellis Horwood Limited, 1989.

8.59 Mizerek, P., Disinfection techniques for water and wastewater. The National Environmental Journal, 22-28, 1996.

8.60 National Fire Protection Association (NFPA 2018). Guide for Venting of Deflagrations. NFPA 68, 2018.

8.61 National Fire Protection Association (NFPA 2014), Explosion Prevention Systems, NFPA 69, 2014.

8.62 National Fire Protection Association (NFPA 2018a). Flammable and Combustible Liquids Code. NFPA 30, 2018.

8.63 National Fire Protection Association (NFPA 2004). Recommended Practice for the Classification of Combustible Dusts and of Hazardous (Classified) Locations for Electrical Installations in Chemical Process Areas, NFPA 499, 2004.

8.64 National Fire Protection Association (NFPA 2006). Standard for the Prevention of Fire and Dust Explosions from the Manufacturing, Processing, and Handling of Combustible Particulate Solids NFPA 654, 2006.

8.65 National Institute for Occupational Safety and Health (NIOSH), DHHS (NIOSH) Publication No. 2005-149, Pocket Guide to Chemical Hazards, 2005.

8.66 National Institute for Occupational Safety and Health (NIOSH), Registry of Toxic Effects of Substances (RTECS); www.cdc.gov/niosh/rtecs/default.html

8.67 American Institute of Chemical Engineers (AIChE), Center Chemical for Process Safety (CCPS), Chemical Reactivity Worksheet 4.0.3 (CRW 4.0.3) (2019)

8.68 Negron, R., Using ultraviolet disinfection in place of chlorination. The National Environmental Journal, 48-50, 1994.

8.69 Norman, D., Turn Signals are the Facial Expressions of Automobiles, Addison-Wesley Publishing Company, 1992.

8.70 US Occupational Safety & Health Administration (US OSHA), 29 CFR 1910.1200, Hazard Communication Standard, 2012.

8.71 Petroski, H. (1985). To Engineer Is Human: The Role of Failure in Successful Design. St. Martin's Press/Vintage Books, 1985/1992.

8.72 Rand, G., Petrocelli, S., Fundamentals of Aquatic Toxicology. Hemisphere Publishing, 1985.

8.73 Sanders, R. E. Management of Change in Chemical Plants Learning From Case Histories, Butterworth-Heinemann, 1993.

8.74 Stankiewicz, A., Re-engineering the Chemical Processing Plant: Process Intensification (pp. 261-308), CRC Press/Marcel Dekker, Inc., 2003.

8.75 Tyler, B., Thomas, A., Doran, P., and Greig, T., A toxicity hazard index, Chemical Health and Safety, 3, 19-25, 1996.

8.76 United Nations, Global Harmonized System of Labeling and Classification of Chemicals (GHS), 4th Edition; United Nations, New York and Geneva, 2011.

8.77 U.S. Chemical Safety and Hazard Investigation Board, Investigative Report, LPG Fire at Valero – McKee Refinery. Washington, DC: Report No. 2007-05-I-TX, 2008.

8.78 U.S. Coast Guard (USCG), CIM 16616.6A, Chemical Data Guide for Bulk Shipment by Water, 1990.

8.79 U.S. Department of Energy (USDOE), AEGLs, ERPGs, or Rev. 21 TEELs for Chemicals of Concern 2005, DKC-05-0002, 2005.

8.80 U.S. Department of Health and Human Services, Agency for Toxic Substances and Disease Registry (ATSDR), www.atsdr.cdc.gov/

8.81 Vernon, R., Cohrssen, B., Patty's Industrial Hygiene and Toxicology, Sixth Edition. John Wiley and Sons, 2010.

8.82 Wells. G., Rose, L., The Art of Chemical Process Design, 266. Elsevier, 1986.

8.83 Yaws, C., Chemical Properties Handbook. McGraw-Hill, 1999.

8.84 Yoshida, T., Wu, J., Hosoya, F., Hatano, H., Matsuzawa, T., and Wata, Y. Hazard evaluation of dibenzoylperoxide (BPO), Proc. Int. Pyrotech. Semin.. 2 (17), 993-98, 1991.

8.85 Zabetakis, M., Bureau of Mines Bulletin 627, Flammability characteristics of combustible gases and vapors. National Technical Information Services, 1965.

9. Inherent Safety and Security

9.1 INTRODUCTION

This chapter is concerned with the application of inherent safety (IS) concepts as a component in addressing chemical security risk. Inherently safer approaches can reduce or eliminate hazards integral to the process and thereby provide benefits to facility security. With increasing worldwide concern over terrorism and the increased potential for intentional attacks on chemical facilities and assets, those responsible for chemical facilities must approach the security of their processes and facilities from a holistic viewpoint. In a security context, an "asset" is a target of potential interest for an intentional attack. As such, an asset may be an individual production unit, a specific piece of equipment, a particular area of the facility, or even the facility as a whole. Increasingly, chemical engineers are being called upon to apply their chemical process expertise and techniques to chemical process *security* management because it:

1. minimizes the risk of harm to the public, employees, and environment from intentional acts against a process facility and;
2. protects the assets within a process facility by maintaining ongoing integrity of operations and preserves the value of the investment.

Inherently safer approaches and concepts identified throughout this book incorporate security principles that reduce perception of the facility as a target, encourage greater safety through facility layout or design, expand buffer zones, substitute or minimize hazardous chemicals, and apply controls to limit release or mitigate consequences. Each IS option has the potential to enhance "defense in depth" and may improve process security so long as application does not conflict with operational requirements or create unintended security or safety exposures.

9.2 CHEMICAL SECURITY RISK

As discussed in previous chapters, process safety risk is defined as a function of the *consequence* of an event and the likelihood, or *probability*, of its occurrence (Ref 9.4 CCPS):

$$\text{Risk(R)}_{\text{Safety}} = f\ [\text{Consequence (C), Likelihood (L)}]$$

In both safety and security assessment, the potential consequences of an event can generally be estimated. Likelihood, however, cannot be estimated for security in the same way it can be estimated for safety.

For unplanned scenarios, the likelihood (probability or frequency) of occurrence may be estimated by relying on statistically valid data, such as historic failure data or an assumption that the event is random and can be described by a normal probability distribution.

Such an approach is *not* appropriate for events potentially arising from intentional acts, such as a terrorist attack. An intentional act is not random and therefore cannot normally be estimated by a probability distribution. In addition, intentional acts may be discrete and independent of other acts – one act does not necessarily contribute to or influence another. With an intelligent and adaptive adversary, terrorists may learn from one another's methods, and those methods may evolve in response to site specific conditions, goals of the terrorist or even target security measures. As a result, a history of intentional acts cannot be combined to create a body of data analogous to equipment failure data or other statistically valid data sets. Therefore, a combination of the following factors must be used as a proxy for likelihood in security risk assessment:

- the *threat* posed by an adversary to an asset;
- the *vulnerability* of an asset to an attack; and,
- an asset's *attractiveness* as a target.

Security risk can therefore be considered a function of consequence (C), threat (T), vulnerability (V) and attractiveness (A) or:

$$\text{Risk (R)}_{\text{Security}} = f\,[\text{Consequence (C), Threat (T), Vulnerability (V), Attractiveness (A)}]$$

With regard to security, it is appropriate to define the terms of "consequence," "threat," "vulnerability," and "attractiveness."

Consequence is the severity of loss or damage that can be expected from a successful attack against an asset. Examples of relevant consequences include:

- Injuries to the public or to workers
- Significant environmental damage (such as contamination of drinking water)
- Direct and indirect significant financial losses to the company
- Disruption to the national, regional, or local operations and economy
- Loss of business viability

The US Department of Homeland Security (DHS) has identified five potential consequence categories or security issues associated with the production, use, storage or distribution of chemicals: (Ref 9.4 CCPS)

1. *Release*. Toxic, flammable, or explosive chemicals or materials that, if released from a facility, could create significant adverse consequences for human life or health.
2. *Theft or Diversion*. Chemicals or materials that, if stolen or diverted, could be used as weapons or easily converted into weapons using simple chemistry, equipment or techniques.
3. *Sabotage or Contamination*. Chemicals or materials that, if mixed with readily available materials, could release poisonous gasses or create other significant adverse consequences for human life or health.
4. *Critical Relationship to Corporate or Government Mission*. Chemicals, materials, or facilities that, if unavailable, could

compromise the government's ability to deliver essential services during an emergency.

5. *Critical Relationship to Corporate or National Economy.* Chemicals, materials or facilities that, if unavailable, could create significant adverse consequences for the corporation's existence or economic well-being, or the national or regional governmental economy.

The first three issues (i.e., release, theft/diversion and sabotage/contamination) relate to the properties of the chemicals and the potential for adverse human health effects. The last two issues–government mission and the economy–reflect the critical uses of some chemicals that may or may not be hazardous.

Inherent Safety (minimization, moderation, or substitution) can affect consequences by either reducing or eliminating the hazard, or by moderating the hazard. Simplification may reduce the opportunities for an event to escalate to a larger degree of consequence.

Threat can be defined as any indication, circumstance, or event with the potential to cause loss of, or damage to, an asset. It is also the intention and capability of an adversary to undertake actions that would be detrimental to valued assets. Threats are manifestations of an adversary's malevolent intent directed at the chemical asset or use of the chemical asset as a means to attack a different target, such as stealing a chemical for use in producing an improvised explosive device. The magnitude of a threat is influenced by the intention and capability of an adversary to undertake actions that would be detrimental to valued assets. Sources of threats may be categorized as terrorists (international or domestic); current or former disgruntled employees or contractors; activists, pressure groups, single-issue zealots; or criminals (i.e., white collar, cyber hacker, organized, opportunists).

Adversaries may be categorized as "insiders" (internal threats), "outsiders" (external threats) or a combination of both insiders and outsiders (internal-external collusion). Government law enforcement and security agencies may provide additional information on potential adversaries, with regard to motives and tactics. Companies may also choose to acquire such information through other channels, including commercial intelligence services.

Inherent Safety can affect threat. For some adversaries, the reduction or absence of the hazards may lessen or eliminate their interest in a target, or their opportunity to commit an attack. However, inherent safety strategies do not necessarily remove the threat, as a determined adversary may not be persuaded, especially if the inherently safer elements result in a safer operation but doesn't completely eliminate the opportunity for them to inflict damage.

Vulnerability is any weakness that can be exploited by an adversary. Vulnerabilities may include, but are not limited to:

1. structural characteristics
2. equipment properties
3. personnel behavior
4. locations of people, equipment and buildings
5. operational, cyber, and personnel practices

Vulnerabilities are estimates of an asset's ability to withstand specific attack scenarios, considering existing security elements. The scenarios are usually derived from a combination of facility brainstorming ("knowing what I know about this asset, this is how I'd bring it down") and intelligence estimates or other information about the motivations and tactics of specific adversaries. The Department of Homeland Security provides specific scenarios to facilities required to conduct Security Vulnerability Analyses (SVAs) under CFATS.

Vulnerability may be less affected by inherent safety than other factors unless sources are consolidated, quantities of dangerous chemicals reduced, or otherwise changed. It is more likely that the vulnerability is the same following the inherent safety application to the hazard itself (1st or 2nd order), but, as mentioned before, the consequences may be different. Inherent safety applied to layers of security may be effective in making the layers robust, but this isn't necessarily reducing the hazard.

Attractiveness is the perceived value of attacking a given asset. The perceived attractiveness of a specific asset depends upon the goals of the attacker, and how well the potential consequences of a successful attack align with those goals. Other factors that may increase an asset's attractiveness include:

- potential "newsworthiness" of a successful attack, or the ability to garner significant publicity from an attack;
- potential for using a target to create collateral damage to an adjoining target through the proximity of an asset to a desirable target; and
- proximity of an asset to a publicly accessible vantage point from which an attack could be launched, such as a public transportation route or highway overpass.

The attractiveness of an asset is heavily influenced by threat, consequence and vulnerability. As a result, targets with similar potential consequences or vulnerabilities may not be equally attractive, depending on the goals and motivations of the adversary. An asset owned or operated by a large, well known global company may be more attractive to some adversaries than facilities with similar or even greater potential consequences or vulnerabilities operated by small or relatively unknown companies. A facility operated by a company known to have a relationship with an adversary's enemy might be a more attractive target than a similar facility operated by a company that is not known to have such relationships. Because of the influence of threats, consequences and vulnerabilities, attractiveness should be considered a dependent, rather than an independent, security risk variable.

Inherent safety may affect attractiveness in a similar manner as it can threat. The adversary may be less attracted to a modified process that doesn't offer the same potential consequences.

9.3 SECURITY STRATEGIES

Security strategies for the process industries are generally based on the application of four key concepts: (Ref 9.4 CCPS)

Deter: A security strategy to prevent or discourage the occurrence of a breach of security by means of fear, doubt, or reduction in facility attractiveness. Physical security systems, such as warning signs, lights, uniformed guards, cameras, and barriers, are examples of systems that provide deterrence.

Detect: A security strategy to identify an adversary attempting to commit a malicious act or other criminal activity in order to provide real-time observation, interception and post-incident analysis of the activities and identity of the adversary.

Delay: A security strategy to provide various barriers to slow the progress of an adversary in penetrating a site to prevent an attack or theft, or in leaving a restricted area to assist in apprehension and prevention of theft.

Respond: The act of reacting to detected criminal activity either immediately following detection, such as notifying local authorities for assistance, or post-incident via surveillance tapes or logs.

Resilience/Resiliency: (Ref 9.8 DHS) The ability to resist, absorb, recover from, or successfully adapt to adversity or a change in conditions. In the context of energy security, resilience is measured in terms of robustness, resourcefulness, and rapid recovery.

A complete security design includes these concepts in "Layers of Protection" or a "Defense in Depth" arrangement (Figure 9.1). Ideally, the most critical assets should be in the center of the conceptual concentric levels of increasingly more stringent security measures. Security scenarios often include direct attacks on or near an asset. For this reason, the spatial relationship or proximity between the location of the target asset and the location of the physical countermeasures is important.

The facility could detect, deter, delay, or respond to the event at multiple points in the layers of security. Optimally, security would deter and detect at the outermost layers to provide sufficient time for responders to suppress or neutralize the adversary prior to initiation of the event. The layers of protection for critical assets may need to be quite robust because the adversaries are intentionally attempting to breach the protective features and may use whatever means are available to help ensure a successful attack. This may include the use of explosives or other destructive events that result in widespread common cause failures. Particularly motivated adversaries may go to extreme lengths, including suicide attacks, to breach the security layers of protection.

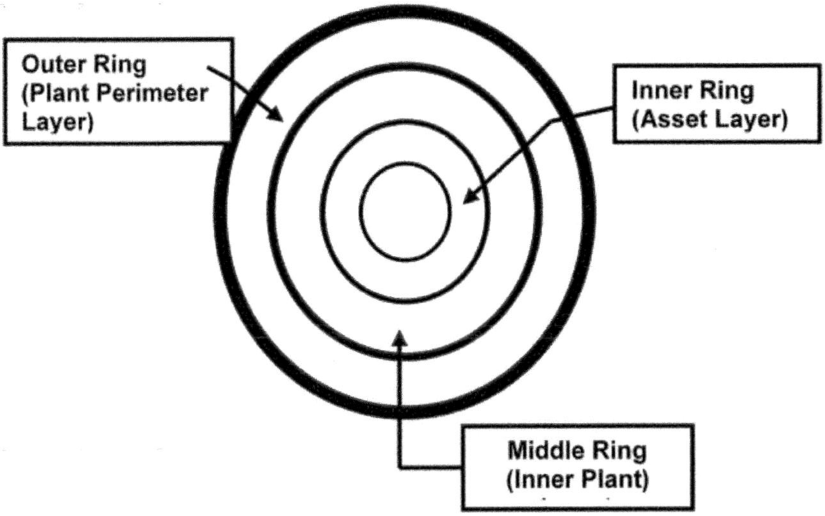

Figure 9.1: Security Layers of Protection or Defense in Depth

9.4 COUNTERMEASURES

Countermeasures are actions taken to reduce or eliminate one or more vulnerabilities. Countermeasures include security officers, security barriers, technical security systems, cyber security, response, security procedures, and administrative controls to address the following strategies:

- Physical Security (Security systems, protective personnel, and architectural features that are intended to improve protection).
- Technical Security (Electronic systems for increased protection or for other security purposes including access control systems, card readers, keypads, electric locks, remote control openers, alarm systems, intrusion detection equipment, annunciating and reporting systems, central stations monitoring, video surveillance equipment, voice communications systems, listening devices, computer security, encryption, data auditing, and scanners).

- Cyber Security (Protection of critical information systems including hardware, software, infrastructure, and data from loss, corruption, theft, or damage).
- Crisis Management and Emergency Response Plans (process by which an organization deals with a major event that threatens to harm the organization, its stakeholders, or the general public).
- Policies and Procedures (policies are the rules that govern how a company conducts business; whereas procedures are a set of steps for administering a process).
- Information Security (the practice of preventing unauthorized access, use, disclosure, disruption, modification, inspection, recording or destruction of information).
- Intelligence (Information to characterize specific or general threats when considering a threat's motivation, capabilities, and activities).
- Inherent Safety (a concept and approach to safety that focuses on eliminating or reducing the hazards associated with materials and operations used in the process where this reduction or elimination is permanent and inseparable).

9.5 ASSESSING SECURITY VULNERABILITIES

Leaders in the chemical industry have recognized the potential for chemical facilities, or chemicals themselves, to be used as weapons by terrorists or other criminals. The industry also understood the need to expand existing security programs to address these new and serious threats. U.S. processing industries built on existing process safety management systems to develop *security* management systems that included requirements to assess and prioritize potential security risks posed by chemical facilities, and to implement measures to address those risks. Examples include the Responsible Care® Security Code, adopted by the Chemical Manufacturers Association in June 2003 (Ref 9.1 ACC) and API's *Security Guidelines for the Petroleum Industry* (Ref 9.3 API)

The American Institute of Chemical Engineers supported the development of such security management systems by publishing *Guidelines for Analyzing and Managing the Security Vulnerabilities of Fixed*

Chemical Sites in 2002. The Guidelines describe a risk-based approach to managing chemical facility security that allows assessment of a wide range of threats, ranging from vandalism to terrorism. With the recognition of threats, consequences, and vulnerabilities, and the evaluation of security event risks, a security management system can be organized to more effectively mitigate the risks identified in the assessment.

Several trade associations, such as the American Petroleum Institute (Ref 9.2 ANSI/API) and states require companies to use security risk assessment (SRA) or security vulnerability assessment (SVA) methodologies that meet the design criteria set forth in the CCPS *Guidelines*. New Jersey allows facilities to use CCPS or CCPS-equivalent SVAs to meet the requirements of Administrative Order 2005-05 (see Chapter 10); Maryland and the City of Baltimore also allow facilities, subject to their chemical security requirements, to use the CCPS or a CCPS-equivalent SVA methodology.

The U.S. Department of Homeland Security also adopted elements of the CCPS methodology as part of the Chemical Facility Anti-Terrorism Standards (CFATS). (Ref 9.5 CFATS) The elements of a typical SVA approach are shown in Figure 9.2, which describes the methodology published by CCPS in *Guidelines for Analyzing and Managing the Security Vulnerabilities of Fixed Chemical Sites* (Ref 9.4 CCPS).

Upon completion of an SVA, a report or other documentation is prepared to communicate its results to management for appropriate action. The report should be carefully protected as company-sensitive information, because it includes descriptions of identified vulnerabilities and related countermeasures recommended to address those vulnerabilities.

9.6 INHERENT SAFETY AND CHEMICAL SECURITY

Application of inherent safety (IS) principles may allow a facility to go beyond traditional *physical* security measures. Physical security includes such considerations as security officers, barriers, and surveillance equipment. However, some assets may lend themselves to security evaluation using IS concepts, in addition to physical, technical, and even cyber security options. A combination of these strategies, that may also

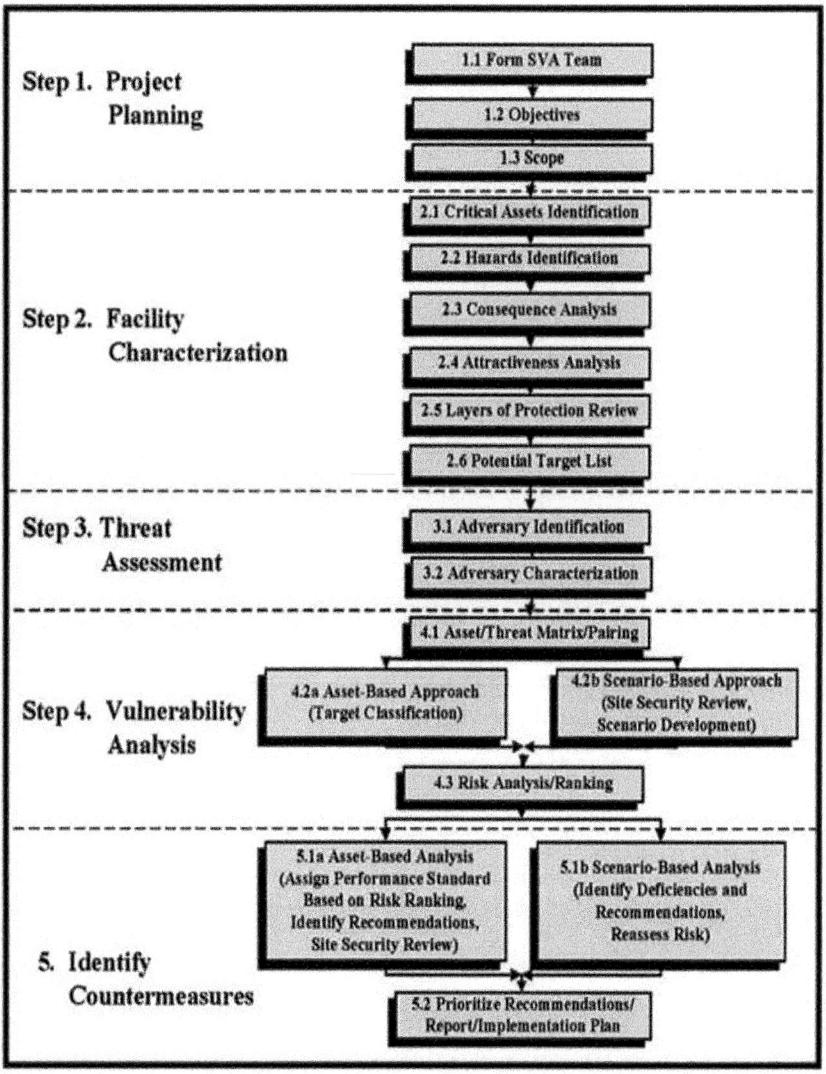

Figure 9.2: Elements in a CCPS Security Vulnerability Assessment (Ref 9.4 CCPS)

include changes in the process itself could be evaluated to create a more robust, effective and individualistic facility security program.

The IS knowledgeable chemical engineer has an opportunity to positively influence security, as well as safety, through the process life cycle. As is true for safety, security issues can be addressed in the concept and design phases of a project, allowing for more cost-effective considerations that could eliminate or greatly reduce security risks. As described in Chapter 5, inherent safety concepts may be applicable to both new and existing/modified processes by answering questions such as:

1. First, could a potential consequence be eliminated altogether? (First order IS measure)
2. Second, could the magnitude or severity of the potential consequence be reduced? (2nd order IS measure)
3. Finally, could the remaining vulnerabilities be reduced through consideration of the hierarchy of passive, active, and procedural measures? (Layers of Security)

Consequence. As described earlier, chemical security consequences typically result from one of five security issues:

- *Release* of a toxic, flammable or explosive chemical
- *Theft or diversion* of a chemical that could be used as or used to produce a chemical weapon
- *Sabotage or contamination* of a shipment to cause a toxic release during transportation
- Loss of a government or corporate *mission critical* chemical
- Loss of a corporate or nationally important economically critical chemical or *economically critical* facility

Inherent safety (IS) principles are more applicable to potential consequences from some of these security issues than others and may be most readily apparent in the context of security release scenarios. If feasible, substituting a less hazardous chemical, reducing the amount of hazardous chemicals used, or operating under more moderate process conditions may contribute to a facility's security posture, as well as provide a viable safety option.

Opportunities to apply First and Second order IS concepts may be limited. There may be no viable alternative for a critical reagent or reducing inventory may make a process difficult or unsafe to operate, such as the need for a surge tank. Moreover, some security issues do not lend themselves to an IS solution. Some chemicals of concern because of potential theft may not be acutely toxic, flammable or otherwise hazardous. In such situations, IS concepts may be more appropriately applied in the consideration of additional layers of protection, facility layout, process safeguards, or buffer zones to reduce the vulnerability of hazardous chemicals to theft, diversion or destruction.

Vulnerability. Similar to the way that layers of protection are used to reduce the probability or frequency of a chemical accident, layers of protection may be implemented to reduce an asset's vulnerability by making the asset less attractive, and by increasing the difficulty of a successful attack. Depending on the security concern associated with an asset, such layers may include:

- Facility or asset perimeter security
- Access controls for persons, vehicles and packages
- Measures to deter, detect and delay an attack
- Measures to secure chemicals and equipment to prevent theft, diversion, contamination or sabotage
- Securing business, process control and other computer systems
- Monitoring, communications and warning systems
- Emergency response plans and recovery

See 6 CFR 27.230 for a list of security performance measures required by the Department of Homeland Security for high risk chemical facilities.

Traditional security measures such as physical security, background checks, administrative controls, access controls, or other preventive or protective security measures may also be considered from an inherent safety perspective. Applying IS concepts to layers of security, particularly minimization and simplification strategies, can make them more robust, reliable, effective, available, and maybe even less expensive to install or maintain. Ideally, layers of protection should be:

- Independent: not sharing common-cause failures (i.e., loss of power, single human operations, or maintenance acts that would cause more than one layer to fail);
- Effective: capable of detecting, delaying, and/or deterring an adversary;
- Auditable: each layer can be checked for its effectiveness;
- Passive: requiring no energy or human input to function properly.

However, passive countermeasures may not be possible for all security elements. Gates will need to open and close, lights and cameras will need power, a human will likely be required to implement a visitor/guest vetting and escort procedure and provide response to interdict and neutralize an adversary.

In some instances, security countermeasures may also increase safety at the facility. For example, a facility may create a buffer zone between the public areas and the plant fence, and another between the plant fence and critical process equipment. These security measures will create:

- *detection zone(s)* if they are free of obstacles and have sufficient depth to allow for adversaries to be detected while attempting unauthorized entry;
- *standoff zone(s)* provided they have sufficient depth to keep adversaries from using explosives or standoff weapons effectively from the perimeter;
- *delay zone(s)* allowing intervening force time to respond, or time for operators to take evasive action before an adversary reaches a target following detection.

Such buffer zones may also reduce potential off-site impacts from an accidental release by increasing the distance between the potential release and an off-site receptor.

9.7 LIMITATIONS TO IMPLEMENTING IS CONCEPTS IN SECURITY MANAGEMENT

As discussed in Chapter 11, identifying and implementing inherently safer measures may entail evaluating risk tradeoffs to help assure that risk is actually being reduced and not shifted elsewhere or converted to a different one. This is especially true in the context of security in which an inadvertent shift in risk from one population to another could have significantly severe consequences.

9.7.1 Balancing competing security or other concerns. Consider a batch process that includes a large storage tank for a hazardous intermediate product to allow the facility to switch among different uses of the intermediate. While the facility could redesign the process to remove the intermediate storage and, thereby remove or reduce the consequence, the manufacturing flexibility provided by the intermediate storage may be necessary for the facility's continued operations. In order to remain in operation, a more viable security strategy would be to establish or enhance other security countermeasures. Continued operation itself could be a security issue if the facility produced a mission or economically-critical chemical, or if the facility itself was critical to the regional economy.

9.7.2 Creating new security concerns. As described in Chapter 4, phosphorous oxychloride could be used in place of phosgene, and 1,1,1-trichloroethane could be used in place of hydrochloric acid gas (Ref 9.6 Kletz). Phosphorus oxychloride and trimethyl phosphite are listed as chemical weapons precursors under the Chemical Weapons Convention, (Ref 9.7 OPCW) and export controls which create additional recordkeeping and reporting burdens for the facilities. However, phosphorus oxychloride and trimethyl phosphate are listed as Chemicals of Interest under the Chemical Facility Anti-Terrorism Standard (CFATS), 6 CFR Part 27. From a security standpoint, these chemicals present a concern for theft or diversion because any of them can be converted to chemical weapons using simple chemistry, equipment and techniques. A facility converting to this alternative process could still be required under CFATS to implement security measures to assure that these chemicals are not stolen from the facility, and that shipments are not misdirected or diverted. Moreover, phosgene is also listed in the CWC and as a CFATS chemical of interest for theft, but little phosgene is shipped, meaning that a CFATS facility with phosgene would be less likely to be required to implement anti-theft/anti-diversion measure.

9.7.3 Shifting security risk from one concern to another. Chlorine may raise two different security concerns, depending on the quantity and container – *release* of chlorine gas from a vessel at a facility could have the potential to create serious off-site consequences, or *theft or diversion* of chlorine in transportation packaging for use as a chemical weapon at an unknown time and location.

A facility using chlorine from 90-ton rail cars, a potential security release concern, could reduce the potential consequences from an attack on the facility by switching to chlorine from smaller cylinders. However, chlorine in smaller containers could create a security theft/diversion concern not only for the facility itself but, by increasing the number of chlorine cylinders, could create a transit threat as well. Increasing the number of cylinders at the site also provides greater opportunity for a shipment to be diverted, or individual cylinders stolen, for nefarious purposes. Alternately, if the process were modified to substitute aqueous chlorine bleach versus gaseous chlorine, the target attractiveness and consequences may be significantly reduced.

A facility with multiple potential targets for terrorists may adopt a security policy to minimize vehicular traffic into or out of the facility due to concerns that a vehicle coming into a facility could be carrying an explosive to attack an asset within the facility or could be used to smuggle weapon or bomb precursors out of the facility. This could contrast with a recommendation to reduce the hazardous inventory on-site by increasing the number of smaller shipments into the facility. Either the reduction in vehicular traffic or the reduction in inventory could be more desirable, depending on other conditions at the site or in the surrounding community (see examples below). Regardless, both would need to be identified and evaluated to assure that the increase in the vulnerability created by the increased vehicular traffic did not create a greater overall security risk than the potential consequence created by the larger inventory.

Shifting security risk from one location to another: As described in Chapter 13, a plant might reduce the size of a hazardous material inventory on the site such that the facility would receive shipments by truck (typically about 30,000-pound shipments) rather than rail (typically about 300,000-pound shipments for many materials) thus increasing the number of shipments by a factor of 10. Not only would the increase in the number of shipments potentially increase the risk of shipment diversion (discussed earlier), but a truck route past a terrorist's target (such as a school or government building) might increase the attractiveness of the shipment itself as a target, particularly if the rail line did not run past iconic or other soft targets. This is not to say that the reduction in inventory should not be considered, only that security factors outside the facility must also be considered before a decision is made.

If the security of alternative storage is less robust than at the facility, the overall security risk would not have been reduced, but merely shifted to another location. Knowing whether moving the inventory reduces or increases security risk would depend on an evaluation of the relative security postures of the facility and the alternatives. If the facility security were more robust, then increasing the amount of hazardous material stored within the facility's security perimeter might reduce the overall security risk to the community, since the CFATS (6 CFR 27) requirements for security at chemical facilities does not extend to chemicals of interest stored in rail yards.

9.8 CONCLUSION

Inherently safer systems can be beneficial for security as well as for safety risk management. Inherent safety can be applied for security as risk assessment determines to reduce the hazards and consequences of misuse.

Inherent safety and process security alternatives can be complex. Inherent safety may or may not be the most effective security strategy, particularly if a new security concern is created or the security risk is simply shifted. It may be unrealistic to expect that all hazards are eliminated by inherent safety. Instead, all options must be examined in a complete context to assess risk reduction benefits and costs and to consider risk/risk tradeoffs that may occur. They also must be weighed against the benefits of the operation, as well as alternatives, such as more traditional security measures, to manage exposures, and whether overall risk is sufficiently managed.

9.9 REFERENCES

9.1 American Chemistry Council (ACC), *Responsible Care*, (www.responsiblecaretoolkit.com/security.asp)

9.2 American National Standards Institute/American Petroleum Institute (ANSI/API); ANSI/API Standard 780 "Security Risk Assessment Methodology for the Petroleum and Petrochemical Industries, First Edition; 2013

9.3 American Petroleum Institute (API), new.api.org/policy/otherissues/upload/SecurityGuideEd3.pdf

9.4 Center for Chemical Process Safety (CCPS), *Guidelines for Managing and Analyzing the Security Vulnerabilities of Fixed Chemical Sites*, American Institute of Chemical Engineers, 2002.

9.5 Chemical Facility Anti-Terrorism Standards (CFATS), 6 CFR Part 27, Interim Final Rule published April 9, 2007 (72 *Fed. Reg.* 17696).

9.6 Kletz, T.A. (1998). *Process Plants: A Handbook for Inherently Safer Design.* Philadelphia, PA: Taylor & Francis.

9.7 Organization for the Prohibition of Chemical Weapons (OPCW), Convention on the Prohibition of The Development, Production, Stockpiling and Use of Chemical Weapons and On Their Destruction, (www.opcw.org/)

9.8 United States Department of Homeland Security (DHS), *Energy Sector Specific Plan*, 2010.

10. Implementing Inherently Safer Design

10.1 INTRODUCTION

To be most effective, implementation of inherently safer designs/technologies requires a systematic management approach, sound technical basis, and management and cultural emphasis on inherent safety as an important organizational value and tool for reducing the risk of process-related incidents. At a corporate level, it first requires management commitment and leadership to provide resources and establish policies and procedures to integrate the use of inherent safety into the framework of the company's overall process safety management program. This may be also driven by regulatory requirements to implement ISD in addition to a company's commitment to reducing risk. As a general philosophy, its concepts should be woven into the way that facilities are designed, constructed, operated and maintained, so that company practitioners can continuously look for and identify ways to maximize the inherent safety of an operation. This is very similar in concept to the current industry emphasis on "Lean Manufacturing," which involves an overall management approach to eliminate waste in over-production, waiting time, transportation, processing, inventory, motion and scrap, and on "Kaizen," a philosophy of continual improvement in business processes to accomplish this objective.

It is now more common for companies to build inherently safer design principles into their process safety management systems. For example, this can be accomplished by incorporating inherently safer design concepts into existing safety and process hazards reviews. Companies may wish to enhance their existing review systems with inherent safety reviews at key points in the process life cycle. This chapter discusses methodologies to conduct inherent safety reviews at three key stages of the life cycle:

1. During product and process development;

2. During conceptual facilities' planning early in the process design phase; and,
3. During various phases of routine operation, including modifications and incident investigations.

Many companies may elect to include the objectives and features of these reviews into their existing process safety management system. If companies decide to incorporate inherent safety into existing systems and reviews, particular attention needs to be given to the subject in R&D for new product and process development. This is the life cycle stage where application of inherently safer design concepts can have the greatest impact. The reality is that retrofitting ISD into an existing process design is much harder and much more expensive than building it into the design originally. Overton and King (Ref 10.19 Overton) note that implementation of inherently safer technology will be an evolutionary process occurring over time. For example, a management system that calls for periodic IS reviews for existing processes, such as PHA revalidations, provides on-going opportunities to identify new technology breakthroughs that can be economically applied to increase the level of inherent safety in a process. Significant risk reduction may be achieved in this manner. An important key to incorporating the tenets of IS into the entire spectrum of process safety management activities is making sure that operations, maintenance, and engineering staffs are aware of its technology and benefits.

10.2 MANAGEMENT SYSTEM APPROACH FOR IS

An effective inherent safety (IS) program requires an integrated management system approach that embodies the "Plan-Do-Check-Act" management cycle. A high level corporate process safety management policy and program that explicitly includes IS as one of the key principles of risk management is generally a starting point. This document would typically outline the management system required, and would include, at a minimum, the following elements within the framework of the process safety management program:

- Development of IS policies, procedures, and design standards
- IS education for affected functions (i.e., research, process design, engineering, operations, maintenance)

- IS reviews of new and existing facilities, including recommendation follow-up tracking (may be incorporated into project design reviews and facility PHAs)
- Implementation of IS concepts in on-going aspects of an operation (i.e., management of change, maintenance, SOPs)
- A systematic management review process, including performance metrics, performance assessment, and review and implementation of improvements

Amyotte, et al. (Ref 10.1 Amyotte) describe how IS concepts can be integrated into each element of a process safety management program, including maintenance and operating procedures, training, management of change, and incident investigation. Dowell (Ref 10.11 Dowell) emphasizes the need to integrate the many environment, safety and health elements, including those involving process safety management, into a comprehensive management system so that they become part of a company's way of doing business, and not separate management programs. IS is unlikely to succeed as a separate program; it needs to be integrated into an overall Process Safety Management (PSM) program. The process hazard analyses (PHAs) will drive understanding of the process hazards/risks, after which IS techniques could be used to control or eliminate the identified hazards/risks.

In this way, *inherent* safety becomes an *inherent* part of the company's system for improving process safety and reducing risk. Application of IS principles to Risk Based Process Safety elements is addressed more comprehensively in Chapter 11.

Like other management systems, an IS management system moves ISD from the conceptual stage to implementation in relevant process safety-related activities. Without a management systems approach, IS will remain as a concept only, and not a functional element of the organization's safety efforts.

10.3 EDUCATION AND AWARENESS

10.3.1 Making IS a Corporate Philosophy

Once management has established its commitment to the principles of inherent safety, including the development of a management system through policies and procedures, training is required for those who have responsibilities that impact the consideration and use of IS in process safety-related activities. Management promotion and adoption of the principles of inherent safety, wherever possible, is necessary to provide a visible demonstration of leadership and commitment to effective process safety management (Ref 10.1 Amyotte). However, this is not sufficient by itself to establish the IS program. Education and awareness of IS concepts and application are also necessary.

As a general philosophy, one of the objectives of an IS education and awareness program is that all employees responsible for design or engineering decisions become aware of inherently safer concepts and "automatically" look for opportunities in their everyday work activities. This is more proactive than any review process. As technologists do their design and operations tasks, they understand the value of IS and implement the concepts where appropriate at all levels of detail, from basic process chemistry selection by research chemists, to the writing of operating procedures by plant personnel.

10.3.2 IS in Education

In recent years, the basic concepts of chemical process safety have increasingly been integrated into the undergraduate chemical engineering curriculum at many colleges and universities. As the profile of inherently safer technology increases, interest in IS is expected to likewise increase. In order to better prepare students for work in the chemical process industries, this education should be extended to both chemistry and technical management programs. Basic chemistry is a key to ISD, and chemists need to be taught to recognize this and look for IS chemistries when conducting research into new or modified processes. IS has a parallel in current efforts to institute the philosophy and practice of "green chemistry," which is the design of chemical products and processes in such a way as to reduce or eliminate the use and generation of hazardous substances, to minimize or prevent pollution and the creation of hazardous waste. Green chemistry, which has inherent safety as a parallel benefit, is an overarching philosophy of chemistry now becoming an important part of the chemistry education curriculum.

Another parallel to IS that exists in the environmental field is the current emphasis on pollution prevention, which calls for designing processes to minimize the amount of waste generated, rather than relying on "end-of-pipe" means to treat wastes generated. This approach also has its basis largely on basic process design and operating conditions, as well as modifying raw materials and, in some cases, products. This philosophy has been adopted by many companies, and has, in fact, been instituted by regulation in many states through requirements for Pollution Prevention plans.

CCPS sponsors SAChE (Safety and Chemical Engineering Education), whose major activities include the development of instructional materials for chemical process safety education in the chemical engineering curriculum, as well as sponsorship of workshops for chemical engineering faculty on process safety topics. SAChE has developed an instructional module on inherently safer processes, as has the Institution of Chemical Engineers in the UK in conjunction with the International Process Safety Group (Ref 10.16 IChemE) (see Section 10.7).

Teaching inherently safer technology concepts in undergraduate chemistry, chemical engineering, and related disciplines will be of great benefit as students move into industry after graduation, and will help further the practice of IS.

10.4 ORGANIZATIONAL CULTURE

An organization's Safety Culture, which is the collection of its values, beliefs, and behaviors—or "the way the organization conducts its business"—flows from visible management commitment and leadership. As an active strategy for eliminating or reducing—as opposed to just managing—risk, inherently safer operations and the associated innovation that is often required must become ingrained in the comprehensive process safety culture, in the same way that the commitment to Total Quality Management (TQM) or Lean Manufacturing has become ingrained—through management leadership and commitment. In most companies today, there must be a well-understood business benefit in addition to the safety benefit in order for management to be convinced of the need for inherently safer operations. This often comes through effective training and communication throughout the organization on the concepts, applications (particularly in process development and early design stages), and benefits of ISD—such as reduced inventories, smaller equipment, and fewer add-on safeguards to be maintained, etc.—as well as its limitations. The evaluation of both initial and on-going costs associated with inherently safer designs can often help enhance this understanding. Many IS improvements can be implemented cost-effectively, particularly incremental improvement in existing plants. Kletz (Ref 10.18 Kletz,) discusses cultural and management barriers to the implementation of inherently safer designs, as well as the actions needed to overcome these barriers. Khan and Amyotte (Ref 10.17 Khan) also provide guidance for making the use of inherently safer design principles more routine.

Turney (Ref 10.22 Turney) lays out five steps, taken from the European Process Safety Centre's Statement of Good Practice, which are necessary for effective adoption of inherent safety within an organization:

- Support by a champion
- Suitable training
- Application from the earliest stage of a project
- Reviews throughout project development
- Recognition and reward of those involved in the project

By incorporating inherently safer design into the organizational culture, inherent safety becomes an on-going way of examining and addressing processes and their hazards, and this philosophy then permeates all aspects of the process safety management systems. For example, once its application is fully understood, the concept of simplification with respect to human factors can become an inherent aspect of the writing of standard operating procedures.

10.4.1 *Multiple Demands of IS in the PSM program*

Dowell (Ref 10.11 Dowell) points out that excellence in operations, which must include personnel and process safety, requires a comprehensive management system approach. The challenge is to integrate the many ESH compliance standards (PSM, RMP, Responsible Care®, RP750, company requirements, etc.) and special program activities (ISO 9000, Sustainability, etc.) into the fabric of the system of making chemicals, which, in turn, is grounded in company culture. These activities must go beyond doing the minimum required to give the appearance of compliance. Figure 10.1 illustrates the required foundation of

Figure 10.1: Management leadership is the foundation for process safety management

management leadership (Ref 10.3 Auger). Managing process safety is based on a corporate mission which management implements. By their words and actions, managers at all levels of the corporation must understand the benefits of IS and show their commitment to such programs in tangible ways. The latter should include providing the necessary resources, such as personnel, time, and funding, for IS-related activities like reviews and evaluation/ implementation of follow-up recommendations that arise.

10.4.2 Incorporating IS into Normal Design Process

As stated previously, IS should be made a part of the normal design process, by incorporating IS thinking into design processes and standards. Only by implementing IS consistently throughout the organization—from Research through Operations—will it become "the way we do business."

Consideration of IS should begin at the start of the process life cycle, which is the initial process research and development. Because the basic level of inherent safety is shaped at this stage, attention to IS at this point can have the greatest impact on process safety—even more so than at design stage reviews where piping and instrumentation drawings (P&IDs) have already been developed. It is imperative that designers incorporate IS into their thinking in the initial process selection, design, and equipment specification, as the ability to modify the design becomes more difficult as the project progresses into the detailed design and construction stages.

When a hazard is identified at any stage of development, the designer should ask the following questions in this order (Ref 10.14 Gupta):

1. Can the hazard be eliminated (1st Order of IS)?
2. If the hazard cannot be eliminated, can the magnitude of the hazard, or potential incidents arising from the hazard, be significantly reduced (2nd Order of IS)?
3. Do the alternative designs identified from the answers to Questions 1 and 2 increase the magnitude of other existing hazards, or introduce any new hazards to the process?
4. If the alternative designs increase other hazards, or introduce new ones, develop an understanding of these hazards and their relative importance.
5. Identify passive, active, and procedural safeguards appropriate for management of all hazards.
6. Use a logical decision process to select the optimum design, considering inherent safety and other safety characteristics, and all other relevant design parameters.

It is important for a designer, upon identifying a hazard, not to skip past the first four questions and go directly to question 5—identifying "add-on" safety features to manage and control a hazard whose existence and magnitude the designer accepts. While it will not always be possible or feasible to eliminate or significantly reduce all hazards, it is certain that it will never happen if the question is never asked. This is the heart of the inherently safer design philosophy – to first ask if hazards can be eliminated from the process, or if they can be significantly reduced. Designing of safety systems to manage and control hazards should not proceed until it has been firmly established that it is not feasible to eliminate or reduce these hazards from the start.

Basic process selection is at the core of inherent safety at the pre-design stage, for this is the point where the basic chemistry and selection of unit operations is established. Where one or more alternative processes for producing the desired product exist, they should be compared in terms of their inherent hazards, such as raw materials used, intermediates and wastes produced, and operating conditions including temperature and pressure. Selection of the inherently safer process is not always straightforward, and a number of conflicts may exist that need to be reconciled to optimize the level of risk against other factors (see Chapter 13). Ankers (Ref 10.2 Ankers) describes a software application for identifying inherently safer process options, with an emphasis on early hazard identification. A number of methods have been developed for measuring and comparing the relative level of inherent safety between two or more processes. These are discussed later in this chapter.

The use of corporate design standards, which incorporate the use of applicable inherently safer design features can also be beneficial in providing guidance for designers of new processes and helping to establish company expectations and decisions regarding risk minimization. Many companies have established internal design guides that incorporate the use of IS designs for new facilities, as well as for major modifications to existing facilities.

Many companies rely on engineering contractors for design work at various levels, from detailed engineering of a process developed by the customer, all the way to "turnkey" licensed technology and plants. Companies should clearly define expectations for IS considerations in the design process, *before* contracts are awarded. Some contractors may not be aware of IS concepts and how to implement them, so providing an engineering standard on how these concepts are to be considered is a crucial factor. The company should provide guidance as to how IS should be considered in layout and residual hazard management, as well as how it will be documented.

The 2015 Control of Major Accident Hazards (COMAH) regulation in the UK requires the consideration and documentation of inherently safer design alternatives during the initial design stage. In addition to the standard IS strategies, it includes plant layout as part of the overall approach, particularly to prevent the potential for domino effects by fire, explosion (pressure wave and missiles) or toxic gas cloud that could cause loss of operations control in another location. Plant layout, which is established during the design stage, is often a compromise between several factors such as:

- The need to keep distances for materials transfer between plant/storage units to a minimum to reduce costs and risks;
- The geographical limitations of the site;
- Interaction with existing or planned facilities on-site, such as existing roadways, drainage and utilities routings;
- Interaction with other plants on-site;
- The need for plant operability and maintainability;
- The need to locate hazardous materials facilities as far as possible from site boundaries and from people living in the local neighborhood;
- The need to prevent confinement where release of flammable substances may occur;
- The need to provide access for emergency services;
- The need to provide emergency escape routes for on-site personnel; and,
- The need to provide acceptable working conditions for operators.

These issues are partially dealt with in the PSM and RMP regulations under Facility Siting, which involves the spatial relationship between process hazards and people but can also be applied to the spatial relationship between hazards and other buildings, equipment, or processes. In its most basic form, using inherently safer strategies for siting calls for locating hazardous operations in remote areas, away from inhabited areas, with sufficient buffer zone for anticipated "urban creep". For this reason, the manufacture of explosives is conducted in unpopulated areas of the country.

Other IS considerations dealt with primarily at the design stage include:

- Designing vessels to withstand the maximum temperature and pressure to which they are likely to be subjected in operation.
- Designing control valves to limit the maximum flow/temperature/pressure that downstream equipment can withstand.
- Use of seal-less pumps and designing pump spillback loops to reduce risk associated with deadheading.
- Selection of materials of construction that are compatible with the process fluids to be handled and the chosen design conditions, in order to minimize corrosion/erosion potential rather than relying on inspection programs to identify and correct.
- Minimizing the size of vessels, piping and other equipment to limit the potential release quantity.
- Minimizing the number of flanged and threaded joints and hoses/flexible connections which present increased leak potential.
- Addressing issues associated with reactive chemicals, such as eliminating the potential for water contamination of the process if a water-reactive material is being used.

In most cases, an assessment of alternatives will involve consideration of cost—both initial and ongoing, such as inspections, testing and maintenance—and other factors, including relative risk reduction using other risk management approaches, such as emergency relief/mitigation system vs. higher vessel design pressure. For complex, technically challenging, or inherently higher risk projects, a formal risk comparison should be carried out using tools such as ranking methodologies or quantitative risk assessment. Ankers (Ref 10.2 Ankers) describes a software application for identifying inherently safer process options, with an emphasis on early hazard identification. Other methods are discussed later in this chapter.

10.5 INHERENT SAFETY REVIEWS

The quantity of literature on company inherent safety policies and procedures has continued to increase in recent years, indicating that awareness of inherent safety principles is improving. Besides presenting an opportunity to reduce risk, IS reviews are mandated by regulation for high-risk facilities in New Jersey and Contra Costa County, California. Recently, California has adopted revised process safety regulations for oil refineries in the state that include a strong IS component. While consideration of IS is strongly implied in UK and EU safety regulations, there are no formal requirements to conduct IS reviews (See Chapter 14). The discussion of IS in relation to reduced security risk, and its consideration for inclusion in chemical security legislation, have also raised the visibility and awareness of IS in the U.S. Companies which have chosen to incorporate inherent safety principles into their existing hazard management programs may want to consider highlighting, or otherwise identifying, inherently safer design aspects of those programs. This will promote an increased awareness of the concepts among a broad population of chemists, engineers, and business leaders. In turn, this heightened awareness will encourage the application of inherently safer design principles as a part of the normal, everyday work processes.

Several companies have published descriptions of their inherent safety review practices:

- Bayer (Ref 10.20 Pilz) uses a procedure based on hazard analysis, focusing on the application of inherent safety principles to reduce or eliminate hazards.
- Dow (Ref 10.21 Scheffler; Ref. 10.13, Gowland) uses the *Dow Fire and Explosion Index* (Ref 10.10 Dow 1994b) and the Dow Chemical Exposure Index (Ref 10.9 Dow 1994a) as measures of inherent safety, while applying inherently safer design principles to reduce hazards.
- Exxon Chemical (Ref 10.12 French; Ref 10.23 Wixom) has developed an inherent safety review process that includes applications at various points in the process life cycle. Its expansion of IS concepts to safety, health, and environmental reviews is documented in *Inherent Safety, Health and Environmental (SHE) Reviews*.

- BP has published *Inherently Safer Design Guidelines for New Projects and Developments*.
- DuPont (Ref 10.6 Clark) describes how ISP (Inherently Safer Processing) is integrated into the overall corporate PSM program, based on a checklist and a corporate training program. A semi-quantitative ISP scoring system is used by corporate R&D to ensure that it is appropriately considered at the earliest stage of a process life cycle.

In addition, the Contra Costa County, CA, Health Services Department has issued a guidance document for the IS review of existing facilities, and new facilities at the chemistry-forming, facilities design scoping and development, and basic project design stages. IS analyses must be performed for all situations where a "major chemical accident or release"—as defined in the standard—could reasonably occur. This document also includes guidance for evaluating the feasibility of recommendations, and for IS review documentation. IS reviews for existing processes can be conducted as part of an initial/five-year PHA or as a separate study, using a checklist or guideword analysis that incorporates IS. See Appendix A for a detailed example of an IST checklist. Table 10.3 that appears later in this document offers sample Guidewords, while Table 10.5 shows a sample Guideword Matrix.

10.5.1 Inherent Safety Review Objectives

The objectives for an inherent safety review are to employ a synergistic team to:

- Understand the hazards.
- Find ways to eliminate or reduce those hazards.

The first major objective for the inherent safety review is the development of a good understanding of the hazards involved in the process. Early understanding of these hazards provides time for the development team to implement recommendations from the inherent safety effort. Hazards associated with flammability, pressure, and temperature are relatively easy to identify. Reactive chemistry hazards are not. They are frequently difficult to identify and understand in the lab and pilot plant. Special calorimetry equipment and expertise are often needed to fully characterize the hazards of runaway reactions and decompositions. Similarly, industrial hygiene and toxicology expertise is desirable to help define and understand health hazards associated with the chemicals employed.

Eliminating and reducing hazards and their associated risks is the second major objective. Applying inherent safety principles early in the product/process development effort provides the greatest opportunity to achieve the objectives of the inherent safety review process for the project at hand. If these principles are applied late in the effort, the results may have to be applied to the "project after next," as the schedule may not permit implementation of the results.

Experience with inherent safety reviews for a new process or project indicates that project investment costs are often reduced as a result of this exercise. Eliminating equipment and reducing the need for safety systems are typically the main contributors to investment reduction. Capturing these potential savings largely depends on the timing of the reviews within the process or project development cycle.

The same two primary objectives apply to an IS review of an existing process. Generally, these reviews are designed to identify either incremental improvements, primarily in the area of simplification, possible major improvements that can be made in the event of a process renovation or major equipment replacement or IS opportunities for future facilities manufacturing the same product. Often, large capital expenditures to improve IS are difficult to justify if the process has been operating without major incident. However, when major projects are planned, IS reviews can identify opportunities for incorporating inherently safer design features.

10.5.2 Good Preparation is Required for Effective Inherent Safety Reviews

Any IS review should begin with a brief overview of IS concepts, as well as the process being studied and the approach to conducting the review (HAZOP, What-If?, Checklist, etc.).

The completeness of the information described in steps 1 to 9 (Chapter 10.5.5) prior to the review will influence the quality of the inherent safety review. The chemist needs to define the desired reactions and develop an understanding of potential side reactions.

Effects of mischarges or process deviations on reaction chemistry should be developed. These process chemistry information requirements are discussed in Section 5.2.

Reaction kinetics are often difficult to develop in the normal experimental laboratory or pilot plant. Specialized reaction calorimeters and reactive chemistry testing, such as differential scanning calorimeter (DSC), Accelerating Rate Calorimeter (ARC□), and a vent sizing package (VSP) are required to obtain thermodynamic and kinetic information (Ref 10.4 CCPS 1995). Some chemical companies have established reactive chemistry laboratories with reaction engineering specialists to assist in developing this information. Specialized analytical laboratory testing firms also provide this expertise and experimental capability.

Industrial hygiene/toxicology professionals need to develop acute and chronic toxicology information on all the materials used and produced in the process. This information should also include the potential products of abnormal reactions. The industrial hygiene member of the team should be prepared to explain the toxicology information on the safety data sheet (SDS) to the review team.

Another necessity is physical property data for the process chemicals. This information includes such items as melting point, boiling point, vapor pressure, water solubility, flammability data, and odor threshold, to name a few. All this information must be organized and coordinated to satisfy the company's process and facilities documentation, and regulatory requirements.

For IS reviews at the design stage, and for existing facilities, accurate P&IDs are also necessary, as well as updated standard operating instructions, and process safety information that includes the size of the equipment and the materials of construction. Operations and maintenance personnel with experience in the process or similar processes should be included in the review, as they often have valuable knowledge and experience to contribute.

10.5.3 Inherent Safety Review Timing

Inherent safety reviews should be considered during appropriate stages of a typical project's life cycle. These may be stand-alone reviews or incorporated with other process hazard analysis or appropriate project review. Suggested timing includes:

- During the chemistry forming (synthesis) stage for product/process research and development to focus on the chemistry and process (refer to Section 8.2).
- During the facilities design scoping and development prior to completion of the design basis to focus on equipment and configuration (refer to Sections 8.3 and 8.4).
- During the basic design phase of the project (i.e., design stage PHA), to ensure that identified IS opportunities are incorporated into the design.
- During regular operations to identify potential improvements in existing facilities, as well as those that can be incorporated into the design of the next plant.

The timing of IS reviews in the first three stages is critical. For each stage, there is a window of opportunity. If carried out too early, essential information may not be available. If carried out too late, the design will be too far advanced, and changes may cause significant expense as well as project delays (Ref 10.18 Kletz).

It is recommended that an IS review be revalidated periodically, like the five-year PHA cycle. The revalidation should include and document the following:

- IS and other process safety improvements made since the last review was conducted or a new method was selected to perform the IS analysis;
- An IS review for all changes that have been made since the last analysis, unless specifically reviewed for IS under Management of Change;
- Review of any significant incidents that occurred in the process since the previous review; and,
- Review of any new and existing technologies not previously reviewed that can potentially make the process inherently safer.
- Review of any new corporate requirements or regulatory changes that necessitate having an inherently safer option for the said unit.

10.5.4 *Inherent Safety Review Team Composition*

The composition of the inherent safety review team will vary depending upon the stages of the development cycle and the nature of the product/process. The team composition is generally four to seven members selected from the typical skill areas checked in Table 10.1. Knowledgeable people, with an appropriate matrix of skills, are required for a successful review effort. The industrial hygienist/toxicologist and chemist play key roles on the team to ensure understanding of the hazards associated with reactions, chemicals, intermediates, and products, and to explain hazards, especially where there may be choices among chemicals or processes.

Often, an organization will strive for the elimination of a specific toxic material from a given process, such as chlorine from a water or wastewater treatment process. Alternatives will also have other hazards and risks that require an informed choice. The industrial hygienist, chemist, and safety engineer play an important role in generating the information needed to select between alternatives.

10.5.5 *Inherent Safety Review Process Overview*

As previously stated, good preparation is very important for an effective inherent safety review, particularly for a new process. Preparation for the review is summarized in Figure 10.2 and includes the following background information:

1. Define the desired product.
2. Describe optional routes to manufacture the desired product (if available), including raw materials, intermediates, and waste streams.
3. Prepare simplified process flow diagram and tentative layout.
    - Include alternative processes
4. Define chemical reactions.
    - Desired and undesired
    - Determine potential for runaway reactions/decompositions.
5. List all chemicals and materials employed
    - Develop a chemical compatibility matrix.
    - Include air, water, rust, byproducts, contaminants, etc.
7. Define physical, chemical, and toxic properties.

- Provide NFPA hazard ratings or equivalent
- Characterize combustible dust hazards (MIE, Kst, etc.)

8. Define process conditions (pressure, temperature, etc.).
9. Estimate quantities used in each process system (tanks, reactors, etc.).
    - State plant capacity basis
    - Estimate quantities of wastes/emissions

Table 10.1: Inherent Safety Review Team Composition

	Product Development	Design Development	Design Stage PHA	Operations
Industrial Hygienist/ Toxicologist	√	√	√	√
Chemist	√	√	√	√
Process Engineer	√	√	√	√
Safety Engineer	√	√	√	√
Process Technology Leader	√	√	√	√
Environmental Scientist/ Engineer	√	√	√	√
Control Engineer		√	√	√

Operator			√	√
Operations Supervisor		√	√	√
Maintenance			√	√

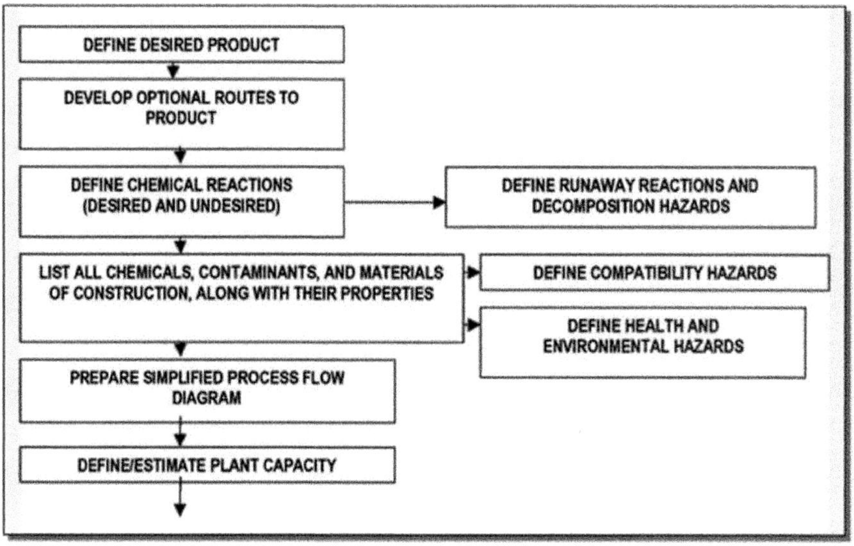

Figure 10.2: Inherent Safety Review Preparation

After the background information is developed, the inherent safety review can be arranged. The review steps are summarized in Figure 10.3. Steps proposed for the review are as follows:

1. Review the background information from steps 1 to 9 above.
2. Define major potential hazards.
3. Systematically review the process flow schematic, looking at each process step and hazardous material to identify creative ways to apply inherent safety principles to eliminate or reduce hazards.

4. During the inherent safety review, at the design development stage, identify potential human factors/ergonomics issues that should be addressed by the design team.
5. Document the review and follow-up items.

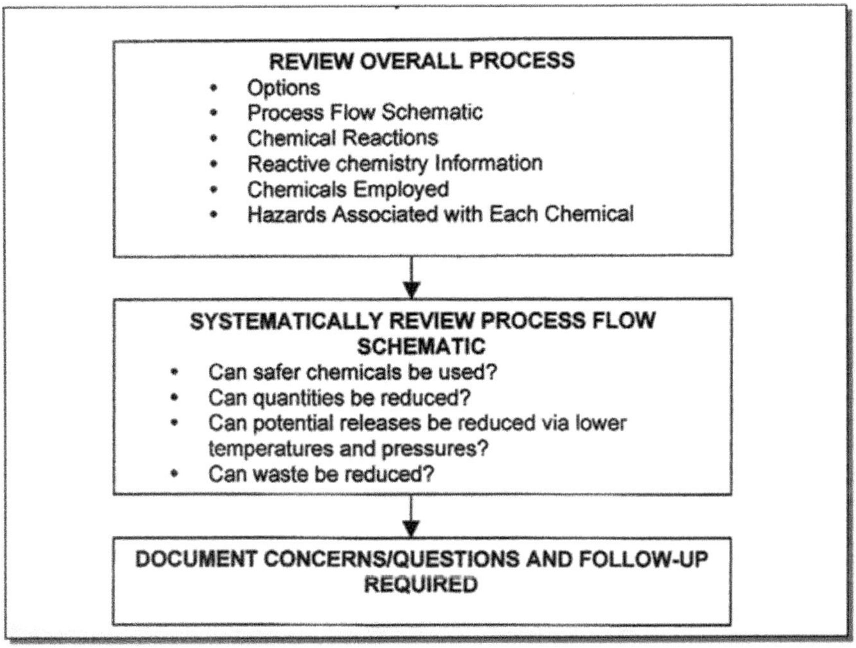

Figure 10.3: Inherent Safety Review Process

During the systematic review of the process flow schematic the team will examine some of the following questions:

- Can safer chemicals be used (i.e., non-toxic/non-flammable or non-volatile reactant)?
- Can quantities be reduced (particularly intermediate requiring storage)?
- Can potential releases be reduced via lower temperatures or pressures, elimination of equipment or by using seal-less pumps?

- Can waste be reduced (i.e., by using a regenerable catalyst or by recycling)?
- What additional information (toxicology information or reactive chemicals data) is required?
- Can the equipment or process be simplified, be made mistake proof, or at least mistake tolerant, by applying ergonomic/human factors principles?
- Have trace materials or contaminants in the process been identified, along with their effects?

It is not unusual to have several inherent safety reviews during the product/process development effort. Early reviews will often not have all of the information required for steps 1 to 9 as depicted in Figure 10.2. The follow-up items will describe what is necessary to obtain the missing information, such as toxicity data on new intermediates or products of undesired reactions.

A more comprehensive IS review process example is provided in Appendix B. This example includes information on pre-study preparation, team training, and IS review conduct and documentation.

10.5.6 Focus of Inherent Safety Reviews at Different Stages

The focus of an inherent safety review will change from the chemistry-forming stage through regular operations. While each review is unique, Table 10.2 shows which areas receive emphasis at a particular stage. In the regular operations review, the team is identifying inherent safety opportunities not only for the existing process, but also for the next major expansion or grass roots facility. This is particularly true if the process was not originally designed with inherent safety in mind. The check marks indicate the relative emphasis each area receives.

In the "chemistry-forming" inherent safety review, the team will cover such things as:

- Understanding the hazards
- Choosing the best route to produce a given chemical or product
- Process improvements
- Reactor types and conditions
- Intermediate storage optimization
- Waste minimization

- Separation technology
- Requirements for additional information

During design scoping, the team will concentrate on minimizing equipment, reducing inventories, simplifying the process, reducing wastes, and optimizing process conditions. Inherent safety concepts should also be considered during process hazards reviews, such as HAZOP, for both new and existing processes. The initial design should be "mistake-proofed," and each safety device and procedure examined to see if there is a way to eliminate the need for it.

When the inherent safety process has been expanded to review regular or routine operation, the team should look at all aspects of inherent safety to provide suggested improvements for both the existing facility and for the next plant. Even if the process was originally designed with inherent safety in mind, these improvements may arise from advances in technology, changes in product specifications or application, or lessons learned from incidents and near-misses, both in the facility being studied or in similar facilities elsewhere.

Table 10.2: Focus of Different Inherent Safety Reviews

Note: The number of check marks indicates the relative importance of the strategy

	Chemistry and Process Selection	Design Scoping	Regular Operation
Minimize • Reduce quantities	√	√√√	√√
Substitute • Use safer materials	√√√	√	√√
Moderate • Use less hazardous conditions	√√	√√	√√

• Reduce wastes			
Simplify	√	√√√	√√

10.5.7 Stage in the Process Life Cycle

The following stages in a process life cycle have been previously defined:

- Process Research and Development
  - Chemistry
- Conceptual Design Phase
  - Process engineering
  - Pilot plant operation
- Detailed engineering
  - Plant design
- Construction and Startup
- Routine operations
  - Operations
  - Inspection
  - Maintenance
- Plant and Process Modification
- Decommissioning

As depicted in Figure 10.4, the types of IST reviews described in this chapter should be conducted at each process life cycle stage, using the most appropriate method.

What follows are examples of the kinds of decisions to which inherently safer design concepts should be applied by the designer or operator of a process at various stages of its life cycle (Ref 10.14 Gupta).

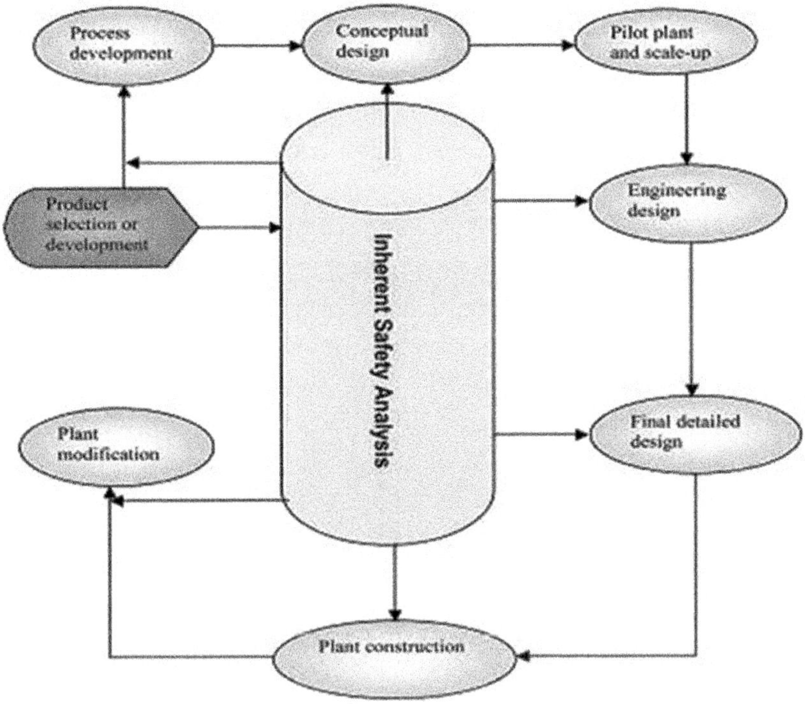

Figure 10.4. Applicability of inherently safer design at various stages of process and plant development.

Examples of Decisions to Which Inherently Safer Design Concepts Should be Applied

CONCEPTUAL PROCESS RESEARCH AND DEVELOPMENT. Selection of basic process technology, raw materials, intermediate products, by-products and waste products, chemical synthesis routes

PROCESS RESEARCH AND DEVELOPMENT. Selection of specific unit operations, types of reactors and other processing equipment, operating conditions, recycle streams, product purification, waste treatment processes

PRELIMINARY PLANT DESIGN. Location of manufacturing facility, relative location of units on a selected site, size and number of production lines, size of raw material, intermediate, and product storage facilities, selection of specific equipment types for the required unit operations, process control philosophy

DETAILED PLANT DESIGN. Size of all equipment, pressure rating and detailed design of all equipment and piping, inventory in processing equipment, location of specific equipment in the plant, length, route, and size of piping, utility design, layout of equipment, detailed control system design, plant-operator interface design

OPERATION. Identification of opportunities to modify plant to enhance inherent safety (reduce inventory, upgrade with more modern equipment, identify opportunities for inherently safer operation based on improved process understanding), consideration of inherently safer design when making modifications and changes, "user-friendly" operating instructions and procedures

New chemical processes will eventually replace the processes of today, and when these new facilities are designed, fundamental changes in chemistry and materials can be made more cost effectively. Process design depends on the physical and chemical properties of the materials being handled; changes can have far reaching implications in the equipment design and are best applied early in the engineering process

(Ref 10.19 Overton). IS reviews at the design stage of a new process are most critical for the implementation of inherently safer alternatives.

IS reviews should be conducted early in the development phase of a new process and then reviewed throughout the different project design phases, including the following, where applicable:

- During the chemistry forming (synthesis) phase for product/process research and development to focus on the chemistry and process
- During facilities design scoping and development, prior to completion of the design basis, to focus on equipment and configuration
- During the basic design phase of the project

However, the emphasis on the importance of early consideration of inherently safer design (ISD) alternatives can lead engineers and managers responsible for the operation of existing facilities to conclude that ISD is not relevant to them. They have an existing plant with established technology that is not easily or inexpensively changed. With existing plants, it is often not economically feasible to apply new inherently safer technology. If the proposed change involves different chemicals or significantly different operating conditions, then it is likely that a major capital expenditure will be required to replace existing equipment (Ref 10.19 Overton).

But ISD is relevant to existing plants and processes. Opportunities often exist to apply ISD principles to an existing plant, making it inherently safer. Furthermore, tools such as release consequence modeling and quantitative risk analysis (QRA) can be beneficial to understanding the potential benefits of ISD options, helping to gain management support for implementing ISD modifications to existing facilities (Ref 10.15 Hendershot).

Additional guidance on conducting IS reviews at the following major process life cycle stages is provided in Section D of the Contra Costa County Industrial Safety Ordinance (ISO) guidance document. See Chapter 14 for a detailed description of this ordinance and its application (Ref 10.7 Contra Costa County):

- Chemistry-forming phase

- Facilities Design Scoping and Development
- Basic Project Design
- Existing Process Units

10.6 REACTIVE CHEMICALS SCREENING

Experience has shown that reactive chemistry hazards are sometimes undetected during bench scale and pilot-plant development of new products and processes. These chemistry hazards must be specifically identified so they can be addressed in the inherent safety review process. Chemists should be encouraged and trained to explore reactive chemistry of abnormal operations. Simple reactive chemicals screening tools, such as the interactions matrix described in Chapter 8.2, can be used by R&D chemists.

The CCPS book, *Essential Practices for Managing Chemical Reactivity Hazards* (Ref 10.5 CCPS 2003), outlines a process for identifying and managing reactive chemistry hazards throughout the life cycle of a facility. Adequately managing chemical reactivity hazards requires routine knowledge of:

- the potential for uncontrolled reaction(s) to take place within the facility.
- how such reactions might be initiated (i.e., heat, contamination, inadvertent mixing, impact, friction, electrical short, lightning).
- how to recognize when an uncontrolled reaction is taking place.
- what the consequences would be if such a reaction took place (i.e., toxic gas release, fire, explosion).

Once the materials to be used in the process have been identified, they must be screened for potential reactive hazards. CCPS (Ref 10.5 CCPS 2003) provides a detailed methodology for reactive chemicals screening, both for intentional and unintentional reactivity issues, as well as guidance on how to manage those hazards. From an IS perspective, opportunities to eliminate or minimize, rather than manage the hazards, are preferred. This includes avoiding inadvertent reagents that can create a hazard through unintentional mixing or contact. For example, not using water in any process involving water-reactive materials is inherently safer than trying to prevent contact between the two. Many of the questions in the IST checklist provided in Appendix A can be adapted for consideration of reactivity hazards, such as:

- Can reactive raw materials inventory be reduced?
- Is it possible to completely eliminate reactive raw materials, process intermediates, or by-products by using an alternative process or chemistry?
- Is it possible to eliminate reactive components (i.e., water, iron, air, etc.) from the process to prevent inadvertent contact?
- Is it possible to design operating conditions such that materials that become unstable at elevated temperatures or freeze at low temperatures will not operate in those ranges?
- Is the equipment designed to totally contain the materials that might be present inside at the maximum attainable process temperature?
- Can equipment be designed such that it is difficult or impossible to create a potential hazardous situation due to an operating or maintenance error?

CCPS (Ref 10.5 CCPS 2003) also offers the following as a typical agenda for an inherent safety review that includes consideration of reactive hazards, either at the concept or development stage of a new facility.

- Review what is known of the chemical reactivity and other hazards that will need to be contained and controlled in the proposed process. This existing level of knowledge might come from past experience, suppliers, literature reviews, incident reports, etc.
- Based on the level of knowledge of chemical reactivity hazards, determine if additional screening will be needed for these hazards. Having reactive functional groups might indicate the need to perform literature searches, access databases or run differential scanning calorimetry.
- Discuss possible process alternatives and their relative hazards, including such topics as alternative solvents and possible incompatibilities to avoid.
- Brainstorm and discuss possible ways to reduce the hazards, such as a checklist that can be used as an aid to the brainstorming process.

- Obtain consensus on significant unknowns that will need to be addressed.
- Document the review, including attendees, scope, approach, and decisions.
- Assign follow-up items with responsibilities, goal completion dates, and a closure mechanism, such as a specific time to reconvene.

An inherently safer technology checklist is provided in Appendix A.

10.7 INHERENT SAFETY REVIEW TRAINING

IS training should include, at minimum, the following topics:

- The hierarchy of risk management strategies, from inherent to passive to active to procedural levels
- Various approaches to applying inherently safer systems, including minimization, substitution, moderation, and simplification
- Company design standards, examples, and other information that provides guidance on evaluation and implementation of IS design features
- Potential IS conflicts

A sample IS training program outline is provided below (Ref 10.15 Hendershot).

- History of Inherently Safer Design
- Basic Concepts
- Chemical Process Safety Strategies
- Inherently Safer Design Strategies- Minimize, Substitute, Moderate, Simplify
- Inherent safety and reliability
- Robust processes
- Inherent safety in other industries
- Inherent safety conflicts and measuring inherent safety
- When to consider inherently safer design
- Inherently safer design tools

Training in IS should also include company requirements for evaluating and implementing inherently safer technology during the life cycle of a process, as well as how this technology should be applied to the various process safety management program elements. As with most elements of safety, including process safety, employee participation is very important, as employees at all levels will be asked to help identify, evaluate, and implement changes to operations that may improve the inherent safety of the process. All relevant functions must have a good understanding of IS and its possible conflicts, so that good judgments can be made on the issue of risk reduction vs. cost, and the potential impact of changes on other important operating parameters, including quality, yields, operating costs, and waste generation, etc. It is also important to emphasize that the proper application of other risk management strategies, like active and procedural controls, can often satisfactorily manage risk at an acceptable level.

Training requirements for inherent safety should be merged into existing management systems. In general, inherent safety training objectives should be established for all personnel, including engineers, scientists, and business leaders, as well as for operations and maintenance personnel. To be truly effective, training should be provided at two levels: "awareness" level training to provide a basic understanding of IS principles and application, and "technical," or specific, training for those whose job function calls for the evaluation and implementation of IS. Awareness training should include the basic concepts of inherent safety, along with how these concepts are to be implemented in the organization's process safety management, new product development, and project execution programs. More specific training should be considered for those people who play key roles in inherently safer processes such as:

- R&D chemists involved in product and process development.
- Process development and process design engineers involved in process scoping, development, and design activities.
- Functional specialists, such as safety engineers and industrial hygienists.
- Method leaders (those personnel who will conduct the IS reviews).

If an organization chooses to include inherent safety reviews in their implementation approach, training should be included on how to conduct such a review.

The Institute of Chemical Engineers (UK) Inherently Safer Process Design (Ref 10.16 IChemE) is an excellent ISD program available on CD-ROM. Updated by the International Process Safety Group, the program provides the training support needed to ensure that the principles of inherent safety are understood and applied to both new plant designs and proposed changes to existing plants, with an emphasis on past experiences and lessons learned. It contains a CD-ROM with trainer's notes, 150 illustrated slides, video and a variety of case studies on best practices. New materials include a checklist section, more information about inherent safety assessment and design tools, and an extended bibliography of more than 400 inherent safety references. The program includes a comprehensive review, management issues, process intensification, sustainability, waste minimization and inherently safer processing, barriers to inherent safety and its relationship to cost and legislation, inherent safety and plant reliability, human factors, chemical reactivity hazards, and case studies.

Companies may wish to develop their own workshops to train potential team members in the inherent safety review process. The workshop can provide background information on inherent safety concepts, the extensive systems required to manage hazardous materials, and the elements of the inherent safety review process. Videos, problems, examples, and team exercises should be included to maximize student interest and learning.

The training management system should also define requirements for initial and refresher training. Definition of these requirements will assist in maintaining the IS culture and skill level by ensuring appropriate training for new employees, transferred and relocated employees and for experienced personnel.

10.8 DOCUMENTATION OF THE INHERENTLY SAFER DESIGN FEATURES OF A PROCESS

An important factor in implementing inherently safer design is the issue of company memory with respect to the management of process safety information. The intent of this component is to ensure that knowledge and information gained from plant experience that is likely to be important for the future safety of a facility is well-documented so that it is not forgotten or overlooked as personnel and organizational changes occur (Ref 10.8 CSChE). Hendershot (Ref 10.15 Hendershot) argues that this is especially critical when dealing with inherent safety and inherently safer design (ISD) features. He gives several examples where ISD features were essentially put at risk because the reasons they were implemented were not clearly and adequately documented. Potentially, this could compromise facility safety when future modifications are made by people who do not understand the original designer's intent, or what is involved in a particular safety feature. Examples include why a certain size feed line for limiting maximum reagent flow is part of the safety design basis, or how the routing of a pipe is intended to minimize consequences of a spill.

ISD features are particularly susceptible to lapses in corporate memory given that, unlike an add-on device such as a high-pressure alarm, they are such a fundamental part of the design that their purpose may not be obvious (Ref 10.15 Hendershot; Ref 10.1 Amyotte). All these inherently safer design features must be documented in original design manuals, and appropriate process safety information, including P&IDs and SOPs, must be readily available. This issue is also pertinent to Management of Change in that a proposed modification should be reviewed carefully to ensure no ISD features are being compromised.

10.8.1 IS Review Documentation

Whichever type of IS analysis is conducted, a report of the review should be generated to document the study. This report should include, at minimum, the following information:

- A summary of the approach used for the IS review (i.e., methodology, checklist used, etc.).
- Names and qualifications of the team facilitator/leader and team makeup, including positions, names, and any relevant experience or training.
- IS alternatives considered, as well as those already implemented or included in the design.
- If an independent inherently safer systems analysis was conducted, documentation should include the method used for the analysis, what inherently safer systems were considered, and the results of each consideration. If an IS checklist was used,

document the reasons why items were not considered, for example, if they were not applicable or had been considered previously.

- Documentation of rationale for rejecting potential IS opportunities (cost, creation of other safety or operability problem, etc.).
- Recommendations/action plans for further evaluation or implementation of IS alternatives identified during the study.

If the IS review was conducted as part of a larger study (i.e., PHA or hazard review), this information should be incorporated into the report of this activity. It is recommended that this information becomes a part of the permanent process safety file and be maintained for the life of the process. Electronic versions in an editable format (i.e., MS Word) should be maintained to facilitate future updates and revalidations.

The rationale for why recommendations from IS reviews were rejected should follow the following guidance, which includes for declining recommendations from incident investigations and process hazards analyses:

- The analysis upon which the recommendations are based contains factual errors.
- The recommendation is not necessary. For example, the safeguards may be inadequate, but the consequences are operational, or the consequence or severity of the scenario would not result in a significant release.
- Another IS alternative would provide a sufficient level of hazard reduction. (NOTE: Implementing only one option to address identified hazards may not be adequate to address the greatest hazard reduction or elimination. However, it is not necessary to implement more than one IS alternative if the implementation of a second IS alternative does not add any significant hazard reduction or has been documented as not feasible.)
- The recommendation is not feasible due to one or more of the reasons listed below:
    - The recommendation is in conflict with existing federal, state, or local laws.

    - The recommendation is in conflict with Recognized and Generally Accepted Good Engineering Practices (RAGAGEP).
- The recommendation is economically impractical, such that the process unit would stop being fiscally feasible. This can include the following factors:
    - Capital investment
    - Product quality
    - Total direct manufacturing costs
    - Operability of the plant
    - Demolition and future clean-up and disposal cost
- The recommendation would have a strong enough negative social impact that the project should not be implemented. Some examples could include an unacceptable visual or noise impact on the community, or increased traffic congestion.
- The recommendation may violate a license agreement that cannot be modified, and so must remain in effect.
- The recommendation may decrease the hazard but would increase the overall risk, or push the risk to a sector where it is less controllable (e.g.,from site to transportation).
- An alternative measure would provide more risk reduction than the IS approach rejected.

If the IS alternative recommended is rejected because it will create more risk, or if other non-IS modifications are made that would reduce the overall risk more than if the IS option were implemented, the rationale of how this determination was made should be documented. A qualitative or quantitative risk assessment could be used as a basis for the risk analysis.

10.8.2 Time Required for an Inherent Safety Review

The time required for IS process reviews depends on the type of review (i.e., new facility, existing facility, modification) as well as the level of complexity. Relatively simple or low-hazard process reviews can be completed in several hours. Reviews of complex projects may require one to two days. One key factor is the extent of evaluation of potentially inherently safer options to be done during the review session(s) vs. that done offline. As with PHAs and other types of hazard reviews, the team should be used to help identify the opportunity or alternative, and then assign it to an individual or subteam. This group will evaluate the alternative off-line, and then report back on the results of their evaluation, which can often involve detailed cost estimates and risk comparisons in order to determine feasibility. IS review teams should not engineer solutions as part of the review. A recommendation to implement a particular IS design feature often involves a judgment on cost vs. degree of risk reduction, as well as resolution of other potential conflicts that may arise during any evaluation of potentially inherently safer options. (See Chapter 13 for a fuller discussion of these potential conflicts.)

The key to an effective and efficient review is good preparation, including updated process drawings and the development of IS review protocols that help the review leader to ask the team pertinent questions, and help ensure that the fullest possible range of IS alternatives is considered. Depending on the outcome of the team review, recommendations for further consideration of alternative designs may need to be resolved, with perhaps additional reviews by all or part of the team.

10.9 SUMMARY

Inherent safety is an exciting field that has caught the attention of researchers, plant designers, management and regulators worldwide. A number of methods for measuring inherently safety have been developed. However, a straightforward, generally accepted approach is needed to compare different process designs for a particular end product.

The best way to invent and build inherently safer processes is to make it a part of the thinking of all engineers and chemists as they go about their work, primarily through education. As they recognize the importance and benefits of hazard elimination, their creativity in inventing new ways to eliminate or reduce hazards as they go about their work activities will go further toward enhancing inherent safety than any particular review methodology could. Once company personnel start applying the principles of ISD to new and existing processes, the obvious benefits would become apparent, prompting further use. As inherently safer design is successfully applied to process industries, it can be adapted by other accident-prone industries such as mining, construction, transportation, and others. (Ref 10.14 Gupta).

10.10 REFERENCES

10.1 Amyotte, P.R. Goraya, A.U, Hendershot, D.C., and Khan, F.I., *Incorporation of inherent safety principles in process safety management.* In Proceedings of 21st Annual International Conference–Process Safety Challenges in a Global Economy, World Dolphin Hotel, Orlando, Florida, April 23-27, 2006 (pp.175-207) New York: American Institute of Chemical Engineers, 2006.

10.2 Ankers, R., *Introducing inherently safer concepts early in process development with PRORA.* In CCPS Inherently Safer Process Workshop, May 17, 1995, Chicago, IL. New York: American Institute of Chemical Engineers, 1995.

10.3 Auger, J.E., *Build a proper PSM program from the ground up.* Chemical Engineering Progress, 91 (1), 47-53, 1995.

10.4 Center for Chemical Process Safety (CCPS), *Guidelines for Chemical Reactivity Evaluation and Application to Process Design*. New York: American Institute of Chemical Engineers, 1995.

10.5 Center for Chemical Process Safety (CCPS) (2003). *Essential Practices for Managing Chemical Reactivity Hazards.* New York: American Institute of Chemical Engineers.

10.6 Clark, D.G., *Applying the limitation of effects ISP strategy when siting and designing facilities.* In Proceedings of the 41st Annual Loss Prevention Symposium, April 22-26, 2007, Houston, TX (Paper 1C). New York: American Institute of Chemical Engineers, 2007.

10.7 Contra Costa County Industrial Safety Ordinance (ISO) guidance document, *cchealth.org/hazmat/pdf/iso/section_d.pdf.*

10.8 CSChE, *Process Safety Management, Third Edition*. Ottawa, ON: Canadian Society for Chemical Engineering, 2002.

10.9 Dow Chemical Company, 1994a, *Dow's Chemical Exposure Index Guide, 1st Edition*. New York: American Institute of Chemical Engineers, 1994.

10.10 Dow Chemical Company, 1994b, Dow's *Fire and Explosion Index Hazard Classification Guide, 7th Edition*. New York: American Institute of Chemical Engineers, 1994.

10.11 Dowell, A.M., Regulations*: Build a system or add layers?* Process Safety Progress, 20 (4).247-242.

10.12 French, R.W., Williams, D.D., and Wixom, E.D., *Inherent safety, health and environmental (SHE) reviews.* Process Safety Progress 15 (1), 48-51, 1996.

10.13 Gowland, R.T., *Putting numbers on inherent safety.* Chemical Engineering 103 (3), 82-86, 1996.

10.14 Gupta, J. and Hendershot, D.C., *Inherently safer design.* In S. Mannan (Ed.) in Lees Loss Prevention in the Process Industries: Hazard Identification, Assessment & Control, 3rd Edition. London, UK: Butterworth-Heinemann, 2005.

10.15 Hendershot, D.C. et al., *Implementing inherently safer design in an existing plant.* Presented at the 7th Biennial Process Plant Safety Symposium, American Institute of Chemical Engineers 2005 Spring National Meeting, Atlanta, GA, April 10-14, 2005.

10.16 The Institution of Chemical Engineers and The International Process Safety Group, *Inherently Safer Process Design*. Rugby, England: The Institution of Chemical Engineers, 1995.

10.17 Khan F.I. and Amyotte, P.R., *How to make inherent safety practice a reality.* The Canadian Journal of Chemical Engineering, 81, 2-16, 2003.

10.18 Kletz, T.A., The constraints on inherently safer design and other innovations. Process Safety Progress, 18 (1), 64, 1999.

10.19 Overton, T.A. and King G., *Inherently safer technology: An evolutionary approach.* Process Safety Progress 25 (2), 116-119, 2006.

10.20 Pilz, V., *Bayer's procedure for the design and operation of safe chemical plants.* Inherently Safer Process Design, 4.54-4.65. Rugby, England: The Institution of Chemical Engineers, 1995.

10.21 Scheffler, N.E., *Inherently safer latex plants.* Process Safety Progress, 15 (1), 11-17, 1996.

10.22 Turney, R.D., *Inherent safety: What can be done to increase the use of the concept?"* In Proceedings of 10th International Symposium on Loss Prevention and Safety Promotion in the Process Industries, Stockholm, Sweden, June 19-21, 2001 (pp. 519–528). Amsterdam, The Netherlands: Elsevier, 2001.

10.23 Wixom, E. D., *Building inherent safety into corporate safety, health and environmental programs.* In CCPS Inherently Safer Process Workshop, May 17, 1995, Chicago, IL, 1995.

11. Inherent Safety & the Elements of a RBPS Program

Chapter 11 will describe the relationship between the four main IS strategies, i.e., *Substitution*, *Minimization*, *Moderation*, and *Simplification* and each element of a PSM/RBPS program. The possible use of the IS strategies in carrying-out the activities of the elements will be described. The interrelationship between inherent safety and PSM programs has been previously described by Kletz and Amyotte (Ref 11.16 Kletz 2010) and Amyotte, Hendershot, et al (Ref 11.5 Amyotte) and their conclusions are summarized herein. Amyotte & Kletz argue that inherent safety is implicitly part of PSM/RBPS programs, that it not be included as a discrete element, but as part of the thought process for the design of these programs and how they are applied and implemented. In their book and in this chapter more explicit applications of the IS strategies are proposed (Ref 11.16 Kletz 2010). For some PSM/RBPS elements the relationship with the IS strategies is direct and relatively strong, e.g., PHA, whereas for other elements the relationship is more indirect, e.g., Measurement and Metrics.

The four IS strategies are usually applied in the context of the chemicals and the hazards/risks that they present, e.g., substituting one chemical with another that presents less of a process safety hazard, or minimizing the inventories of the chemicals that can create process safety hazards. Traditionally, IS has been regarded as an intrinsic part of the design of the process, i.e., part of the "hardware" aspect of the process or as part of the properties of the process with respect to the hazards presented by the hazardous materials used in the process. In this chapter, the application of the same thought processes of IS can be applied to aspects of the process that are largely programmatic, i.e., the design and execution of the PSM/RBPS program policies, practices, and procedures. This is a new application of IS concepts. For example, *Minimization* has traditionally been associated with reducing the inventory of hazardous materials, which is a physical aspect of the process. However, we can pose the question: Can/should shift schedules and their working times be *minimized* to avoid fatigue? This is a possible application of *minimization* to a non-physical or programmatic/procedural aspect of the process. Another example: Can a different type of non-destructive testing be *substituted* to increase the likelihood of detecting a particular type of corrosion. This is an example of how the IS strategies can be used in the planning of a mechanical integrity program. Another possible use of the four IS strategies is their use in guidewords, deviations, checklists, and other process safety analytical and assessment tools. In particular, the HIRA/PHA, MOC, and Incident Investigation elements are examples of such possible use.

While the use of the four main IS strategies in this non-traditional manner represents a novel use of the terms, taking credit for only the programmatic use of the IS strategies might not be considered adequate, Also, proposed the use of IS strategies in programmatic applications in this chapter is not intended to supplant their possible use in traditional, physical aspects of the process.

Another variation of the IS strategy of *Moderation* is segregation. This refers to the spatial relationship between where the hazards are located and where the people onsite are located, as well as the avoidance of the spread of the effects of process safety incidents, or domino/cascade effects. In the expanded definition of *Moderation*, this is sometimes also referred to as the limitation of effects (Ref 11.16 Kletz 2010). Also see Chapters 2 and 5 for a discussion of the concept of segregation as part of the IS strategy of *Moderation*.

Although it is not an explicit element in the CCPS RBPS model, the applicability of process safety regulations are defined by formal requirements, usually the presence at a facility of certain chemicals above specified threshold amounts. By adjusting facility operations or storage procedures to always be below these threshold quantities so as to not trigger the applicability process safety regulations is an example of *Minimization*. Many organizations have used this strategy to lower the allowable inventories of chemicals covered by regulation onsite to avoid the application of the regulations at their facilities (Ref 11.16 Kletz 2010). These facilities are also lowering the process safety risk at the same time.

In each of the sections of this chapter that follow, the possible use of the four IS strategies in the elements of a PSM/RBPS program are described. For similarity of use and clarity, several of the elements have been combined for this purpose.

11.1 PROCESS SAFETY CULTURE

The application of the IS strategies in both the traditional and non-traditional manners is a manifestation of placing process safety at the same priority as production and is an indicator of strong leadership and a positive process safety culture. Although the four IS strategies are not directly applied in assessing or modifying the process safety culture in an organization, the prevailing culture should not preclude their consideration in any element of a PSM/RBPS program and the voluntary application of IS principles is also an indicator of a commitment to and sound process safety culture and effective PSM/RBPS program management.

In explaining the importance of leadership and culture to process safety, Amyotte and Kletz state:

> *"...The absence of any commitment to the hierarchy of controls is another manifestation of the priority of platforms over people [20]. Paraphrasing this statement, it could be said that the absence of a commitment to inherent safety is a manifestation of placing priority on production over safety. This can be expressed in a more positive, encouraging manner by saying that management promotion and adoption of the principles of inherent safety wherever possible is a visible demonstration of leadership and commitment to effective process safety management."* (Ref 11.16 Kletz 2010)

The importance of process safety culture is well documented in the post incident investigation reports for major accidents over the past 25 years. These are summarized, and the entire issue of process safety culture is explored by CCPS (Ref 11.9 CCPS Culture). The relationship between process safety culture and inherent safety is close. This is because IS concepts and strategies are not currently mandatory. Therefore, the implementation of IS in any given organization will require a conscious decision to support the concept of IS and intentionally apply it to upgrade the PSM/RBPS program. This will require firm and far-sighted leadership and a sound process safety culture throughout the organization, else the effort will be wasted.

11.2 COMPLIANCE WITH STANDARDS

IS strategies are indirectly built-into many codes, standards, and other Recognized and Generally Accepted Good Engineering Practices (RAGAGEPs) by establishing design limits for process parameters, establishing the criteria for dimensional characteristics (e.g., the minimum thickness of pressure retaining equipment) and design margins, and setting allowable practices for choosing the materials of construction. These design principles are examples of *Moderation*.

API-751 (Ref 11.4 API-751), which provides mandatory (i.e., "shall" provisions) and non-mandatory (i.e., "should" provisions) practices for the design, operation, and maintenance of hydrogen fluoride (HF) alkylation units in refineries, includes several examples of the use of the IS strategies *Minimization* and *Moderation*. These include: vessel design to minimize HF inventory to levels adequate for unit operation (taking into account the time required to order and receive HF and the number of HF deliveries), vessel baffles to segment HF inventory and limit the amount of HF that can be released should one side of the vessel be compromised, minimizing piping dead-legs, and flanges as potential leak sources, using dual seal or seal-less pumps, and using additives to reduce the vapor pressure of the HF. Other detailed design considerations are included in API-751, such as minimizing the area exposed to localized high water concentrations of HF. This will minimize corrosion in the exposed equipment, e.g., a condenser which cools the incoming vapor rapidly will minimize the tube area exposed to the first droplets of condensate (the water-rich azeotrope). There are similar standards in industry which provide similar practices and guidance for other types of operations with significant process safety risks, e.g., ANSI K61.1 (Ref 11.1 ANSI K61.1) for anhydrous ammonia, the Chlorine Institute Pamphlets (e.g., Pamphlet #6 for chlorine piping (Ref 11.12 CI #6) for chlorine and hydrochloric acid.

The internal process safety, engineering, and project procedures and standards of organizations should mandate the use or consideration of these RAGAGEPS. For example, many companies and/or facilities having piping specifications that have been customized for their given materials being processed and the process conditions. However, nearly all of these individual piping specifications refer to an industry consensus RAGAGEP for piping design (ASME B31.3, the Process Piping Code, in many parts of the world). The provisions of the RAGAGEPs and the safety margins contained in them provide a minimum level of *Moderation* to the piping design where they are followed.

11.3 WORKFORCE INVOLVEMENT

Communicating with onsite stakeholders regarding the risks is an important aspect of a process safety program and including workers in the development and implementation of the process safety program is a regulatory requirement in some regions. As with the communications to offsite stakeholders, the communication itself does not involve a direct application of the four IS strategies. However, informing employees that IS has been used and how it has been used will enhance the workforce involvement program (or employee participation as it also known). Hopefully, this will result in a more positive outcome when these communications are undertaken and also result in a more positive relationship with the employees. Organizations should encourage IS thinking via training and education programs. An example of successfully doing this is increased efforts by employees to eliminate risks rather than just reducing the risk when performing Job Safety Analyses (JSA) for operational and maintenance tasks.

11.4 PROCESS KNOWLEDGE MANAGEMENT

Although the Process Knowledge Management element of PSM/RBPS (or Process Safety Information as it is sometimes known) documents the design, construction, and operations of a process, it can also serve as a useful checklist of issues where the four IS strategies can be applied. For example, the requirement that the chemical, physical, and hazards properties of the chemicals be documented may be used to help guide efforts in *Substitution* and *Minimization*. The requirement that the safe upper and lower limits of the processes be documented can be used to help guide efforts in *Moderation*.

Also relevant, from a different perspective, is the company memory of process safety information, or the management of that information.

This is as important as the identification/creation of the information. The objective is to ensure that knowledge and information gained from plant experience, and which is likely to be important for the future safety of a facility, is well-documented so it is not forgotten or overlooked as personnel and organizational changes occur. That is, the information should become institutionalized. This is especially critical when dealing with inherent safety and inherently safer design features. IS features that have been implemented can be put at risk because the reasons they were implemented are not clearly and adequately documented. This results in compromising of facility safety when future modifications are made by people who do not understand the intent of the original designer. Thus, this also becomes an important aspect of management of change. IS features are particularly susceptible to lapses in corporate memory given that they are such a fundamental part of the design that their purpose may not be obvious, unlike an add-on device such as a high-pressure alarm. The use of the IS strategies should be clearly documented in the process safety information so that they are not inadvertently negated when changes are being contemplated later. (Ref 11.16 Kletz 2010) (Ref 11.15 Hendershot)

Another aspect of *Simplification* with respect to engineering, project, and other technical records that are associated with process safety programs is the increasing use of electronic data management systems (EDMS) to create, issue, store, and maintain these records. While EDMSs simplify the overall management of records significantly, care should be exercised to ensure that these electronic systems are not complicated, i.e., access is difficult or restricted, navigation is difficult and not intuitive, support functions such as printing, file management, document security, etc. are difficult to use. Such problems tend to negate the advantages of EDMSs.

11.5 HAZARD IDENTIFICATION AND RISK ANALYSIS

PHA/HIRA is the PSM/RBPS element with the most direct relation to the four IS strategies and the element where IS has been used the most in the PSM community (other than in the security of chemical/processing facilities). The routine use of IS checklists in PHAs, the use of IS strategies in the formulation of guidewords/deviations or What-If questions during PHAs should be considered. An example of such a checklist is shown in

Table 11.1 (Ref 11.16 Kletz 2010). Others have been published in the literature (Ref 11.7 Bollinger) (Ref 11.10 CCPS 1998). A more extensive version is included in Appendix A of this book.

In addition to the general use of IS checklists in PHAs/HIRAs, reviews devoted to only IS are sometimes used where the four IS strategies are examined explicitly to determine if there are feasible ways to use them. Such IS-only reviews are currently required in two United States jurisdictions: Contra Costa County, California, and New Jersey (See Chapter 14). Often the format and study approach for these IS-only reviews is PHA-like, i.e., worksheets, guidewords/deviations, and similar techniques are often used for conducting them. Contra Costa Health Services, in their IS guidance document includes analytical activities on how IS should be addressed for new vs. existing processes (Ref 14.13 CCHS). See Chapter 14.

Amyotte and Kletz also state that PHAs for capital project reviews, as well as other PHAs that are performed should include the following purposes:

- to identify hazards, hazardous situations and specific events that could produce undesirable consequences,
- to examine the currently available safety measures to deal with the identified hazards and events, and
- to suggest alternatives for risk reduction based on inherent, as well as engineered (add-on) and procedural, safety. (Ref 11.16 Kletz 2010)

In Appendix B of this book, an extensive example of IS-specific reviews using a PHA-like approach is presented.

Facility siting is usually considered part of the PHA/HIRA element of PSM. Facility siting is examined qualitatively during PHAs/HIRAs using checklists typically. It is also addressed quantitatively in most facilities as a separate analytical activity. The vulnerability to the effects of explosions of the administration and control buildings at Flixborough and the contractor trailers at the BP Texas City refinery are often cited as examples of the importance of this topic. Additionally, the use of segregation in the hierarchy of controls and the use of distance in facility siting are examples of *Moderation*.

Amyotte and Kletz offer another important consideration regarding the use of IS strategies on engineered projects. It is a common practice for operating companies in the chemical/process industry to outsource or contract engineering services, especially for capital projects. This is often the case on large projects where partnerships and joint ventures are formed but may also apply to smaller projects through the awarding of subcontracts. The degree of commonality in risk management practices among the different parties may vary widely and there may not be a common set of expectations for safety performance and risk-awareness. By extension, differences in approach to inherent safety and the corporate value placed on this concept can also vary widely, resulting in an uneven application of IS strategies on any given project. This problem can be partially solved by including detailed requirements for IS reviews and the use of the four main strategies in the project technical specifications. However, engineering contractors may have a less, or differently, developed culture and experience of applying IS in their work than the operating companies they work for (of course, the opposite can also be true, and the engineering company may propose IS solutions to hazard/risk reduction that were not specified by the operating company). (Ref 11.16 Kletz 2010)

Table 11.1 Inherent Safety Checklist for PHAs

Guideword	Checklist Question
Minimize	• Is the storage of all hazardous gases, liquids and solids minimized?
	• Are just in time deliveries used when dealing with hazardous materials?
	• Are all hazardous materials removed or properly disposed of when they are no longer needed or not needed in the next X days?
	• Are all tasks involving working at heights done in as few trips as possible?

	• Are all tripping hazards minimized and all walking surfaces tractional during all weather conditions?
	• Are low noise equipment and machinery taken into consideration when making new purchases?
	• Is shift rotation optimized to avoid fatigue?
	• Are awkward positions and repetitive motions minimized?
	• Are attempts made to completely eliminate raw materials, process intermediates or by-products?
	• Are elbows, bends, and joints in piping minimized?
Substitute	• Can a less toxic, flammable or reactive material be substituted for use?
	• Is there an alternate way of moving product or equipment as to eliminate human strain?
	• Can a water-based product be used in place of a solvent or oil-based product?
	• Are all allergenic materials, products and equipment replaced with non- allergenic materials, products and equipment when possible?
Moderate	• Can potential releases be reduced via lower temperatures or pressures, or elimination of equipment?
	• Are all hazardous gases, liquids and solids stored as far away as possible to eliminate disruption to people, property, production and environment in the event of an incident?

	• When purchasing new equipment, are acceptable models available that operate at lower speeds, pressures, temperatures or volumes?
	• Are workplaces designed such that employee seclusion is minimized?
	• Are food and beverages offered at the workplace appropriate to minimize negative impact and maximize positive impact?
	• Are all power tools de-energized when not in use for extended periods?
Simplify	• Are all lights, sprinklers, accesses/exits, ventilation ducts, electrical installations and machines clear of piled materials?
	• Are equipment and procedures designed such that they cannot be operated incorrectly or carried out incorrectly?
	• Are machine controls located to prevent unintentional activation while allowing easy access for stopping the machine?
	• Are all machines, equipment and electrical installations easily isolated of all sources of power?
	• Is the workplace designed for consideration of human factors (i.e. an ergonomically designed workplace)?

In most PSM/RBPS programs human factors is addressed in the PHA element by analyzing both human error potential in the formulation of PHA causes and by addressing human factors engineering issues in the equipment and operations. Human factors are also an important aspect of the Operating procedures and Conduct of Operations elements. The use of human factors checklists, particularly to analyze human factors engineering issues, should include the four IS strategies, particularly *Minimization* and *Simplification* in the design of the man-machine interface. An example of the importance of human factors is demonstrated by the following incident that occurred at a refinery:

A plug valve is used primarily for on/off, and some throttling services. It controls flow by means of a cylindrical or tapered plug with a hole in the center that lines up with the flow path of the valve to permit flow. The valve is opened or closed with the use of a valve wrench. A quarter turn of the valve blocks the flow path. The typical arrangement for such a valve is to have the valve operator be in-line with the flow path when the valve is open, and to have the valve operator be perpendicular with the flow path when the valve is in the closed position. In preparing the pump for maintenance, an operator relied on the visual position of valve wrench to determine that the suction valve was open. He moved the wrench to what he believed was the closed position with the wrench perpendicular to the flow of product. The valve was actually open (Figure 11.1). (Ref 11.17 CSB, 2005)

Figure 11.1 Plug Valve Operator

This incident demonstrated not only an important human factors issue, but also a weakness in the MOC program at the refinery where it occurred. Allowing a valve operator to be installed in the opposite manner that convention would require for this type of valve invited an incident to occur. When the valve was re-configured the MOC should have revealed this serious (and obvious) error but did not. Additionally, in this situation facilities should only procure and install quarter-turn ball/plug valves (a common valve application) where the valve is designed so that the valve operator can only be installed in the normal convention. Internal engineering specifications for valves should reflect this convention (see Compliance with Standards, section 11.2 above).

Another example of the importance of human factors in process safety is demonstrated by the fire at another refinery. This incident also involved valve operators. In this case, a gearbox on a pump suction in hydrocarbon service was malfunctioning, which did not allow the valve to be opened. To correct the malfunctioning gearbox the device was partially disassembled in accordance with refinery maintenance procedures. However, this particular valve was an older design (15 of 500 similar valves in the refinery) where a part called a top-cap was part of the pressure retaining boundary of the valve and the support bracket for the gearbox also held the top-cap in place. On newer design valve gearboxes this was not the case. When the bolts were loosened on the top-cap and valve stem operated with a pipe wrench, a large release of hydrocarbon resulted, which ignited due to a welding machine nearby. The fire seriously injured four personnel in the area. Figure 11.2 shows the old and new designs and how the removal of the support bracket helped precipitate the incident (Ref 11.18 CSB, 2016)

In the past 10-15 years there has been a significant amount of research into developing quantitative methods for analyzing inherent safety. Amyotte and Kletz have summarized efforts made by different organizations to develop these quantitative inherent safety evaluation tools. These tools allow relative comparisons of the hazards and the ensuing risk from different processing options. See Chapter 12 for a description of these quantitative IS tools.

Figure 11.2 Valve Gearbox Designs. Ref 11.18 CSB, 2016

11.6 SAFE WORK PRACTICES

The reduction of ignition sources, which is partially a design issue and partially a safe work practice issue is an example of the use of

Minimization. Various means are used to minimize the presence of ignition sources, including Recognized and Generally Accepted Good Engineering Practices (RAGAGEPs) specifying the electrical classification of certain areas/zones of the facility and the electrical design practices of fixed, portable, and mobile equipment that must be used in each area/zone (NFPA 497/499 in the U.S. and ATEX in the EU), hot work procedures and permits, and no-smoking policies (now almost a universal policy in industrial settings and in most work places). Additionally, most facilities have safe work practices or policies for the following situations: control of vehicles in areas with flammable materials, lightning protection, energy emitted from equipment (e.g., hot surfaces, radiant heat, frictional heat or sparks), control of portable electronic devices in areas with flammable materials, static electricity and stray currents, and control of equipment with open flames (e.g., ovens, furnaces, fired heaters, boilers, etc.). This also applies to segregation of storage of different materials and the use of buffer zones to help limit the spread of fires should they start (*Moderation*).

The technologies used for various SWP practices can also benefit from the application of IS strategies, e.g., Substitution of better techniques in a given situation, and Simplification of techniques where possible. For example, for energy isolation, is double block and bleed always the right approach for the isolation of pressure from fluids? Isolation valves are subject to leakage. Perhaps another approach is better given the hazards involved, e.g., using blinds. The physical configuration of the equipment involved will also play a role in how energy isolation is accomplished.

As with operating procedures, *Simplification* can be employed to make safe work practice procedures and permits simpler to initiate and administer. Complicated safe work permit forms should be avoided if possible. The design of these documents, particularly the permit forms, is an important human factors issue. There is a tendency to try to squeeze as many items onto a single sheet permit form as possible. This makes the form difficult to use and increases the likelihood that something will be missed. Moving lesser used or applicable items to checklists in the safe work procedure(s) that the permit is part of with a single item on the permit form refers to the checklist may be a better way to present the information needed and check that the appropriate safety measures have been verified before work begins. Additionally, the increasing practice of combining all safe work issues on one single permit form, i.e., basic job information, authorization, and closure; hot work; confined space; equipment opening; lockout/tagout; PPE; vehicle entry; etc. has made these all-in-one permit forms very complicated. While this practice may reduce the number of separate permits required for a given job, it makes the all-purpose permit form very complex and more difficult to use.

Finally, SWP permitting processes should not be allowed to evolve into "check-the-box" activities. This can occur when complicated permits are used in a manner that does not give due consideration to each item on the permit and they are rapidly filled-out. This can sometimes be detected when each item is not individually addressed, i.e., a line with a down arrow is used to quickly answer a list of questions. If the importance of having a completed permit document seems to outweigh the thought process associated with carefully recognizing and addressing the hazards of each individual job, then the permit processes may have become a "check-the-box" activity. Unfortunately, this may only become apparent when hazards are not addressed properly and incidents or near misses occur.

11.7 ASSET INTEGRITY AND RELIABILITY

Asset integrity (sometimes referred to as Mechanical Integrity) includes a broad range of technical activities in a PSM/RBPS program, including:

- Identification of equipment to be included in the asset integrity program.
- Written maintenance procedures.
- Inspection, testing, and preventive maintenance (ITPM) programs
- Maintenance personnel training and qualification.
- Management of asset integrity deficiencies.
- Asset integrity quality assurance (QA), which involves the design, fabrication, and installation of the equipment, as well as the replacement of materials.

Each of these sub-areas of asset integrity can involve consideration of the four IS strategies. For example, *Substitution*, *Moderation*, and

Simplification can be applied in the selection of the materials of construction when designing and fabricating equipment, *Minimization* can be applied in developing and implementing risk-based inspection programs where ITPM frequencies and schedules are optimized. *Minimization* can also be employed in the development of ITPM programs, e.g., the *minimization* of dead/ low flow legs in piping systems, which will reduce the number of locations that are susceptible to galvanic corrosion in no flow/low flow areas of the processes. Also, *Substitution* can be used by employing new, or different, and more effective ITPM methods, such as the use of improved non-destructive testing (NDT) techniques.

The QA portion of an asset integrity program is often where the traditional uses of IS strategies are employed because it is the when equipment is designed. The preliminary or conceptual design stage of a project is when the basic chemistry is determined, and this offers the best opportunity to use the *Substitution* and *Moderation* strategies. It is also during this phase of a project when the basic equipment sizing is determined based on the desired throughput of the process and Minimization can be used in this phase.

The opportunity also exists during the preliminary or conceptual design stage to make decisions about the capacity of the equipment to inherently withstand certain upset conditions, particularly the robustness of the pressure retaining boundary, i.e., the primary containment of the process. For example, if the maximum pressure achievable in the process given abnormal events in with respect to the chemistry, and other abnormal events such as external fires is 200 psig. If the primary containment is designed to mechanically withstand 300 psig without relying on any active safeguard, this can be regarded as a use of *Moderation*. Some would regard this as a design choice rather than an IS option. However, those choices, when they are intentionally made to avoid a certain hazard(s) without relying on any other active or passive safeguard can be regarded as inherent in nature. Another example might be the intentional choice of certain materials of construction during preliminary design. If the materials of construction are selected to specifically make the process immune to certain types of corrosion or damage mechanisms that would affect other materials of construction, that can also be regarded as an inherent solution using Substitution.

Another example would be to adjust the operating temperature of a process (either higher or lower) to avoid the temperature where corrosion under insulation (CUI) can occur, given the materials of construction. For some materials of construction CUI is possible at or around ambient temperatures, so the only option is to increase the operating temperature. This is novel, and perhaps counterintuitive use of the IS strategy of Moderation, which usually involves lowering temperatures.

Following the appropriate RAGAGEPs also make use of the IS strategies as they are indirectly built-into these codes, standards, and other guidance by establishing design limits for process parameters, establishing the criteria for dimensional characteristics (e.g., the minimum thickness of pressure retaining equipment) and setting the design margins for the process/equipment, and establishing allowable practices for choosing the materials of construction. By specifying known design margins for pressure, temperature, and other process conditions, the process will have a known level of robustness in its design. Therefore, the application of the design principles contained in the governing RAGAGEPs are examples of *Moderation*. This also includes internal design standards, which often provide even more inherent protection in the design of processes/equipment. Also see Compliance with Standards in Section 11.2.

11.8 CONTRACTOR MANAGEMENT

In recent years many organizations in the process safety community have begun to use third party companies to help fulfill their mandatory contractor management requirements. This is a novel application of the IS strategy of *Simplification*. It does not necessarily reduce the requirements, which are usually established via regulation. However, it does simplify the contractor management activities and recordkeeping for the host facility. It also makes information about contractors and their status with respect to the facility's and regulatory requirements available online which is also simpler than manual records. These third-party organizations consist of both global online services (e.g., ISNetWorld, PICS - now Avetta, etc.) that manage the pre-qualification process for contractors, perform periodic audits, and maintain required records, as well as regional or local consortia (e.g., the Houston Area Safety Council, Delaware Valley Safety Council, etc.) that provide and record the training and administration of contractor personnel before they arrive onsite for work. Host facilities typically employ one or both types of services.

11.9 TRAINING AND PERFORMANCE ASSURANCE / PROCESS SAFETY COMPETENCY

Training should be conducted on inherent safety itself. This should vary from simple awareness level training for those who will not be involved directly in attempting to implement the IS strategies to more detailed IS training for those who will be required to participate in formal IS reviews or engineering personnel who will be reviewing the state-of-the-art in process designs to weigh IS options. The more detailed IS training should include how IS reviews are to be performed and documented, how IS guidewords and deviations should be formulated and used in PHAs, and how the feasibility of IS options is to be evaluated so that it is performed consistently.

The use of training, particularly practical training is a strong component in the *Simplification* of processes. Where available, simulators represent a superior method of training. While this can be regarded as *Substitution* for on-the-job practical training in the field, it actually supplements practical training because it allows operators to practice their response to abnormal and upset events while seeing the actual process parameter indications and alarms that the process conditions and their response actions would create, and by "feeling" a small amount of the stress that these conditions impose. This is invaluable training feedback that is not possible without creating the actual upset conditions in the process. The nuclear and airline businesses (as well as the military) have recognized the value of simulator training and require it periodically as a condition of holding a reactor operator's or pilot's license. However, simulator training is not mandatory in general industry, including the chemical/processing industry. In this way, simulators actually represent an indirect example of *Simplification* as they make the practiced response of operators to abnormal and upset conditions easier and more consistent.

Employing the state-of-the-art in the design of the processing technology is also a use of *Substitution* and *Moderation*, both in the process engineering, as well as in the engineering of the equipment. Keeping up-to-date on the state-of-the-art is an aspect of process safety competency as it requires proactive activities to obtain and maintain knowledge and expertise in the research, development, and engineering of the type of process technology in use. Also, in a basic sense, process *Simplification* should result in less required training as layers of complexity are removed from the process, less human interface is required, and the human interface that is required is simpler, easier to understand, and easier to operate.

11.10 MANAGEMENT OF CHANGE / OPERATIONAL READINESS

Management of Change and Operational Readiness (or Pre-startup Safety Review as it is often referred to) are often combined elements by organizations in the same overall MOC procedure in PSM programs, with the OR/PSSR one of the last steps in the MOC process. These elements offer several ways to incorporate the four IS strategies. First and foremost, MOC is required to be used whenever IS changes to existing equipment are being proposed. However, the reverse should also be true, that is, whenever a change is being contemplated to existing equipment, the MOC process should include IS considerations. This is just as important as looking for IS opportunities while processing a MOC for other reasons. The use and consideration of IS strategies should be incorporated into the identification of prospective changes and their technical basis, the MOC safety and hazards reviews or PHAs conducted to examine the potential process safety impacts of prospective changes, and the modification of operating and other procedures pursuant to a change. In particular, the safety review of the proposed change can employ checklists with IS questions similar to what has been described for PHA/HIRA above. The four IS strategies should also be included in MOC/PSSR procedure checklists. The overall MOC process and where IS guidewords and checklists can be used in it is shown in Figure 11.3 (Ref 11.16 Kletz 2010). Table 11.2 shows a list of MOC questions that can be used when reviewing proposed changes. As with Process Knowledge Management, the use of IS strategies and concepts in a design of a process should be clearly documented because they are inherent and thus not obvious to those who did not participate in the weighing of design options and therefore the design could be compromised, and the process safety risk increased by later changes where the rationale for the original design is unknown.

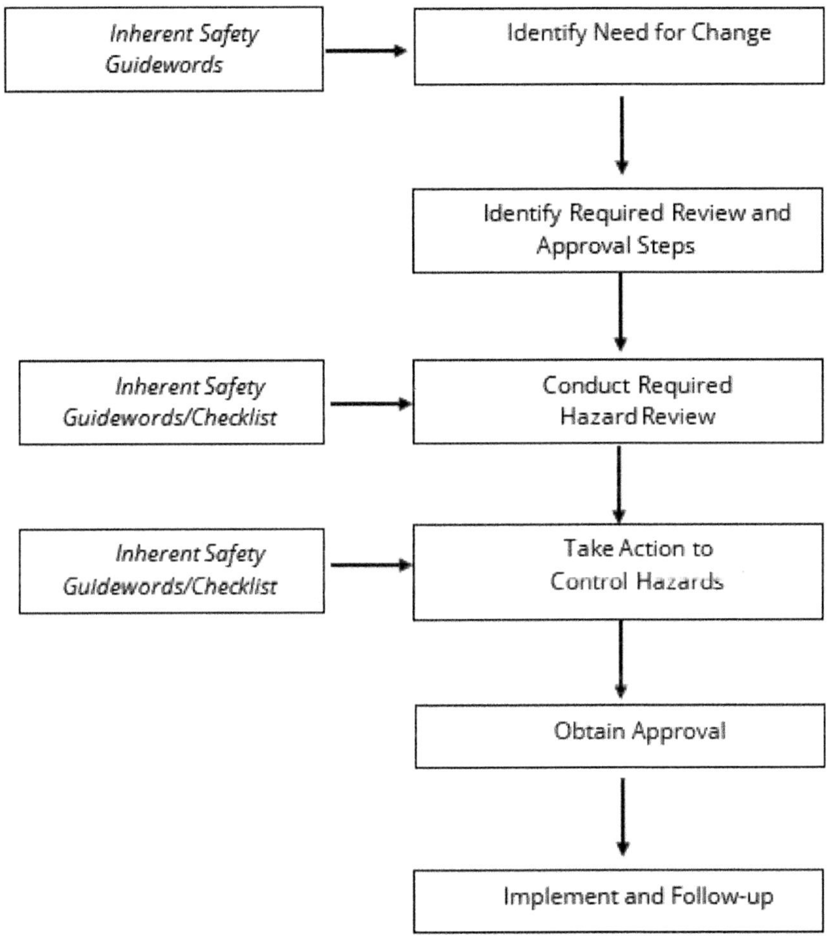

Figure 11.3 Inherent Safety-Based Management of Change Process

Table 11.2 MOC IS Checklist

IS Strategy	MOC IS Checklist Question
Minimize	• Have the inventories of hazardous materials and materials included within the scope of the PSM program been minimized in the design of the proposed change?
	• Have piping dead legs and low flow portions of piping been minimized as the design of the proposed change? Can additional piping dead legs and low flow portions of piping be eliminated while implementing the change?
	• Can just-in-time deliveries be used when dealing with hazardous materials in the modified process? If so, has the possible risk transfer to transportation sectors been evaluated?
	• Have ignition sources been eliminated or minimized to the extent possible in the modified process?
	• Are the human interfaces with the altered process been evaluated so reduce fatigue? Will the existing shift rotation require modification as a result of the change?
	• Has the opportunity to completely eliminate hazardous raw materials, process intermediates or by-products been evaluated as part of the change even if the change was proposed for a different reason?
	• Has the opportunity to eliminate or minimize ignition sources been evaluated as part of the change even if the change was proposed for a different reason?

	• Has the opportunity to reduce inventories of hazardous materials and materials included within the scope of the PSM program been evaluated in the equipment that is appurtenant to or connected to the equipment being changed (even if the change being proposed was for a different reason)?
Substitute	• In the review of the proposed change can a less toxic, flammable or reactive material be substituted for use?
	• Can alternative materials of construction be substituted that will offer higher resistance to (or even immunity from) known corrosion and damage mechanisms in the equipment being changed? If so, is it possible to extend the alternative materials of construction be extended to additional portions of the process at the same time?
	• Can water-based materials be used in place of a solvent or oil-based product?
Moderate	• Can lower temperatures, pressures, speeds or other process parameters be used to achieve the desired reactions in the modified process?
	• In reviewing the proposed change are all hazardous gases, liquids and solids stored as far away as possible to eliminate disruption to people, property, production and environment in the event of a release?
Simplify	• Have all procedures and other written guidance for operating, maintaining, or performing other human interface with the modified process been evaluated to make these documents as simple and easy-to-use as possible, particularly upset/emergency procedures, including emergency shutdown procedures.

	• Are equipment designed such that they cannot be operated incorrectly or carried out incorrectly, e.g., are valve operators on quarter-turn manual valves configured so that when the valve operator is in-line with the flow direction the valve is open and when the valve operator is perpendicular to the flow direction the valve is closed?
	• Have DCS and other control panels been designed following well-known conventions for displays, colors, and other characteristics that will make operations intuitive and easy-to-understand and therefore reduce human error?
	• Have DCS and other control panels been evaluated with respect to alarm management principles and conventions?
	• Can rotating equipment and other machinery be stopped locally and from central control rooms or locations?
	• Have other design and operational conventions been followed so as to obviate human errors?
	• Has the opportunity to simplify procedures associated with the process, particularly emergency procedures, been evaluated to simplify the procedures, even if the change was proposed for a different reason?

11.11 CONDUCT OF OPERATIONS / OPERATING PROCEDURES

Conduct of Operations (COO) is a PSM/RBPS element that covers a broad range of process safety program activities. It is a very important aspect of process safety as this is element of the program where operating discipline and adherence to written and approved procedures as well as adherence to strong operating principles is established. It is also a process safety program element that is very sensitive to the prevailing culture in an organization. The four IS strategies are or can be part of the COO element and its related activities as follows (Ref 11.8 CCPS COO):

11.11.1 Minimization

Alarm management. In recent years modern control systems, particularly integrated distributed control systems (DCS) have come under increasing review for the number, ranges, and human factors aspects of their process alarms. Problems in this area have been contributors in several process safety incidents. Other human factors considerations have received increased attention, including: eliminating nuisance alarms; adjusting alarm setpoints so that multiple simultaneous alarms during upset or abnormal conditions are minimized or avoided; selecting alarm displays; choosing how alarms are silenced; and choosing how alarms are recorded. The modification of the controls systems to make improvements has resulted. This entire set of activities is often referred to as alarm management. An industry standard devoted to this topic has been issued (Ref 11.2 ANSI). Alarm management is an important application of the IS strategy of Minimization.

Fatigue management. The CCPS RBPS Guidelines and Conduct of Operations and Operational Discipline concept book includes fatigue as a COO issue (Ref 11.11 RBPS) (Ref 11.8 CCPS COO). Overtime can be a sensitive subject in many facilities. Sometimes there is pressure to work excessive overtime from facility or company management to help solve more basic staffing problems. Sometimes there is peer or family pressure in favor of excessive overtime for the financial benefits that result. Regardless of where the pressure originates, when it exists a culture issue also exists. Overtime can also be an organization work rule or a collective bargaining agreement issue, in which case the solution to the problem will require changing policies and procedures, or re-negotiation of the labor contract. The Baker Panel (Ref 11.6 Baker) noted that fatigue was a contributor to the BP Texas City incident. Also, this problem was not limited to hourly employees and contractors, as salaried employees, including supervisors also worked excessive amounts of hours.

Industry has recognized that fatigue is an important issue, and API RP-755 (Ref 11.3 API 755) has been published to provide guidance for facilities to detect and counter fatigue in those parts of the workforce working night shifts, rotating shifts, extended shifts/days, or call-outs and where their job requires process safety actions. This reference establishes a fatigue risk management system so that the process of managing fatigue among those positions may be sensitive to it is a continuous and sustainable process. However, if there are underlying pressure being brought to bear on personnel to work excessive overtime, then there is a safety/process safety culture issue that must be solved first.

Fitness for duty. Fitness-for-duty refers to a range of possible influences on worker performance, such as impairment due to drug or alcohol, fatigue (addressed above), illness, distraction due to personal issues, and mental condition (Ref 11.8 CCPS COO). No external influences on performance should be allowed for facility personnel. This includes the use of alcohol or drugs while on duty, horseplay, harassment, and other aberrant behaviors. No tolerance for these behaviors goes beyond simply forbidding them in written policies and stating these prohibitions during training/orientation sessions. It means living them at each level of the organization, including self-policing within peer groups and during off shifts. Minimizing the external negative influences on personnel behavior and performance and establishing the minimum requirements for fitness for duty expectations is an important activity.

Shift differences and turnover. The culture of different operating shifts within a given facility sometimes varies widely. What would never be tolerated on one shift might be tolerated on another. This often represents the difference between staffing levels during normal working hours (i.e., day/weekday and off-hours (i.e., night/weekend/holiday shifts), and the presence and attention of technical staff and management during the day shift of normal weekdays. Most of this difference can be attributed to the attitudes and beliefs of supervisors, foremen, and other mid-level management assigned to off shifts. Having to accomplish the same level of production with less resources sometimes fosters the attitude that shortcuts and other deviations from approved procedures are acceptable in order to achieve those goals.

Sometimes this situation fosters a "can-do" attitude where independently finding any solution to a problem and getting around a production delay/issue quickly is seen as a positive attribute. Yet the process conditions and the potential hazards they present are exactly the same regardless of the time of day or day of the week. One of the definitions of the process safety culture is doing the right thing even when nobody else is watching. This definition captures the conditions that exist in many facilities during off shifts, but not in all facilities.

The root causes of some major process safety incidents have included inadequate shift turnover (Ref 11.6 Baker). Some parts of the shift turnover issue are procedural, and some are cultural. Personnel need to understand both the importance of this activity and that it must be formally transacted. In recent years there has been a recognition of the importance of shift turnover and widespread improvement in this activity. Many facilities now have written procedures that specify the minimum requirements for shift turnover, i.e., how and where is to be accomplished, including its documentation. Minimizing the differences between the culture and operating practices between shifts and any differences in shift turnover will help reduce incidents.

Emergency Shutdown & Stop Work Authority. The emergency shutdown of a process or unit (a rare but critical event) is an extremely important aspect of operational discipline. This action can be a controversial activity, or at least regarded as controversial, but should not be. At its core, this is mostly a process safety culture problem, rather than an operations problem. A contributing factor in many serious incidents is facility personnel not realizing that they could have or should have stopped themselves or others from continuing with an operation or task that was becoming dangerous or had the potential to become dangerous. In most companies, high level statements of safety or process safety policy encourage personnel to exercise their good judgement and be proactive about stopping an operation or task from proceeding when they recognize an error or other conditions that could make the situation worse. However, many employees are hesitant to actually exercise their judgement and stop work, especially when the work of others is involved. Minimizing the hesitance to activate an emergency shutdown when warranted and minimizing the hesitance to exercise stop work authority are important process safety cultural issues.

Normalization of Deviation and Loss of a Sense of Vulnerability. The normalization of deviation and loss of a sense of vulnerability are important core principles of process safety culture. Deviations from approved rules, procedures, codes/standards, or other formal guidance that are tolerated so that they do not arouse curiosity, are not questioned or challenged, and become part of the everyday way of operating the facility then become accepted or "normalized.". These eventually become latent failures in a process safety incident scenario. Sustaining a healthy level of respect for the hazards and risk of the facility and its operations, particularly over extended periods of time maintains a sense of vulnerability. A reduced sense of vulnerability results from complacency over time in the absence of incidents or near misses. This is particularly true when high consequence – low frequency events occur that are in the risk profile of the facility or company. The low frequency of such events, which includes serious process safety events results in few, or likely no personnel working in such facilities ever having experienced such an event. Therefore, over a lengthy period of time, they tend to "forget" that such events are possible. This tendency is also characterized by an erosion of the respect that this type of risk should constantly command. Minimizing the occurrence of these cultural situations is an important aspect of operational discipline.

11.11.2 Simplification

Operating Procedures. Operating procedures should be subject to periodic review to 1) ensure that they are current and accurate, and 2) to the extent possible be as simple as possible (*Simplification*). Operating procedures that are written in as simple a style and format as possible will be easier to use and will be used more often by the operators. Operating procedures that are unnecessarily lengthy, complex, and written in a language that is difficult for the intended users to follow will not be used as readily. Another consideration for writing operating procedures is the reading comprehension level of the intended users. Vocabulary and syntax can be selected to match a desired reading comprehension level. This is an important consideration as engineers tend to write the initial drafts of many operating procedures. However, simplicity should not be achieved at the sacrifice of procedure clarity. For example, "open valve slowly" can mean something different to different users because "slowly" is a general term. If the process is sensitive to degree of valve position, then perhaps more detailed directions would be appropriate, e.g., "open valve 10% initially and then 10% every 30 seconds until you reach 70%, then fully open." Operating procedures prepared with IS principles in mind tend to be easier to use and preclude the possibility of operators taking short cuts. This section has been written in the context of operating procedures. However, the same concepts of *Simplification* apply to any PSM-related procedure, e.g., maintenance procedures, emergency response plans, etc.

Operating Limits. The operating limits and limiting conditions for operation that are included in the operating procedures should be subject to *Moderation*, however, this will occur not because the procedures are modified but because the operating conditions in the process are modified (i.e., moderated) and the published values in the procedures will follow suit. Also see Process Knowledge Management/Process Safety Information.

Human factors. The design of control systems displays, and other ergonomic issues that allow operators to accomplish what the operating procedures require easily. Sometimes attention has to be paid to these types of design issues in order to make it possible to accomplish the what the procedures require. For example, if an emergency operating procedure requires that four steps be performed rapidly, and an operator would have only a few minutes to accomplish them, then the process design must be simplified to reduce the time necessary for the operator to get from one location to another quickly, else the function in question should be automated. Simplifying the human-machine interface is an important consideration.

Control of Work/SWP processes. As described above the control of work/SWP procedures and their permits should be as simple as possible.

Bypass/removal of safety or protective features. As with operating procedures and control of work/SWP procedures, a procedure to manage and control the bypass or removal of safety or protective features, including alarms, should be as simple and straightforward as possible, including any permits associated with them.

11.12 EMERGENCY MANAGEMENT

In recent years the scope, complexity, and shear length of emergency response plans at many facilities has grown substantially. The reason for this growth is a combination of several factors, including the consolidation of separate emergency plans required by different regulations, e.g., process safety incidents, environmental release events such as oil pollution, hazardous waste operations, naturally occurring events such as tornadoes, hurricanes, and other severe weather events, transportation events such as train derailments. Also, security-related events such as bomb threats and physical/cyber security breaches have been added to many emergency response plans. Additionally, supplemental procedures related to emergency response, such as crisis management, business continuity, and event recovery have sometimes been made part of the same plans. Non-mandatory federal government guidance has contributed to this growth, with the wide-spread use of the federal "One Plan" guidance published in 1996 (Ref 11.14 Fed Reg).

As with operating procedures, reducing the complexity of emergency response and emergency action plans where possible will make these vital plans easier to use, particularly during upset/emergency situations where the tension is higher. This is an important application of *Simplification*.

The review of modification of emergency response plans is also an opportunity to challenge and possibly eliminate/minimize some of the hazards addressed in the plans. For example, review of the fire fighting and response procedures (if the facility responds to fires) is an opportunity to employ the IS strategy of *Minimization* to review and possibly eliminate ignition sources, or to use distance to the maximum extent and segregate fire hazards if possible. It is also an opportunity to review the fixed and mobile equipment that supports the emergency response plan. For example, can the materials of construction of the fire main be upgraded from cast iron to more modern materials to resist corrosion (i.e., *Substitution*), can explosion barriers be used, or occupied building structures be upgraded to withstand higher pressures (i.e., Moderation), or can the response equipment be simplified or storage locations improved to increase the speed of access (*Simplification*).

11.13 INCIDENT INVESTIGATION

As with PHA and MOC, the use of IS checklists and guidewords based on the four IS strategies can be used in investigation activities, root cause analysis, and in the formulation of recommendations. The overall incident investigation process and where IS guidewords and checklists can be used in it is shown in Figure 11.4 (Ref 11.16 Kletz 2010).

The formal investigation of incidents and near misses also represents an opportunity to challenge and possibly eliminate the hazards that caused the events in the first place. This is the goal of root cause analysis, which is a common practice during an incident/near miss investigation. The elimination or reduction of ignition sources or potential leak points as a result of an investigation, as described for emergency response plans is an example of the use of the IS strategy of Minimization. Another example of this IS strategy is as follows: if an investigation is conducted pursuant to an injury caused during a sampling operation, this is an opportunity to question why the sample is required in the first place.

11.14 MEASUREMENTS AND METRICS/AUDITING/MANAGEMENT REVIEW AND CONTINUOUS IMPROVEMENT

The three elements of PSM/RBPS where the process safety program are measured and formally evaluated represent opportunities to use IS strategies.

Minimization of a large list of process safety metrics to a small and more manageable list that is easier to collect and analyze is a preferred practice. This will allow a focus on the key issues that affect the process safety risk rather than diluting attention and time with a large number of KPIs. Also, a metric could be created that provides a measure of how many times (divided by the number of opportunities) that IS strategies were used in both the design and operation of the facility (including its equipment and its chemicals), and in the modification of the PSM/RBPS program policies, practices, and procedures. It is not probably necessary to report this KPI at the same frequency as other PSM/RBPS program metrics, but having the PSM Manager keep track of these IS uses and opportunities will provide some useful data on how the facility and its PSM/RBPS program reflect the state-of-the-art.

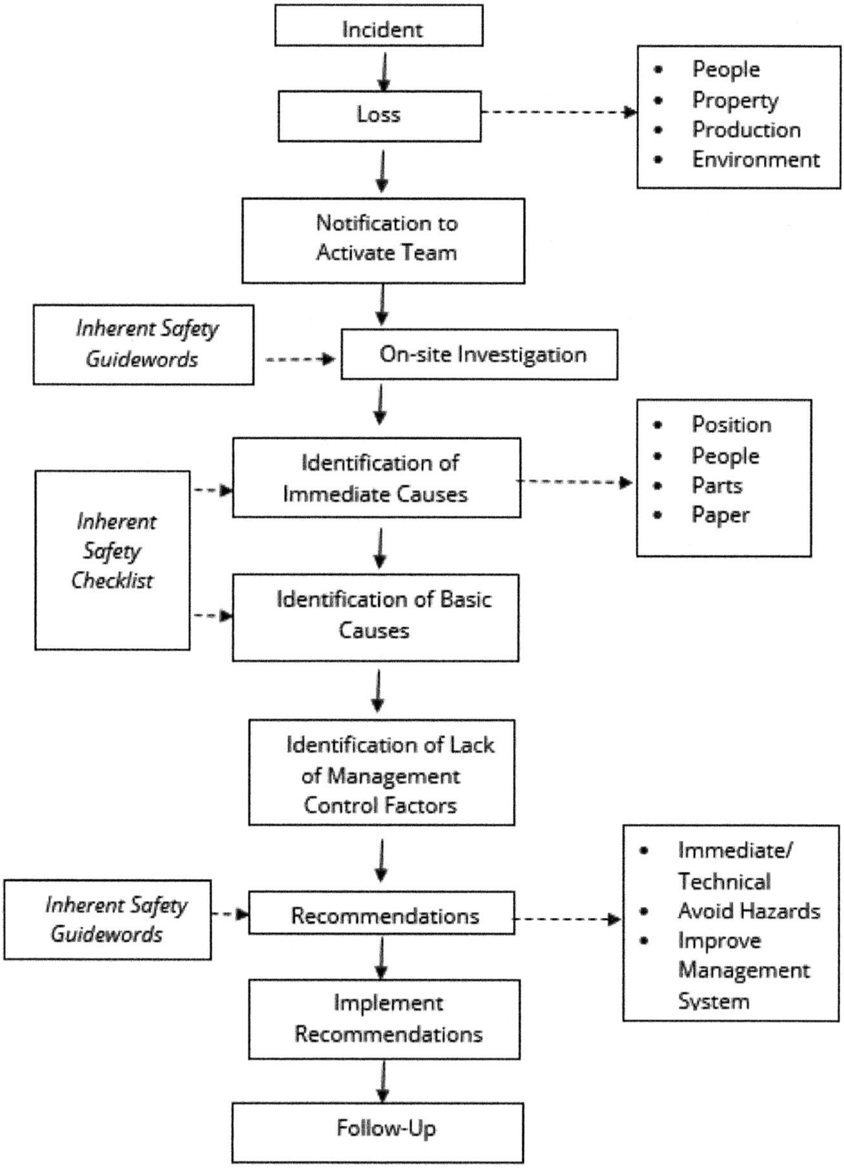

Figure 11.4 Inherent Safety-Based Incident Investigation Methodology

If inherent safety has been adopted by the organization then the process safety audits should include questions and tests for whether the PSM/RBPS program has incorporated the four IS strategies. This should be reflected in the protocol to be used for these audits. As with metrics and audits, PSM program management reviews should study how the IS strategies have been used and how successful this has been. These are not direct applications of the four IS strategies but an examination of whether they have been used in other elements of the PSM/RBPS program.

11.15 SUMMARY

The four IS strategies are usually applied in the context of the chemicals and the hazards/risks that they present, e.g., substituting one chemical with another that presents less of a process safety hazard, or minimizing the inventories of the chemicals that can create process safety hazards. Traditionally, IS has been regarded as an intrinsic part of the design of the process. However, the application of the same thought processes of IS can be applied to aspects of the process safety program that are largely programmatic, i.e., the design and execution of the PSM/RBPS program policies, practices, and procedures. Most of the PSM/RBPS program elements can make use of *Substitution*, *Moderation*, *Minimization*, or *Simplification* in different ways. The proposed use of IS strategies in programmatic applications in this chapter is not intended to supplant their possible use in traditional, physical aspects of the process.

11.16 REFERENCES

11.1 American National Standards institute (ANSI)/Compressed Gas Association, *Requirements for the Storage and Handling of Anhydrous Ammonia, 6th Ed.*, ANSI K61.1/CGA G-2.1, 2014.

11.2 American National Standards institute (ANSI), ANSI/ISA–18.2–2009, *Management of Alarm Systems for the Process Industries*, 2009.

11.3 American Petroleum Institute (API), Fatigue Risk Management Systems for Personnel in the Refining and Petrochemical Industries, API RP-755, 2010.

11.4 American Petroleum Institute (API), *Safe Operation of Hydrofluoric Acid Alkylation Units, 4th Ed.*, API RP-751, 2013.

11.5 Amyotte, P, Goraya, A, Hendershot, D, Khan, F., *Incorporation of Inherent Safety Principles in Process Safety Management*, Proceedings of 21st Annual International Conference – Process Safety Challenges in a Global Economy, Orlando, FL, April 2006.

11.6 Baker, J. et al., The Report of BP U.S. Refineries Independent Safety Review Panel, January 2007 (Baker Panel Report).

11.7 Bollinger, R.E., Clark, D.G., Dowell III, R.M., Ewbank, C., Hendershot, D.C., Lutz, W.K., Meszaros, S.I., Park, D.E. and Wixom, E.D., *Inherently Safer Chemical Processes: A Life Cycle Approach*, American Institute of Chemical Engineers, 1996.

11.8 Center for Chemical Process Safety, *Conduct of Operations and Operational Discipline*, American Institute of Chemical Engineers, 2011.

11.9 Center for Chemical Process Safety, *Essential Practices for Creating Strengthening, & Sustaining Process Safety Culture,* American Institute of Chemical Engineers, draft.

11.10 Center for Chemical Process Safety, *Guidelines for Design Solutions to Process Equipment Failures* American Institute of Chemical Engineers, 1998.

11.11 Center for Chemical Process Safety (CCPS), *Guidelines for Risk Based Process Safety*, American Institute of Chemical Engineers, 2007.

11.12 Chlorine Institute, Pamphlet #6, - Piping Systems for Dry Chlorine, 16th Ed., 2013.

11.13 Contra Costa Health Services (CCHS), California, *Industrial Safety Ordinance Guidance Document, Section D*, June 2011.

11.14 Federal Register, *The National Response Team's Integrated Contingency Plan Guidance*, Vol. 61, No. 109, Fed. Reg. 28642, June 5, 1996.

11.15 Hendershot, D.C., *Tell Me Why*, 8th Annual International Symposium, Mary Kay O'Connor Process Safety Center, Texas A&M University, October 2005.

11.16 Kletz, T., Amyotte, P., Process Plants – A Handbook for Inherently Safer Design, 2nd Ed., CRC Press, 2010.

11.17 U.S. Chemical Safety Board (CSB), *Oil Refinery Fire and Explosion*, Case Study 2004-08-I-NM, October 2005.

11.18 U.S. Chemical Safety Board., Key Lessons from the ExxonMobil Baton Rouge Refinery Isobutane Release and Fire, Final Safety Bulletin, November 2016.

12. Tools for IS Implementation

12.1 IS REVIEW METHODS – OVERVIEW

12.1.1 Three Approaches

Inherent safety reviews of new and existing processes form the foundation of an IS management program. IS reviews for new processes or modifications to existing processes present the best opportunity to identify possible design features that are inherently safer than those proposed. For existing processes, these opportunities are more limited, due to the cost and feasibility involved in making major modifications. However, whenever a change or even an in-kind replacement is planned for an existing facility, an IS review should be conducted to identify any potential IS alternatives that can be incorporated into the design.

Many of the more qualitative hazard analysis tools can be adapted to incorporate the principles of inherent safety, and three basic approaches to conducting IS reviews have emerged in the chemical process industries. These are similar to PHA approaches that have been in use for many years:

- HAZOP
- What-If?
- Checklist

The Checklist approach can be used in combination with both the HAZOP and What-If methodologies. Each of these techniques is described below.

Because of the importance of considering inherent safety early in the design sequence when changes can most readily be made, inherent safety considerations are particularly important in conducting hazard reviews (i.e., a preliminary hazard analysis). The above methods can be applied to these reviews.

12.1.2 Formal IS Reviews

Multiple Approaches for Inherently Safer Review. Although, in most cases, a separate, formal IS review of new or existing processes is the best way to identify opportunities to improve the level of inherent safety, there are a number of valid ways to consider IS in the design of a process.

The INSIDE Project (Ref 12.17 Rogers; Ref 12.13 Mansfield) which is discussed later in this chapter, has developed the Inherent SHE (Safety, Health, Environment) Evaluation Toolkit (INSET) to identify and evaluate inherently safer design options throughout the life of a process.

An outline for conducting an IS review is provided in Appendix B, including pre-study preparation, training of the team on IS, dividing the process up into sections, and conducting and documenting the review.

Factors to Consider in Selecting an Approach. The factors used to select an IS review method are similar to those for selecting a PHA methodology. Each method has its benefits and drawbacks, and selection is based largely on the process life cycle stage that is being reviewed, along with the purpose, scope and objectives of the study. Factors that influence the choice of IS review method include the following:

- Process size
- Process complexity
- Stage in life cycle
- Process operating/incident history
- Phase of process development
- Degree of risk
- Review team's familiarity with techniques
- Opportunities for risk reduction

In general, the larger, more complex, and more hazardous the process (and the more experienced the team with IS), the more detailed the analysis, using perhaps the HAZOP or HAZOP/checklist approach. For smaller, simpler, and less hazardous processes, consider using an IST checklist, such as the one in Appendix A. Early in the design stage, a high-level HAZOP approach using the matrix shown in Table 12.1 can be used in a Preliminary Hazards Analysis-like approach. If the review is being conducted as part of the PHA for the process, of course that method would be used, with the IS guidewords (Table 12.1) used as deviations for each node or subsystem. In general, combining the use of HAZOP, or What-If? Methodology with a checklist provides for creative brainstorming as well as a detailed means to ensure that most issues have been covered. It should be pointed out that no checklist is perfect and there may be opportunities not identified in the checklist that can only be discovered through a more subjective analysis.

12.1.3 IS Review Methods

The following review methods can be used to ensure that inherent safety is considered and documented for hazardous processes:

- **An independent IS analysis done in addition to a PHA, either in tandem or separately**. This analysis should review the process for ways to eliminate or reduce hazards present in the covered process and may be achieved using an IST checklist (Appendix A) or guideword analysis (Table 12.1).
- **An IS analysis that is incorporated into the existing PHA review process**. In most cases, an initial stand-alone IS analysis should be conducted for the entire process to ensure it receives adequate attention. Again, this may be achieved using a checklist (Attachment A) or guideword (Table 12.1) approach. This type of analysis would review the processes for ways to eliminate or reduce hazards, as well as to reduce risks using the other risk management strategies (passive, active, and procedural).

12.1.4 Research & Development Application

There are significant benefits to applying IS concepts and methodologies within Research and Development (R&D)/laboratory and pilot plant operations. IS requirements should be mandated both when a process hazards review is done, and when a project progresses toward full development:

- A process safety review should be required for each new or significantly modified R&D "process" (a semi-works, laboratory experiment, etc.). This review should include hazards identification and evaluation, facility siting, consequence analysis, human factors, and, importantly, inherently safer design. The emphasis should be on the *safety* of the pilot plant operation or experiment being proposed. Thus, the IS review is primarily geared toward the actual performing of the experiment or running of the pilot plant test.

- The maturation of a new or modified "process," from concept to experiment to pilot plant to production facility, can be done using a series of "gate keeping steps." Each gate should be designed to ensure that important issues are resolved *before* progressing to the next step, so as to resolve problems as early as possible. IS issues are no exception and, in fact, are paramount.

The evaluation should address material factors, quantity factors, and process factors. *Material factors* are related to the toxicity, flammability, and reactivity of the materials to be used and created. *Quantity factors* are related to the inventories of chemicals that would be present in the production facility. *Process factors* are related to the process variables, such as pressure, temperature, and ease of operation.

The IS evaluation should include the identification of alternative process technologies to reach the same or similar end product. For each alternative, the researcher should be asked to evaluate the material, quantity, and process factors. The end result would be a list of alternatives offering various degrees of IS implementation and ranked in order of that degree. It is recognized that additional risks of personal injury, environmental harm, lost production, and loss of "right to operate" may be involved. It is also recognized that the most inherently safe processes may not be economically feasible from a business perspective. The evaluation should include the identification of the most significant benefits and barriers to the adoption of the inherently safer process alternatives, and of a less than optimum inherently safe alternative, and the identification of alternative risk management strategies (i.e., passive, active, procedural) that may, in some way, mitigate concerns and reduce risk, albeit perhaps not in strictly inherently safer ways.

12.1.5 PHA - Incorporation into HAZOP or other PHA Techniques

HAZOP. Inherently Safer reviews can be conducted as separate studies, using a HAZOP-like technique, or incorporated into initial HAZOP studies or revalidations. This involves the application of inherent safety guidewords as "deviations," using, for example, the Guideword Matrix provided in Table 12.1. Recommended inherent safety guidewords and their descriptions are shown in Table 12.2 (Ref 12.7 Goraya). As previously mentioned, these guidewords are simply the four most general and widely applicable principles of inherent safety. The description for each is purposely brief and is focused on materials, process routes, equipment and procedures. Use of these guidewords as "mind triggers" for each section of a process, as shown in Table 12.1, or during a particular process safety activity, such as management of change or incident investigation, as subsequently described, will help ensure that the concepts of inherent safety are visible within the process or activity. These guidewords are intended as a supplement to existing tools that may already be in use within a specific process safety protocol.

As with the use of HAZOP methods for conducting regular PHAs, it can be used at any level of subdivision of the process, including node-by-node. While thorough, this can also increase the amount of time required to conduct the study. It also requires a degree of brainstorming by a team that understands IS concepts well and can apply them to a process based on the P&ID and knowledge of the operation. Appendix C provides an example of a combination HAZOP/checklist worksheet that includes review of potential IS opportunities applicable to the particular node and its hazardous materials.

Table 12.1: Guideword Matrix

(Contra Costa County, CA, Industrial Safety Ordinance Guidance Document, Attachment D, June 15, 2011)

Minimization	Substitute/ Eliminate	Moderate	Simplify
Raw Material			
In-process Storage			
Product Inventory			

Process Chemistry			
Process Controls			
Process Piping			
Process Equipment			
Process Conditions			
Maintenance			
Siting			
Transportation			
Misc. Material (i.e., solvents)			

Documentation for incorporating inherent safety review guidewords into a Hazard and Operability Study should be consistent with the documentation used during any HAZOP. As for any other inherently safer system analysis, the documentation should include IS already implemented, IS alternatives considered, the alternatives implemented and not implemented, and the rationale for rejection.

12.1.6 "What-If?" Method

Use of the "What-If" method for IS reviews has the same benefits and pitfalls as its use as a PHA technique. "What-If" analysis provides an opportunity for a team to brainstorm possible process deviation scenarios in order to help identify safeguards, or in this case, IS improvements that can reduce or eliminate the potential for the scenario to develop in the process, such as through elimination of equipment, rather than by adding additional safeguards.

Table 12.2: Inherent Safety Guidewords (Ref 12.1 Amyotte)

Guideword	Description
Substitute	Replace a substance with a less hazardous material, or a processing route with one that does not involve hazardous material. Replace a procedure with one that eliminates or presents less of a hazard.
Minimize	Use smaller quantities of hazardous materials when the use of such materials cannot be avoided. Perform a hazardous procedure as few times as possible when the procedure is unavoidable.
Moderate	Use hazardous materials in their least hazardous forms or identify processing options that involve less severe processing conditions.
Simplify	Design processes, processing equipment, and procedures to eliminate opportunities for errors, i.e., by removing the possibility that two chemicals that may be incompatible under certain conditions can react unfavorably, such that the chemicals will not be together in the process. Designing equipment that cannot be over pressured by process conditions.

12.1.7 Checklist Method

Checklists can be used either independently or in combination with the What-If (or HAZOP) methodologies. In this manner, the checklist analysis is combined with the What-If (or HAZOP) analysis to yield a method that combines the creative brainstorming features of What-If with the systematic, rigorous features of a checklist. In addition, the inherent safety HAZOP guideword/checklist approach helps to ensure that inherent safety considerations are explicitly evaluated in identifying both hazards and safety measures (Ref 12.1 Amyotte).

Appendix A contains an IST checklist based on CCPS (Ref 12.3 CCPS 1998) and additional items from Hendershot (Ref 12.11 Hendershot 2000). This checklist has been incorporated into guidance for conducting IS reviews under the Contra Costa County, CA Industrial Safety Ordinance (ISO). The ISO requires IS evaluations for both new and existing covered processes, as well as for certain modifications where a Major Accident risk has been identified. This checklist is similar to one provided in the CCPS book, *Essential Practices for Managing Chemical Reactivity Hazards* (Ref 12.4 CCPS 2003).

In addition, a set of excellent checklists for a number of specific types of chemical processing units (vessels, reactors, heat transfer equipment, mass transfer equipment, etc.) can be found in CCPS Guidelines for Design Solutions to Process Equipment Failures (Ref 12.3 CCPS 1998). These checklists include suggestions for inherent, passive, active, and procedural approaches to risk management for a number of incident scenarios. An example for reactors is provided in Figure 12.1.

Operational Deviations	Failure Scenarios	Potential Design Solutions		
		Inherently Safer/Passive	**Active**	**Procedural**
Overpressure (Batch, Semi-batch and CSTR Reactors)	Loss of agitation resulting in runaway reaction, or hot bearing/seals causing ignition of flammables in vapor space.	Vessel design accommodating maximum expected pressure Use different type of reactor (plug flow) Alternative agitation methods (e.g., external circulation eliminates shaft seal as a source of ignition in vapor space)	Agitator power consumption or rotation indication interlocked to cutoff feed of reactants or catalyst, or to activate emergency cooling Uninterrupted power supply backup to motor Emergency relief device Pressure or temperature sensors activating bottom discharge valve to drop batch into a dump tank with diluent, poison, or short-stopping agent, or to an emergency containment area Inerting of vapor space Provide nitrogen buffer zone around seal using seal enclosure Automatic agitator trip on low agitation (velocity) sensor, low seal fluid, or low shaft speed	Operators to visually check mechanical seal fluid on regular basis In-vessel agitation (velocity) sensor with alarm Mechanical seal fluid reservoir low level sensor, with alarm Speed or vibration sensor with alarm Manual activation of bottom discharge valve to drop batch into dump tank with diluent, poison or short-stopping agent, or into an emergency containment area Manual activation of inert gas sparging of reactor liquid to effect mixing

Figure 12. 1 Example of Potential Design Solutions for Reactor Failure (Ref 12.3 CCPS 1998)

Other checklists useful for ISD evaluations can be found in such references, as the CCPS book, *Human Factors Methods for Improving Performance in the Process Industries* (Ref 12.5 CCPS 2006).

12.1.8 Consequence-Based Methods

A number of consequence-based methods have been devised to measure the level of inherent safety in a process, based on potential incident consequences. These include the following:

- Rohm and Haas Major Accident Prevention Program
- *Dow Fire and Explosion Index*
- Dow Chemical Exposure Index
- Mond Index (ICI)
- US EPA RMP modeling

The *Dow Fire and Explosion Index* (Ref 12.6 Dow) and the *Mond Index* (Ref 12.12 ICI) are used primarily for determining the hazard level of a process plant. However, their use requires that the process design be fully in place since much of the information required is based on this design. Therefore, they are unsuitable for measuring the level of inherent safety at the conceptual design and preliminary process development stages, where the use of such indices is most useful (Ref 12.9 Gupta 2005).

Shah, Fischer and Hungerbuhler (Ref 12.19 Shah) outline a hierarchical approach in which a chemical process is divided into different layers (substance, reactivity, equipment and safety technology), each representing a different level of analysis. Using an automated tool, these layers are successively, and also possibly repeatedly, assessed for non-idealities of the process with regard to inherent safety. In the case of non-idealities, a worst-case scenario is determined and analyzed with the help of available data, and recommendations are made for possible preventive and protective measures. The assessment of different process aspects—substances, reactivity of substances, operating conditions and worst-case scenarios in equipment units—and the selection of safety technologies for a given chemical process within one single framework provides a clear qualitative and quantitative picture to design inherently safer chemical processes and to prevent accidents.

12.1.9 Other Methods

Traditionally, consequence modeling has been used to estimate the potential impact of potential accidents based on worst-case scenarios (US EPA RMP) or other criteria. A number of programs, both proprietary and public, have been developed for this purpose (PHAST, ALOHA, DEGADIS, CAMEO, RMP*Comp, etc.). These methods are the most direct way of measuring the level of inherent safety, as they do not consider likelihood of the event. Using these programs, plant personnel can estimate the impact of reduced quantity and changes in some process conditions, such as pressure, on potential consequences.

The concepts of inherent safety can also be applied to health, to eliminate or reduce risk of personnel exposure, and environmental goals, to eliminate or reduce risk associated with routine emissions or generation of waste. Measurement tools which can be quickly and easily used early in the life cycle of the process design are particularly important, because that is the time at which the designer has the greatest opportunity to change the basic process technology. In recent years, a number of tools have been developed to measure combined inherent SHE characteristics of processes, and the applicability of several existing tools, such as the Dow Indices, to the understanding of inherent SHE has been recognized (Ref 12.10 Hendershot 1997).

One major chemical company decided that special emphasis should be placed on reducing the risk of low probability, high consequence events, and developed the Major Accident Prevention Program (MAPP). This program systematically examines the handling and processing systems for hazardous chemicals that could present toxic or explosive vapor cloud threats to employees or neighboring communities.

Traditionally, chemical systems with the potential for high consequence events have been studied using hazard analysis techniques that concentrate on reducing the frequency of such events. In contrast, MAPP places the emphasis on reducing the consequences. An advantage of this approach is that damage reduction techniques are passive in nature and lead to inherently safer systems. Risk reductions that reduce the frequency of an event tend to be active systems that require devices that must be operated and maintained to reduce risk.

MAPP follows an established protocol to identify hazards, understand potential consequences, evaluate means to reduce the potential consequences and manage risks appropriately (Ref 12.18 Rohm and Haas).

Analyzing the potential consequences of an accident is a useful way to understand the relative inherent safety of process alternatives. These consequences might consider, for example, the distance to a benchmark level of damage resulting from a fire, explosion, or toxic material release. The Major Accident Prevention Program (MAPP) (Ref 12.16 Renshaw) encourages inherently safer process development by requiring accident consequence analysis of a standard list of potential chemical process accidents for a specified list of high hazard chemicals. Accident consequence analysis is of particular value in understanding the benefits of minimization, moderation, and limitation of effects. Table 12.3 illustrates the use of consequence analysis as a metric for understanding the benefits of a number of inherently safer process options (Ref 12.10 Hendershot 1997).

Table 12.3: Examples of Potential Accident Consequence Analysis as a Measure of Inherent Safety

Inherent Strategy	Description	Consequence
Minimize	Chlorine transfer line rupture — distance to 20 ppm atmospheric concentration D atmospheric stability 3.4 mph wind speed	2-inch line - 3.4 miles *To* *1-inch line - 1.2 miles*
	Explosion overpressure 100 feet from a reactor	3000-gallon batch reactor - 1.1 psig *To* *50-gallon continuous reactor - 0.3 psig*
Substitute	Concentration in the atmosphere 500 feet downwind from a large spill from a storage tank D atmospheric stability 3.4 mph wind speed	Methanol - 1000 ppm *To* *Butanol - 130 ppm*
Moderate	Distance to 500 ppm atmospheric concentration for a large spill of Monomethylamine D atmospheric stability, 3.4 mph wind speed	Stored at 10°C - 1.2 miles To Stored at 3°C — 0.7 miles *To* *Stored at -6°C — 0.4 miles*
	Concentration in the atmosphere 500 feet downwind from a large spill from a storage tank containing methyl acrylate D atmospheric stability, 3.4 mph wind speed	2500 sq. ft. concrete containment dike — 1100 ppm *To* *100 sq. ft. concrete containment pit —* *830 ppm*

The INSIDE (INherent SHE In DEsign) Project is a European government/industry project sponsored by the Commission of the European Community, to encourage and promote inherently safer chemical processes and plants (Ref 12.17 Rogers). The INSIDE Project expands the inherent safety concept to include inherent approaches to Safety, Health, and Environmental (SHE) aspects of chemical processes. The project has developed a set of tools, the INSET Toolkit, to identify and evaluate inherently safer design options throughout the life of a process. Table 12.4 summarizes these tools, and they are described in more detail by the INSIDE Project (Ref 12.17 Rogers).

The tools of particular interest with regard to metrics for measuring the inherent safety of chemical processes are the ISHE Performance Indices listed in Table 12.4 (Tools I.1 through I.11). The ISHE Performance Indices are relatively simple, intended for hand or simple calculator computation, so that a large number of process options can be rapidly evaluated. The various inherent safety, health, and environmental aspects of a process are evaluated using separate indices, and no attempt is made to combine the indices into a single overall measure.

Table 12.4: Summary of INSET Tools (Ref 12.13 Mansfield)

Tool	Description
A	Detailed Constraints and Objectives Analysis
B	Process Option Generation
C	Preliminary Chemistry Route Options Record
D	Preliminary Chemistry Route Rapid ISHE Evaluation Method
E	Preliminary Chemistry Route Detailed ISHE Evaluation Method
F	Chemistry Route Block Diagram Record
G	Chemical Hazards Classification Method

H	Record of Foreseeable Hazards
I	ISHE Performance Indices
I.1	Fire and explosion hazard index
I.2	Acute toxic hazard index
I.3	Inherent health hazard index
I.4	Acute environmental hazard index
I.5	Transport hazard index
I.6	Gaseous/atmospheric emissions environmental index
I.7	Aqueous emissions environmental index
I.8	Solid emissions environmental index
I.9	Energy consumption index
I.10	Reaction hazard index
I.11	Process complexity index
J	Multi-Attribute ISHE Comparative Evaluation
K	Rapid ISHE Screening Method
L	Chemical Reaction Reactivity - Stability Evaluation
M	Process SHE Analysis - Process Hazards Analysis, Ranking Method
N	Equipment Inventory Functional Analysis Method
O	Equipment Simplification Guide
P	Hazards Range Assessment for Gaseous Releases
Q	Siting and Plant Layout Assessment

R	Designing for Operation

The INSET Toolkit instead recommends a multi-attribute decision analysis technique to evaluate the overall inherent SHE aspects of the various process options. The INSET Toolkit is particularly interesting as an Inherent SHE measurement tool because it represents the combined consensus expertise of a number of companies and organizations, and because it is intended to consider safety, health, and environmental factors in one set of tools (Ref 12.10 Hendershot 1997).

Gupta and Edwards (Ref 12.8 Gupta 2003) also propose an ISD measurement procedure for use in differentiating between two or more processes for the same end product. Palaniappan, et al. (Ref 12.15 Palaniappan 2002; Ref 12.14 Palaniappan 2001) have developed an i-Safe index, which includes five supplementary indices: Hazardous Chemical Index (HCI), Hazardous Reaction Index (HRI), Total Chemical Index (TCI), Worst Chemical Index (WCI) and Worst Reaction Index (WRI). Lastly, CCPS (Ref 12.2 CCPS 1995) presents a large range of decision aids for risk analysis and decision making in industry, including cost-benefit analysis, voting methods, weighted scoring methods and decision analysis, so that critical risk decisions can be made in a more consistent, logical, and rigorous manner.

12.2 SUMMARY

This chapter outlines how various tools can be used to support the implementation of inherent safety. Inherent safety considerations are particularly important in conducting hazard reviews (i.e., a preliminary hazard analysis), because of the importance of considering inherent safety early in the design sequence when changes can most readily be made. Qualitative PHA techniques, such as the HAZOP, What-If, and Checklist methodologies, as well as consequence-based methods such as the *Dow Fire and Explosion Index*, can be adapted for use in IS reviews. In addition, consideration of the human-machine interface should be made as early as possible in the design phase of a process. In the next chapter, the discussion will turn to conflicts which may arise during the application of inherently safer design.

12.3 REFERENCES

12.1 Amyotte, P.R. Goraya, A.U, Hendershot, D.C., and Khan, F.I., *Incorporation of inherent safety principles in process safety management.* In Proceedings of 21st Annual International Conference–Process Safety Challenges in a Global Economy, World Dolphin Hotel, Orlando, Florida, April 23-27, 2006 (pp.175-207) New York: American Institute of Chemical Engineers, 2006.

12.2 Center for Chemical Process Safety (CCPS), *Tools for Making Acute Risk Decisions With Chemical Process Safety Applications*. New York: American Institute of Chemical Engineers, 1995.

12.3 Center for Chemical Process Safety (CCPS), Guidelines *for Design Solutions to Process Equipment Failures.* New York: American Institute of Chemical Engineers, 1998.

12.4 Center for Chemical Process Safety (CCPS), *Essential Practices for Managing Chemical Reactivity Hazards.* New York: American Institute of Chemical Engineers, 2003.

12.5 Center for Chemical Process Safety (CCPS), Human *Factors Methods for Improving Performance in the Process Industries.* New York: American Institute of Chemical Engineers, 2006.

12.6 Dow Chemical Company, Dow's *Fire and Explosion Index Hazard Classification Guide, 7th Edition*. New York: American Institute of Chemical Engineers, 1994.

12.7 Goraya, A., Amyotte, P.R. and Khan, F.I., *An inherent safety-based incident investigation methodology.* Process Safety Progress, 23 (3), 197-205, 2004.

12.8 Gupta, J. and Edwards D.W., A *simple graphical method for measuring inherent safety.* Journal of Hazardous Materials, *104*, 15-30, 2003.

12.9 Gupta, J. and Hendershot, D.C., *Inherently safer design*. In S. Mannan (Ed.) in Lees Loss Prevention in the Process Industries: Hazard Identification, Assessment & Control, 3rd Edition. London, UK: Butterworth-Heinemann, 2005.

12.10 Hendershot, D.C., Safety *through design in the chemical process industry: Inherently safer process design.* Presented at the Benchmarks for World Class Safety Through Design Symposium, sponsored by the Institute for Safety Through Design, National Safety Council, Bloomingdale, IL, August 19-20, 1997.

12.11 Hendershot, D.C., *Process minimization: Making plants safer*. Chemical Engineering Progress, 30 (1), 35-40, 2000.

12.12 Imperial Chemical Industries (ICI), *The Mond Index, Second Edition*. Winnington, Northwich, Chesire, U. K.: Imperial Chemical Industries PLC, 1985.

12.13 Mansfield, D., Malmen, Y. and Suokas, E., "*The Development of an Integrated Toolkit for Inherent SHE.*" International Conference and Workshop on Process Safety Management and Inherently Safer Processes, October 8-11, 1996, Orlando, FL (pp. 103-117). New York: American Institute of Chemical Engineers, 1996.

12.14 Palaniappan, C., *Expert System for Design of Inherently Safer Chemical Processes*. Singapore: National University of Singapore, 2001.

12.15 Palaniappan, C., Srinivisan, R., and Tan, R.B.H., *Expert System for Design of Inherently Safer Processes,* Ind.Eng. Chem. Res., **41**, 6698-6722, 2002.

12.16 Renshaw, F. M., A *major accident prevention program.* Plant/Operations Progress, 9 (3), 194-197, 1990.

12.17 Rogers, R.L., Mansfield, D.P., Malmen, Y., Turney, R.D., and Verwoerd, M. *The INSIDE Project: Integrating inherent safety in chemical process development and plant design.* In G.A. Melhem and H.G. Fisher (Eds.). International Symposium on Runaway Reactions and Pressure Relief Design, August 2-4, 1995, Boston, MA (pp. 668-689). New York: American Institute of Chemical Engineers, 1995.

12.18 Rohm and Haas, *2000 EHS and Sustainability Annual Report*, Revised December 2002.

12.19 Shah, S., Fischer, U. and Hungerbühler, K., A hierarchical approach for the evaluation of chemical process aspects from the perspective of inherent Safety. Trans IChemE, 81, Part B, November 2003.

13. Inherently Safer Design Conflicts

13.1 INTRODUCTION

In several cases the inherent safety advantages of one process are clear when compared with the alternatives. One or more hazards may be significantly reduced, while others are unaffected or only marginally increased. Aqueous latex paints, for example, are inherently safer than solvent-based paints, although there are applications where the increased performance of solvent-based paints justifies their use with the appropriate layers of protection.

Unfortunately, many times it is not clear which of several process alternatives is inherently safer. Because nearly all chemical processes have many hazards associated with them, an alternative which reduces one hazard may increase a different hazard. Consider the following example that requires a choice between several possible processes. Process A uses flammable materials of low toxicity; process B uses non-combustible, but volatile and moderately toxic materials; while process C uses non-combustible and non-toxic materials but operates at high pressure. Which process is inherently safer? The answer to this question will depend on the specific details of the process options. In addressing these questions, it is essential that all significant hazards be identified and understood, including:

- Acute toxicity
- Chronic toxicity
- Flammability
- Reactivity
- Instability
- Extreme conditions (temperature or pressure)

- Environmental hazards, including:
    - Air pollution
    - Water pollution
    - Groundwater contamination
    - Waste disposal.

The hazards associated with normal plant operations, such as stack and fugitive emissions, as well as those resulting from specific incidents, such as spills, leaks, fires and explosions, must also be considered. Lastly, a designer needs to also consider business and economic factors in making a process selection. These include:

- Capital investment
- Product quality
- Total manufacturing costs
- Operability of the plant
- Demolition, future clean-up and disposal cost

Design strategies that result in an inherently safer design may also tend to improve process economics. For example, minimizing the size of equipment or simplifying a process by eliminating equipment may reduce capital investment and reduce operating and maintenance costs. However, overall process economics are very complex and impacted by many factors, so it may not always be true that an inherently safer process is also economically more attractive.

An inherently safer process offers greater safety potential, often at a lower process life-cycle cost. However, selection of an inherently safer technology does not guarantee that the actual *implementation* of that technology will result in a safer operation than an alternate process in actual operation. A classic example is alternative modes of transportation. Table 13.1 compares some inherent safety characteristics of airplane and automobile travel. It is apparent that automobile travel, *per se*, is inherently safer, since a car carries fewer people, travels on the ground at a lower speed, contains a smaller quantity of flammable fuel, and operates in an external environment usually capable of supporting life. However, for all trips for which air travel is a practical alternative, it is, in fact safer (Ref 13.11 Evans; Ref 13.28 Sivak). This is because the hazards inherent in air travel are very effectively managed by passive, active, and procedural risk management systems, including, extensive redundancy in the construction of the aircraft, rigorous training and qualification of all personnel, sophisticated management of airplane movement through the air traffic control system, rigorous maintenance of all equipment, and many other systems. These systems all cost a lot of money to organize, maintain, and operate, but it is worth doing this because of the benefits of air transport—the ability to move people and goods long distances in a short time. Society may make similar choices in other technologies, including chemical manufacturing (Ref 13.17 Hendershot 2006). That said, recent increases in air traffic accidents can be traced to breakdowns in these risk management systems.

Table 13.1: Comparison of ISD attributes of air travel and automobile travel (Ref 13.17 Hendershot 2006)

ISD Characteristic	Air Travel	Automobile Travel
"Inventory" of passengers	Up to 800 for Airbus A380	A few people
Elevation above ground	Typically, several kilometers	On the ground
Control requirement	Three dimensions (forward-backward; right-left; up-down)	Two dimensions (forward-backward; right-left)
Speed	~ 900 km/h	~ 100 km/h
Flammable fuel inventory	Thousands of liters	Tens of liters
Passenger compartment	Pressure vessel	May be a convertible (i.e., the outside environment is

		usually capable of supporting life)

Another example is a significant risk in the home: falling down stairs. Recognizing this, Kletz has suggested that, *with respect to the hazard of falling down the stairs*, living in a one-floor ranch house with no stairs is inherently safer. However, McQuaid (Ref 13.22 McQuaid) points out that only "falling down the stairs" accidents are prevented by living in a one-floor house. He gives an example of a situation in which a two-story house might be considered inherently safer. In the 1970s, in a town near Antwerp, people woke up one morning to find all of their domestic animals were dead. A dense, ground hugging cloud of chlorine had been released from a nearby chemical plant during the night. The animals on the ground floor were killed, while the people sleeping in second story bedrooms were unharmed.

One can envision other hazards against which a two-story house might be considered inherently safer. For example, in an area prone to flooding, a second floor provides an escape from rising waters. A drive through many beach resort communities shows that people have adopted this design philosophy, building houses on stilts to avoid damage from high water. An even more inherently safe solution would be to not build in flood prone areas at all, but to build houses on high ground (Ref 13.16 Hendershot 1995).

The traditional strategy of providing layers of protection for a less inherently safe process can be quite effective, although the expenditure of resources to install and maintain the layers may be very large. In some cases, the benefits of the more hazardous technology will be sufficient to justify the costs required to reduce risk to a tolerable level. However, even when we determine that the benefits of an inherently less safe technology justify its use, we should always continue to look for inherently safer alternatives. Technology continues to evolve and advance, and inherently safer alternatives which are not economically attractive today may be very attractive in the future. The development of new inherently safer technology offers the promise of meeting process safety goals more reliably and more economically.

Inherent safety is assessed relative to a particular hazard, or perhaps a group of hazards, but essentially never relative to all hazards. A chemical process is a complex, interconnected organism in which a change in one area of the system can impact the rest of the system, with effects cascading throughout the process. These interactions must be understood and evaluated. Similarly, the chemical industry can be viewed as an ecosystem with complex interactions, interconnections, and dependencies. Understanding these relationships is necessary in order to reach a well-balanced resolution when technological options conflict.

13.2 EXAMPLES OF INHERENT SAFETY CONFLICTS

13.2.1 Continuous vs. batch reactor

The use of continuous, rather than batch reactors is a strategy that is often proposed for improving the inherent safety of a chemical process (Ref 13.4 CCPS 1993). This modification generally succeeds because a continuous reactor is usually much smaller, reducing the material and energy inventory of the process, increasing heat transfer per unit of reaction mass, and improving mixing. However, batch reactors may also have safety advantages, and, under the right circumstances, may be judged to be inherently safer.

Consider a simple reaction:

$$\mathrm{A + B} \xrightarrow[\textit{Solvent S}]{\textit{Catalyst C}} \mathrm{D + E + HEAT}$$

The reaction is exothermic and proceeds virtually instantaneously to complete conversion in the presence of Catalyst C. The process hazard of concern is that the reaction mass becomes extremely unstable if Reactant B is overcharged, or Catalyst C is left out. The resulting buildup of unreacted Reactant B may result in a potentially explosive reaction if Reactant B exceeds a known critical concentration. Two processes are proposed for this reaction, a batch process (Figure 13.1) and a continuous process (Figure 13.2).

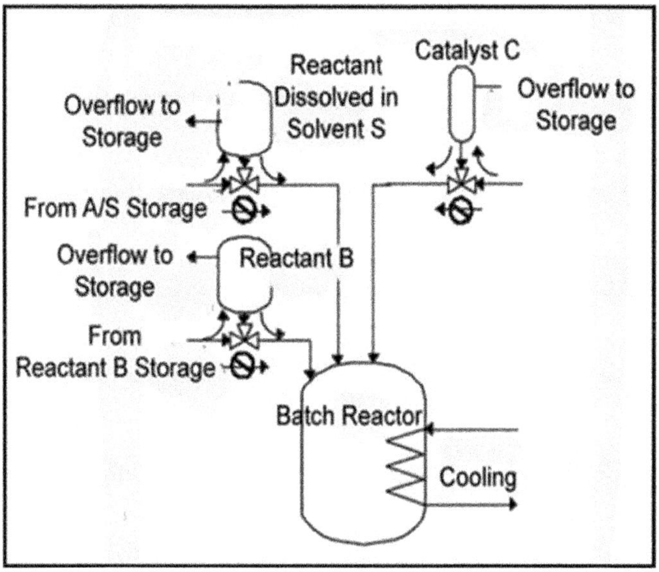

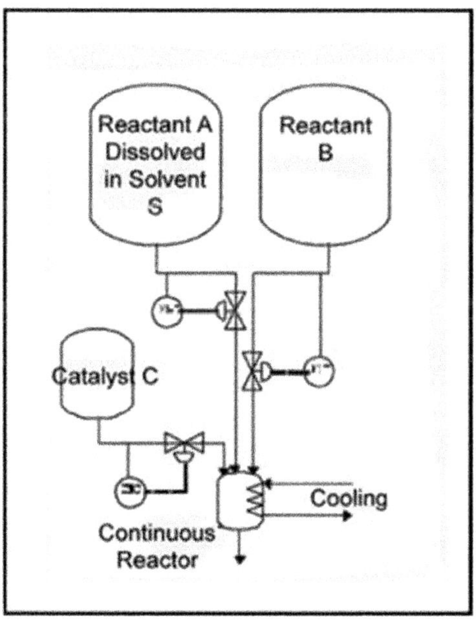

Figures 13.1 and 13.2: Examples of batch (top) and continuous (bottom) process approaches to the reaction described above.

In the batch process, the correct amount of Reactant A dissolved in Solvent S is fed to the reactor from a weigh tank sized to hold exactly the correct charge for one batch. The apparatus features three-way valves which do not allow flow directly from the storage tank to the reactor and will return any excess material fed to each charge tank to a tank overflow to ensure the correct charge. Catalyst C is then added, and an exotherm created by the heat of mixing confirms that the catalyst charge has been made. Reactant B is then fed at a controlled rate to maintain the desired reaction temperature. The feed tank is sized for the correct charge for one batch. Even if the entire charge of Reactant B is somehow fed without reacting, the size is such that it is not possible to reach the critical concentration of Reactant B unless the operators filled and emptied the charge tank several times. The safety advantage of the batch process is that it is very difficult to reach a hazardous condition. But, the disadvantage is that the reactor is very large, so the consequence of the potentially explosive reaction would also be large in the event that the critical concentration was somehow reached.

The continuous reactor, on the other hand, can be sized much smaller–perhaps 1/10 the size of the batch reactor for the same production volume. Therefore, the consequence of the potentially explosive reaction is much smaller in case the critical concentration of Reactant B is reached. However, to keep the process running, the continuous process must be connected to a large feed inventory of Reactant A/Solvent S mixture, Catalyst C, and Reactant B. It is possible, with various feed flow rate ratios, to produce almost any concentration of Reactant B in the reactor. The continuous process relies on instrumentation, such as flow meters, ratio controllers, and control valves, to ensure that the materials are fed in the proper ratio. The control instrumentation, logic, and other hardware could fail in a way that results in a concentration of Reactant B in the reactor in excess of the critical concentration. While the control instrumentation could be made highly redundant and reliable, the continuous process still relies on instrumentation to prevent the Reactant B concentration from exceeding the critical value and should therefore be considered inherently less safe. In addition, with the continuous process operating at steady state and the reactor temperature controlled by cooling the reactor jacket, there is no positive feedback signal to verify that Catalyst

C is being fed. As a result, the continuous process has both an inherent safety advantage—the reactor is significantly smaller—and a disadvantage—it relies on instrumentation and control to ensure that the critical Reactant B concentration is not exceeded.

There is no general answer to the question of which of these two proposed processes is inherently safer. The conclusion will depend on the specifics of real process and reaction kinetics, including the specific consequence of exceeding the critical concentration of Reactant B, the size of the various material charges, and other process-specific details. The example illustrates the point that process alternatives can have both inherent safety advantages and disadvantages, and they must be carefully considered before deciding between the two options. Beyond that, many potential variations that can further enhance inherent safety also need to be explored. For example, for the continuous process, one might use metering pumps with a limited range of flow rates—perhaps driven by a single motor and drive shaft—to control the feed ratios, rather than flow sensors, controllers and valves (Ref 13.15 Hendershot 1994).

13.2.2 Reduced toxicity vs. reactive hazard

For many processes, a designer can choose from several reaction solvents. These solvents will sometimes differ significantly in their hazardous characteristics, such as toxicity, flammability, and volatility. The CCPS *Guidelines for Engineering Design for Process Safety* (Ref 13.4 CCPS 1993) provides an example. For an exothermic batch reaction, there may be a choice between a toxic, but non-volatile solvent, and a non-toxic, but volatile solvent (perhaps water). Each solvent has inherent safety advantages and disadvantages, as summarized in Table 13.2 (Ref 13.15 Hendershot 1994).

The additional factors that must be considered in order to reach a decision on these solvent options include such issues as: environmental impacts; economic considerations, such as capital costs versus operating costs; solvent availability (for example, for materials that may become difficult to procure due to a changing regulatory landscape). Social and ethical considerations, such as materials that may affect the ozone layer, also enter the decision making.

Table 13.2: Some inherent safety advantages and disadvantages of alternative process solvents

Solvent	Inherent Safety Advantages	Inherent Safety Disadvantages
Non-toxic, volatile	Solvent is non-toxic, reducing hazards resulting from normal handling, and hazards resulting from a discharge due to a runaway reaction. The volatile solvent limits temperature rise in case of a runaway reaction due to the "tempering" effect when the solvent boils.	High vapor pressure of solvent results in potential for high pressure in the reactor in case of a runaway exothermic reaction.
Toxic, non-volatile	Runaway reaction exotherm may not be sufficient to raise the reaction mixture temperature to its boiling point, so there is no risk of overpressurizing the reactor.	Potential exposure of personnel to toxic solvent; environmental damage in case of a spill.

13.2.3 Reduced inventory vs. dynamic stability

If the strategy of minimization is employed in a chemical process, there could be a tendency to reduce the slack inventory to a minimum in order to reduce the total potential release volume in the event of a containment breach. One inherent safety conflict with that concept is that a process may then actually be *less* tolerant of operational upsets. For example, assume there is a process that ordinarily has a surge drum with an inventory that allows for compensation of a loss of an upstream process by providing continued feed to the downstream process until the system is restored. In the interest of IS, the surge drum is minimized or eliminated, making the process less stable. If the results of a process upset could lead to a major catastrophic outcome, the apparent IS move may have been a factor in an increased overall risk of catastrophic loss.

13.2.4 Risk transfer vs. risk reduction

When discussing the issue of risk transfer versus risk reduction, one's point of view plays a large part in deciding how a particular risk decision is resolved. If no one is tasked with taking into consideration the overall risks to society or taking a "30,000-foot view" of an issue, then decisions may be made that, while reducing the risk for a portion or subset of parties, results in an increased level of risk for society as a whole.

A population of potentially impacted people may perceive the inherent safety of technology options differently. For example, for a process that requires relatively small quantities of chlorine gas, a plant may have a choice between supply in 1-ton (907 kg.) cylinders or 90-ton (81.6 tonne) railroad tank cars. Neighbors located several miles away from the plant would consider the 1-ton (907 kg.) cylinder supply to be inherently safer because it is unlikely that a leak would affect them at that distance. On the other hand, the plant operators would have to connect and disconnect cylinders 90 times, as opposed to once for a railroad car. They would consider the railroad car to be safer because they are more likely to be exposed to potentially negative impacts by any release, even a small one, and face a much higher frequency of relatively high-risk operations by connecting and disconnecting so many hoses that could potentially contain chlorine. Of course, these hazards can be managed with procedures, personal protective equipment, and other safety management systems, but these are not inherent. Both the neighbor and the operator are correct in their perception of the inherent safety characteristics of the chlorine supply options, but they are concerned about different kinds of incidents. The challenge for the design team of the system is to understand these conflicting requirements and make an intelligent choice, including consideration of the entire risk management system (inherent, passive, active, and procedural) (Ref 13.17 Hendershot 2006).

It is important to consider whether an inherently safer option actually *reduces* risk or merely *transfers* it somewhere else. A plant might reduce the size of a hazardous material storage tank, thereby reducing inventory and site risk. Use of the smaller tank, however, may require a change in how material is shipped to the plant from railroad tank cars (typically about 180,000 to 200,000 lb. (82,000 to 91,000 kg.) shipments for many materials) to trucks (typically about 30,000 to 40,000 lb. (14,000 to 18,000 kg.) shipments) because the smaller tank cannot contain more than a truck load of material. Now, the plant will receive 10 times as many shipments, and they will come by road rather than by rail. Depending on the particular location, road shipments may be more hazardous. Even though the site risk is reduced, the overall risk to society may actually be increased (Ref 13.17 Hendershot 2006).

Another example of transferred risk is when a plant considers producing chlorine on site versus using purchased chlorine via tank cars which are shipped across country. There may be a larger inventory of chlorine when produced on site, which if released may have a greater impact on residences close to the facility. This is contrasted with the shipping of chlorine tank cars over long distances and in some cases through populated areas, presenting a potentially greater (and transferred) hazard. There is also less security of tank cars while being transported.

Minimizing the size of a raw material storage tank will reduce the potential impact of a catastrophic release in the process area, but if the tank is too small to unload an entire tank car of material at once, the potential risk for less than catastrophic spill incidents may increase. If the receiving tank is smaller than the shipping container, the risk of overflowing the tank may increase (Ref 13.26 Overton).

The alkylation process in oil refineries is typically accomplished using either hydrofluoric acid (HF) or sulfuric acid, although other alkylation processes such as ionic liquids are being developed that are considered inherently safer than HF or sulfuric acid. Depending on the concentration, hydrofluoric acid is more toxic and, therefore, inherently more hazardous than sulfuric acid. If released from the alkylation unit, HF would form a toxic cloud, while a release of sulfuric acid would more likely create a pool of liquid. If refineries switched from hydrofluoric acid to sulfuric acid alkylation, it would reduce the process hazard level by use of 2nd order IS (use of less toxic material) at the refinery. But, it may or may not reduce the risk to the broader society. The increased use of sulfuric acid could have a number of negative consequences for safety and the environment:

- HF is regenerated within the alkylation unit, while sulfuric acid must be replaced with waste sulfuric acid regenerated in a separate acid-burning process unit that may be located on or off-site. On average, 0.1 pounds (0.045 kg.) of HF are required per barrel (159 liters) of alkylate produced compared to 42 pounds (19 kg.) of sulfuric acid per barrel (159 liters) of alkylate (Ref 13.23 Myers).
- Because of the difference in acid usage, the HF alkylation process requires about 12 tank truck deliveries per year, while the sulfuric acid process may require 14 to 15 tank trucks per day (Ref 13.23 Myers).
- Regenerating the sulfuric acid in a separate acid burning process unit produces sulfur dioxide and sulfur trioxide. These may be regulated.
- Spent sulfuric acid stored prior to regeneration (if the regeneration unit is located off-site) exists as a mixture with flammable hydrocarbons, creating a flammable waste material (Ref 13.29 US CSB).
- The total volume of chemical waste produced increases.

13.2.5 *Inherent safety and security conflicts*

Inherent safety conflicts may arise as sites assess their security posture (see Chapter 9) if the facility does not consider the potential for shifting risk when making security decisions. For example, a chemical plant may choose to minimize the number of chlorine rail cars it maintains on site; however, this would transfer the risk from the chemical facility to a local rail yard or other storage area, unless security for the railcars at these storage sites was more rigorous than that at the facility. The overall risk to the community could increase if the chemical site had a strong security stance, while the rail yard had open gates and no means of controlling unwanted traffic. In this instance, the chemical company and rail yard would need to work together, and with local law enforcement and emergency responders, to ensure risk is adequately managed and not transferred or increased.

13.3 INHERENT SAFETY – ENVIRONMENTAL HAZARDS

13.3.1 PCBs

Polychlorinated biphenyls (PCBs) were originally introduced in the 1930s as non-flammable cooling and insulating oils for electrical transformers. PCB manufacture was banned in May 1979 due to environmental concerns (Ref 13.1 Boykin). This is an example of how new data and information led to a change in the use of a material due to an improved understanding of its hazards and a reevaluation of the relative importance of different types of hazard (Ref 13.16 Hendershot 1995).

13.3.2 CFCs

With current concerns about the adverse environmental effects of chlorofluorocarbons (CFCs), it is easy to forget that these materials were originally introduced as inherently safer replacements for more hazardous refrigerants then in use. These included ammonia, light hydrocarbons, such as isobutene, ethyl chloride, methyl chloride, methylene chloride and sulfur dioxide (Ref 13.18 Jarabek). These materials are flammable, acutely toxic, or both. A release of one of these substances in the home potentially causes an immediate fire or toxic exposure hazard.

Thomas Midgley, Jr. dramatically introduced CFCs in a lecture to the American Chemical Society in 1937. Midgley filled his lungs with CFC vapors, and then exhaled, extinguishing a candle. This graphically illustrated that CFCs are not flammable or acutely toxic (Ref 13.20 Kauffman). Now, after many decades of use, we have discovered that CFCs cause environmental damage by depleting the stratospheric ozone layer, and their use is being phased out. Since it is unlikely that our society will give up refrigeration or air conditioning, substitute refrigerants are needed. Perhaps hydrofluorocarbons (HFCs) and hydrochlorofluorocarbons (HCFCs) emerged as safe and environmentally acceptable replacements (Ref 13.30 Wallington). However, in some cases, we are going back to the refrigerants which CFCs originally replaced. For example, home refrigerators using isobutene refrigerant are now available in Europe. The manufacturers are not producing "frost-free" versions of these refrigerators because of concerns associated with ignition hazards because of the small heater needed for the defrost cycle of a frost-free refrigerator (Ref 13.24 Chemistry and Industry). People must recognize that they are making a tradeoff when they replace CFCs with other materials. While the alternative materials are safer with respect to long-term environmental damage, they are often more hazardous with respect to flammability and acute toxicity (Ref 13.16 Hendershot 1995).

13.4 INHERENT SAFETY AND HEALTH CONFLICTS

13.4.1 Water Disinfection

Substituting bleach for chlorine in drinking water and wastewater treatment facilities can reduce risk at the water treatment plant but may increase the amount of chlorine required at the bleach manufacturing site. The amount of chlorine needed at the treatment plant—whether from chlorine gas or bleach—will depend on the amount of water to be treated, so the total amount of elemental chlorine required will remain the same. The difference is the way in which the facility receives the chlorine and whether a change from elemental chlorine to bleach will reduce the overall risk, or just shift the risk from one place to another. This well-publicized inherently safer modification can be used to highlight the site-specific challenges to identifying IS opportunities.

Converting from elemental chlorine to bleach will reduce the hazard associated with a release of the material to the population around the water treatment plant, whether from the chlorine/bleach storage vessel or the process of connecting and disconnecting the transported chlorine or bleach to the water treatment process. Due to the economics of sewer and water distribution, such plants are generally located in close proximity to the populations they serve. Because the same amount of chlorine will be needed to treat a given quantity of water, and there is less chlorine in a container of bleach than in the same size container of chlorine gas, more containers of bleach will be required. The increased probability of release due to the increased number of connections/disconnections necessitated by an increased number of shipments must be balanced against the reduced potential consequence from the release of a less hazardous material. The reduction in hazard (in this case, 2nd order IS) could require an increase in the layers of protection surrounding the hazard to reduce the increased probability of release. The goal is for the facility to make sure that the overall risk is reduced - that the increased probability of a release of less hazardous materials presents a lower risk than a lower probability of release of a higher hazard material.

Another example of potential shifts, rather than reductions in risk is the question of whether converting from chlorine gas to bleach shifts risk from the population around the water treatment facility to the facility producing the bleach. If the bleach supplier also supplies chlorine gas by taking large quantities of chlorine and repackaging them into smaller containers, then the facility may be able to readjust the amount of chlorine from repackaging to bleach production. However, a bleach supplier that does not repackage chlorine may be required to increase the amount of elemental chlorine used at that facility in order to meet the increased demand for bleach. If the bleach supplier is also in a more densely populated area, the increased chlorine needed at the facility could increase the risk to that new population. Again, the question of whether overall risk is reduced or shifted will depend on the specifics of an individual water treatment plant, as well as the specifics of the treatment plant's supplier (Ref 13.26 Overton).

13.5 INHERENT SAFETY AND ECONOMIC CONFLICTS

13.5.1 Existing plants – operational vs. re-investment economics in a capital-intensive industry

The following example illustrates the selection of an inherently safer design solution for an existing process.

The design problem was to avoid a significant leak in several water-cooled heat exchangers. These exchangers used material on the process side that reacted violently with water, producing corrosive and toxic by-products. Alternative solutions considered included combinations of passive (double tube sheet or falling film exchangers), active (multiple sensor leak detection with automated isolation), and procedural (a variety of nondestructive testing/ inspection techniques, periodic leak testing with inert gas, improved cleaning procedures) strategies. All of these design alternatives resulted in a lower risk level than the original design. However, none was totally acceptable (see Table 13.3). When management studied the effort and commitment of resources necessary to maintain a less than satisfactory risk level, they chose a design that used a compatible heat transfer fluid (an inherently safer design). Alternatively, a design engineer might choose one of the design solutions offered or choose to combine solutions from more than one category. For example, a combination of reduced inventory (inherently safer) and double-tubesheet construction (passive) might produce the optimal risk reduction alternative. Ultimately, design engineers should make decisions based on the prevailing risk tolerability and cost criteria, and their understanding of the operations and maintenance requirements for the design.

Table 13.3: Process Safety System Design Solutions for a Heat Exchanger Failure Scenario

Design Basis Failure Scenario Tube to tubesheet failure results in mixing of incompatible fluids, resulting in a system over-pressure and/or the formation and release of a toxic material.	
Design Solution Type	Description
1. Inherently safer	A heat transfer fluid compatible with the process fluid
2. Passive	Double tubesheet construction
3. Active	Pressure relief system with discharge to safe location Multiple sensor leak detection with automated isolation
4. Procedural	Periodic manual sampling of the lower pressure fluid Periodic leak detection with inert gas Improved cleaning procedures

Historically, an overemphasis on minimizing initial capital investment, and on time constraints, which often favor active or procedural systems, has resulted in underutilization of inherently safer solutions. Instead, there is an increased dependency on alarms and SISs to reach acceptable risk levels. Economic analyses in the initial design stages often fail to take into consideration the cost of maintaining and proof-testing these systems, which can be significant for large process facilities. When comparing inherently safer design solutions to other solutions, designers should include the total life cycle cost of each alternative before reaching a decision. For example, Noronha, et al. (Re. 13.25 Noronha) describe the use of deflagration pressure containment design in preference to using deflagration suppression or other means of explosion prevention based on life cycle cost and reliability considerations (Ref 13.6 CCPS 1998).

13.5.2 Often more economical, but not necessarily

Figure 13.4 presents a comparison of the four categories of design solutions with respect to several cost and functional parameters, for a heat exchanger failure scenario. Inherently safer/passive solutions (such as exotic metallurgy) tend to have higher associated initial capital outlays; however, operating costs are usually lower than those for the other design solutions. For active solutions (such as on-line monitoring and instrumentation), as compared to inherently safer/passive solutions, reliability is typically lower, and complexity is greater. Operating costs are also likely to be the greatest for active solutions . While procedural solutions are tempting due to their initial very low capital cost and typically lower complexity, they are often also the least reliable, and should be considered only after other solutions have been explored. (Ref 13.6 CCPS 1998).

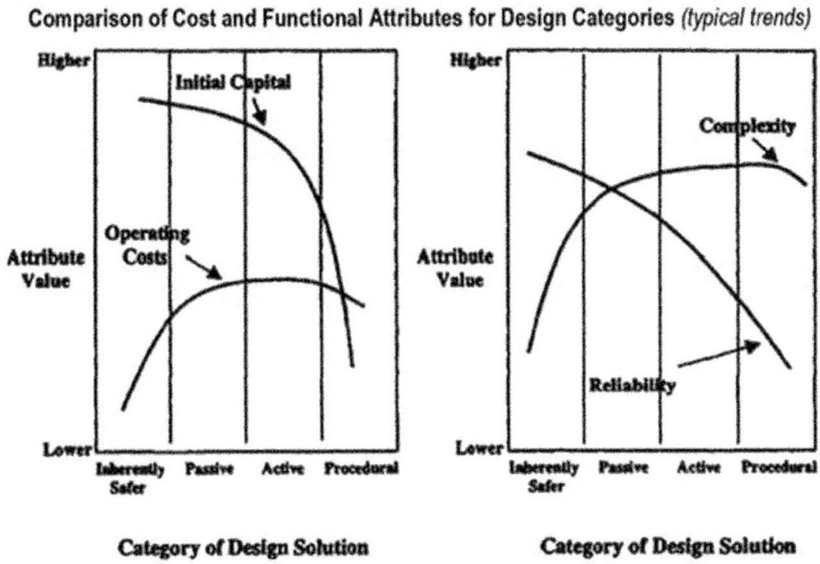

Figure 13.4 Process safety system design solutions for a heat exchanger failure scenario

13.6 TOOLS FOR UNDERSTANDING AND RESOLVING CONFLICTS

Deciding among a number of process options having inherent safety advantages and disadvantages with respect to different hazards can be quite difficult. The first step is to thoroughly understand all hazards associated with the process options. Process hazard analysis and evaluation techniques are appropriate tools (Ref 13.3 CCPS 1992). These include:

- Past history and experience
- Interaction matrices
- "What-if"
- Checklists
- "What-if" Checklists
- Hazard and operability (HAZOP) studies

The hazard identification step is perhaps the most important, because any hazard not identified will not be considered in the decision process. For example, the impact of chlorofluorocarbons on atmospheric ozone was unknown for much of the period of their use, and this potential hazard was not considered until recent years. Process hazards analysis (PHA) tools are addressed in Chapter 12. Any change requires a re-evaluation of the entire system to ensure that any new hazards introduced are properly assessed. Changes should be addressed by a Management of Change (MOC) system that assesses the technical basis for, and the impact of, the change on safety and health. Use of an MOC system to manage changes may be mandated by regulation in certain jurisdictions, depending on the chemicals involved.

Sometimes the consequences of all hazardous incidents can be expressed by a single common measure, such as dollar value of property damage, total economic loss, risk of immediate fatality due to fire, explosion or toxic material exposure. If all consequences can be measured on a common scale, the techniques of quantitative risk analysis (Ref 13.2 CCPS 1989; Ref 13.5 CCPS 1995) may be useful in assessing the relative magnitude of various hazards, and in understanding and ranking total risk of process options.

In many cases, it is not readily apparent how the potential impacts from different hazards can be translated into some common scale or measure. For example, how do you compare long-term environmental damage and health risks from use of CFC refrigerants to the immediate risk of fatality from the fire, explosion, and toxicity hazards associated with many alternative refrigerants? This question does not have a "right" answer. It is not really a scientific question, but instead it is a question of values. Individuals, companies, and society must determine how to value different kinds of risks relative to each other, and base decisions on this evaluation.

Once the hazards have been identified, the process options can be ranked in terms of inherent safety with respect to all identified hazards (see Chapter 12.7). This ranking can be qualitative, placing hazards into consequence and likelihood categories, based on experience and engineering judgment (Ref 13.3 CCPS 1992). More quantitative systems can also be used to rank certain specific types of hazard, for example, the *Dow Fire and Explosion Index* (Ref 13.8 Dow 1994b; Ref 13.12 Gowland) and the Dow Chemical Exposure Index (Ref 13.7 Dow 1994a). Unfortunately, none of these indices consider the full range of hazards.

To get an overall assessment of the process options, it is necessary to use a variety of indices and qualitative techniques and then combine the results. A number of quantitative tools for evaluating inherent safety have been developed or are in the process of being developed. These include Khan and Amyotte (Ref 13.19 Khan), Heikkilä (Ref 13.14 Heikkilä 1999), and Edwards, et al. (Ref 13.10 Edwards 1996).

13.6.1 Tools for understanding and resolving conflicts

The CCPS book, *Tools for Making Acute Risk Decisions with Chemical Process Safety Applications* (Ref 13.5 CCPS 1995), presents a list of factors that, in addition to cost and risk, should be taken into consideration:

- Alternatives available for reducing or eliminating the risks
- Availability of capital
- Codes, standards and regulations and good industry practices
- Company and/or personal liabilities
- Company image
- Costs of implementing available alternatives
- Economic impact of the activity on the local community
- Employment opportunities provided by the activity
- Frequency level(s) of the risk
- Inequities in how the risks and benefits are distributed among members of society
- Number of people at risk
- Perceived benefit of the activity and its impact on the public and/or stockholder image
- Profitability of the activity
- Societal component of the risk, such as the maximum number of people impacted by a single event
- Strategic importance of the activity to the company's growth and survival
- Type(s) of risk, such as human fatality, injury, and acute environmental damage

It is impractical to develop an explicit algorithm for selecting a decision aid. But, decision aids can be flexibly applied, and most can be adapted to an organization's needs and to a wide range of problems. However, an understanding of the key characteristics of the problem, and of the decision aids, is basic to making an appropriate selection.

A key step in selecting a decision aid is to understand the aspects of the problem to be addressed. These aspects include:

- Resource Availability: The time and analytic resources to address the problem
- Problem Complexity: The number of alternatives, the complexity of the system, and the amount of uncertainty in the problem
- Importance/Scrutiny: How sensitive decision makers and stakeholders are to the decision, and how extensively the decision will be reviewed
- Group Involvement: The desire of the organization to involve multiple decision makers or incorporate input from multiple stakeholders in the decision
- Need for Quantification: The desire to be able to point to a quantitative basis for choosing one alternative over another.

In addition to these five problem aspects, there may be constraints that affect the selection of a decision aid. Constraints can include organizational guidelines on the type of analysis to be performed and the types of issues to be addressed.

Specific criteria can be used to evaluate the appropriateness of various decision aids to a given problem. Some key characteristics that differentiate decision aids include:

- *Resource Requirements*: The time, budget, and effort required to use the decision aid
- *Depth of Analysis/Complexity*: The detail and explicitness with which important aspects of the problem are addressed, and the complexity of doing a complete and thorough analysis
- *Logical Rigor*: The mathematical soundness and logical rigor of the analysis
- *Group Focus*: The ability to incorporate group opinions, handle problems with more than one decision maker, and address competing objectives
- *Quantitativeness*: The ability to provide a quantitative basis for the decision, accommodate sensitivity analyses, and address inherently quantitative decisions such as resource allocation problems

- *Track Record*: How long the decision aid has been available and, preferably, demonstrated on actual risk decisions.

Each decision aid can be evaluated against these criteria to determine how well it addresses these concerns. Depending on the distinguishing aspects of a problem, some of these criteria may be more important than others. The final step in selecting a decision aid is to consider the relative importance of these characteristics, and thereby identify decision aids that are best suited to a problem.

There may not be a "right" answer for the decision-making process. Different people will put different values on the relative importance of different kinds of desired and undesired outcomes. Negotiation and consensus building are required in order reach a decision about what the "right" answer is for a particular situation. In order to reach a decision, agreements must also be reached on how the system boundaries will be drawn. Will you consider transportation? Impact on upstream or downstream technology? Markets? Impact of changes in the product on other parts of the entire system?

In many cases, formal tools for decision making can be useful, particularly if the hazards vary greatly in type of consequence or impact. Many of these tools introduce additional rigor, consistency, and logic into the decision process. Some available methods include:

- Weighted scoring methods, such as Kepner-Tregoe Decision Analysis and the Analytical Hierarchy Process (AHP)
- Cost-benefit analysis
- Payoff matrix analysis
- Decision analysis
- Multi-attribute utility analysis

The chemical process industry is beginning to use these techniques in making safety, health, and environmental decisions. In CCPS (Ref 13.5 CCPS 1995), the reader will find a review of these decision aids and others, with special emphasis on how they are employed in making chemical process safety decisions. Hendershot uses a weighted scoring technique based on Kepner-Tregoe Decision Analysis (Ref 13.21 Kepner); Hendershot (Ref 13.16 Hendershot 1995) as illustrated for a generic process in Table 13.4. Reid and Christensen (Ref 13.27 Reid) describe the use of the Analytical Hierarchy Process to evaluate three alternative technologies considered for a metal fabrication application, with the overall goal of minimizing waste from the process.

The following stepwise procedure can be used for decision matrix calculation:

- Assign a weighting factor (1 to 10) to the various parameters based on your judgment of the relative importance of this Safety, Health and Environmental, or other issue.
- For each option, assign a performance factor from 1 to 10, based on the relative performance of that option with respect to the particular parameter. This can be based on judgment or scaled based on some kind of quantitative analysis.
- Multiply the weighting factor by the performance factor for each parameter and process option combination.
- Sum the products for each process option (highest total is most desirable).

See Kepner and Tregoe (Ref 13.21 Kepner) or CCPS (Ref 13.5 CCPS 1995) for additional discussion, particularly on how potential negative consequences may impact the scoring matrix. Note that the weighting factors in this table are for purposes of illustrating the methodology only, and do not represent recommendations on the relative importance of the factors listed.

Table 13.4: An example of a Weighted Scoring Decision Matrix

Parameter	**Weighting Factors**	**Process Options**			
		#1	**#2**	**#3**	**#4**
	Performance Factor >	x 2 =	x 9 =	x 10 =	x 1 =
COST	9	18	81	90	9

	Performance Factor >	x 10 =	x 5 =	x 3 =	x 1 =
SAFETY	10	100	50	30	10
	Performance Factor >	x 3 =	x 5 =	x 1 =	x 10 =
ENVIRONMENT	7	21	35	7	70
	Performance Factor >	x 3 =	x 10 =	x 2 =	x 1 =
OPERABILITY	5	15	50	10	5
	Performance Factor >	x 1 =	x 9 =	x 10 =	x 3 =
DESIGN	3	3	27	30	9
	Performance Factor >	x 7 =	x 5 =	x 10 =	x 1 =
OTHER	3	21	15	30	3
	SUM	**178**	**258**	**197**	**106**

13.7 MEASURING INHERENT SAFETY CHARACTERISTICS

Current efforts to measure the inherent safety of processes and operations are still in development stage and are primarily focused on basic process technology and route selection. These measures have been used occasionally in industry and there is, as yet, little data that relates the application of these indices to process safety outcomes in chemical processing.

13.7.1 Dow Fire and Explosion Index (Ref 13.14 Heikkilä 1999)

The Dow Fire and Explosion Hazard Index was developed initially for use within company operations and was later disseminated in a book published by the American Institute of Chemical Engineers (Ref 13.8 Dow 1994b). The index quantifies the damage from potential fire and explosion incidents and identifies equipment that would be likely to contribute to the creation or escalation of an incident.

The Dow Index is the product of a *Unit Hazard Factor* and a *Material Factor*. The material factor for a process unit is based on the most hazardous substances or mixtures present that would result in a worst-case scenario, and it quantifies the amount of energy released. A material factor is determined for each process unit. The unit hazard factor for a process unit is the product of both general and specific process hazards. General process hazards address such issues as exothermic chemical reactions, endothermic processes, material handling and transfer, enclosed or indoor process units, access, and drainage and spill containment. Special process hazards include factors for toxic materials, operation in or near flammable range, dust explosions, relief pressure, low temperature operation, corrosion and erosion, joint and packing leaks, fired heaters and hot oil systems, and rotating equipment.

13.7.2 Dow Chemical Exposure Index

Like the Fire and Explosion Index, the Chemical Exposure Index was developed by Dow to help its employees design and operate safer facilities. In 1998, an AIChE volume brought the index—which by then was considered a seminal tool for rating the relative acute health hazard potential of a chemical release to workers and the neighboring community—to the entire process industry. The newest edition uses a new methodology for estimating airborne quantity released, which allows for more sophisticated process analyses (Ref. 13.7 Dow 1994a).

13.7.3 Mond Index

The Mond Index was developed by Imperial Chemical Industries (now ICI, an Akzo Nobel Company) after the Flixborough incident, and is based on the *Dow Fire and Explosion Index*. The Dow Index was modified to address: 1) a wider range of processes and storage installations; 2) processing of chemicals having explosive properties; 3) improved hazard consideration for hydrogen; 4) additional special process hazards; and, 5) inclusion of toxicity in assessment.

The Mond Index divides the plant into individual units and takes into consideration plant layout and the creation of separating barriers between units. The hazard potential is initially expressed in terms of a set of indices for fire, explosion, and toxicity. The hazard indices are then reviewed to determine if design changes reduce the hazard, and the revised values. Factors for preventative and protective features are applied, and then final values of the indices are calculated.

13.7.4 Proposed Inherent Safety indices

The *Integrated Inherent Safety Index* (I2SI), developed by Khan and Amyotte (Ref 13.19 Khan) , addresses the economic evaluation and hazard potential identification for each option within the process life cycle. I2SI is comprised of sub-indices accounting for hazard potential, inherent safety potential, add-on control requirements, and the economic aspects of the options. The two main sub-indices are a hazard index and an inherent safety potential index. The hazard index measures the damage potential of the process, considering the process and hazard control measures. The inherent safety potential index addresses the applicability of inherent safety principles to the process. The two sub-indices are combined to produce the I2SI value.

The *Prototype Index of Inherent Safety* (PIIS) for process design was developed by Edwards and Lawrence (Ref 13.9 Edwards 1993). The PIIS is based on a chemical score and a process score. The chemical score takes into consideration inventory, flammability, explosiveness and toxicity. The process score addresses parameters, such as temperature and pressure.

When using an inherent safety index, the user should take the necessary steps to ensure that he/she understands the basis of the index. The developers of the indices use their own judgment and experience in deciding what factors are analyzed and in determining the weighting—sometimes transparently, sometimes hidden—of those factors and how they are combined. The user must be sure that these subjective decisions are in line with their organization's philosophy and goals.

Heikkilä (Ref 13.13 Heikkilä 1996) developed an *Inherent Safety Index* to classify process alternatives during the process synthesis phase. This method assigns scores to chemical factors (heat of main and side reactions, flammability, explosiveness, toxicity, corrosiveness, chemical interaction) and process factors (inventory, process temperature and pressure, equipment safety, safety of process structure). The final index is a summation of the chemical and process factors.

13.8 SUMMARY

It is not always clear which of several process alternatives is inherently safer. Because nearly all chemical processes have a number of hazards associated with them, an alternative which reduces one hazard may increase a different hazard. It is also necessary to consider business and economic factors in making a process selection. Design strategies that result in an inherently safer design may also tend to improve process economics. For example, minimizing the size of equipment or simplifying a process by eliminating equipment will usually reduce capital investment and reduce operating costs. However, overall process economics are very complex and are impacted by many factors, and it may not always be true that an inherently safer process is also economically more attractive.

An inherently safer process offers greater safety potential, often at a lower cost. However, selection of an inherently safer technology does not guarantee that the actual implementation of that technology will result in a safer operation than an alternate inherently safe process. As previously stated in Chapter 2, section 2.8, an inherently safer design also does not necessarily ensure that the technology will be safer in actual operation.

Inherent safety is assessed relative to a particular hazard, or perhaps a group of hazards, but essentially never relative to all hazards. A chemical process is a complex, interconnected organism in which a change in one area of the system can impact the rest, with effects cascading throughout the process. These interactions must be understood and evaluated. Similarly, the chemical industry can be viewed as an ecosystem with complex interactions, interconnections, and dependencies. Understanding these relationships is necessary in order to reach a well-balanced resolution when technological options conflict.

Various means for measuring inherent safety characteristics are available. These include the Dow Fire and Explosion and Chemical Exposure Indices and the Mond Index. Several inherent safety indices have also been developed by Khan and Amyotte, Heikkilä, and Edwards and Lawrence.

13.9 REFERENCES

13.1 Boykin, R.F., Kazarians, M. and Freeman, R.A. (1986). *Comparative fire risk study of PCB transformers,* Risk Analysis, 6 (4), 477-488.

13.2 Center for Chemical Process Safety (CCPS 1989). *Guidelines for Chemical Process Quantitative Risk Analysis*. New York: American Institute of Chemical Engineers, 1989.

13.3 Center for Chemical Process Safety (CCPS 1992). *Guidelines for Hazard Evaluation Procedures, Second Edition With Worked Examples*. New York: American Institute of Chemical Engineers, 1992.

13.4 Center for Chemical Process Safety (CCPS 1993). *Guidelines for Engineering Design for Process Safety*. New York: American Institute of Chemical Engineers, 1993.

13.5 Center for Chemical Process Safety (CCPS 1995). *Tools for Making Acute Risk Decisions With Chemical Process Safety Applications*. New York: American Institute of Chemical Engineers, 1995.

13.6 Center for Chemical Process Safety (CCPS 1998). *Guidelines for Design Solutions to Process Equipment Failures.* New York: American Institute of Chemical Engineers, 1998.

13.7 Dow Chemical Company (1994a). *Dow's Chemical Exposure Index Guide, 1st Edition*. New York: American Institute of Chemical Engineers, 1994.

13.8 Dow Chemical Company (1994b). *Dow's Fire and Explosion Index Hazard Classification Guide, 7th Edition*. New York: American Institute of Chemical Engineers, 1994.

13.9 Edwards, D.W. & Lawrence, D. , (1993) Assessing the inherent safety of chemical process routes: Is there a relation between plant costs and inherent safety? Trans. IChemE. 71, Part B, 252-258, 1993.

13.10 Edwards, D.W., Lawrence, D., and Rushton, A.G. (1996). *Quantifying the inherent safety of chemical process routes.* In 5th World Congress of Chemical Engineering, July 14-18, 1996, San Diego, CA (Paper 52d).

13.11 Evans, L., Frick, M.C., and Schwing, R.C. *Is it safer to fly or drive?* Risk Analysis, 10, 239–246, 1990.

13.12 Gowland, R.T. *Putting numbers on inherent safety.* Chemical Engineering 103 (3), 82-86, 1996.

13.13 Heikkilä, A.-M., Hurme, M. & Järveläinen, M. (1996). *Safety considerations in process synthesis.* Computers Chem. Engng, 20(Suppl. A), S115-S120, 1996.

13.14, Heikkilä, A.-M. (1999). *Inherent safety in process plant design.* VTT Publications 384. Technical Research Centre of Finland, Espoo. (D Tech Thesis for the Helsinki University of Technology), 1999.

13.15 Hendershot, D.C. (1994). *Chemistry—The Key to Inherently Safer Manufacturing Processes.* In 208th American Chemical Society National Meeting, August 21-25, 1994, Washington, DC (Paper No. ENVR-135).

13.16 Hendershot, D.C. (1995). *Conflicts and decisions in the search for inherently safer process options.* Process Safety Progress, 14 (1), 52-56, 1995.

13.17 Hendershot, D.C. (2006). *An overview of inherently safer design.* Process Safety Progress, 25 (2), 103-107, 2006

13.18 Jarabek, A.M., et. al., Mechanistic insights aid the search for CFC substitutes: Risk assessment of HCFC-123 as example. Risk Analysis, 14 (3) 231-250, 1994.

13.19 Khan, F.I. and Amyotte, P.R. *I2SI: A comprehensive quantitative tool for inherent safety and cost evaluation.* Journal of Loss Prevention in the Process Industries, 18, 310-326, 2005.

13.20 Kauffman, G. B., *CFCs, TEL and "Midge."* Chemistry and Industry, 143, 21 February 1994.

13.21 Kepner, C.H., and Tregoe, B.B., *The New Rational Manager.* Princeton, NJ: Princeton Research Press, 1981.

13.22 McQuaid, J., *Know your enemy: The science of accident prevention.* Trans. IchemE, Part B. 69, 9-19, 1991.

13.23 Myers, P.; Mudan, K. and Hachmuth, H. , *The risks of HF and sulfuric acid alkylation.* In International Conference on Hazard Identification and Risk Analysis, Human Factors and Human Reliability in Process Safety. New York: AIChE/CCPS, Health & Safety Executive, U.K. and European Federation of Chemical Engineering, 1992.

13.24 *New look fridges take off without HFCs,* Chemistry and Industry, 130, 21 February 1994.

13.25 Noronha, J.A., Merry, J.T., and Reid, W.C., *Deflagration pressure containment for vessel safety design,* Plant/Operations Progress, 1 (1), 1-6, 1982.

13.26 Overton, T.A. and King G., *Inherently safer technology: An evolutionary approach.* Process Safety Progress 25 (2), 116-119, 2006.

13.27 Reid, R.A., and Christensen, D.C., *Evaluate decision criteria systematically.* Chemical Engineering Progress, 90 (7), 44-49, 1994.

13.28 Sivak, M. and Flannagan, M.J., *Flying and driving after the September 11 attacks.* Am Scientist 91, 6–8, 2003.

13.29 U.S. Chemical Safety and Hazard Investigation Board (US CSB), *Investigative Report, Refinery Incident: Motiva Enterprises.* Washington, DC: Report No. 2001-05-I-DE, 2002.

13.30 Wallington, T.J., et al. *The environmental impact of CFC replacements – HFCs and HCFCs.* Environmental Science and Technology, 28 (7), 320-325, 1994.

14. Inherent Safety Regulatory Initiatives

14.1 INHERENT SAFETY REGULATORY DEVELOPMENTS AND ISSUES

In the past decade regulators and legislators in jurisdictions across the globe have recognized the risk reduction potential in IS and have debated as to whether encouraging or mandating these approaches through regulation could improve overall process safety or security results. Debated options have ranged from requiring that facilities "consider" IS as one of several choices, to actual mandates that IS be implemented, the latter option backed up by giving agencies the authority to override facility determinations and require installation of specific IS elements.

Several key obstacles or misperceptions have resulted in IS being underutilized by industry, including the following:

- A *perception* that IS is technically and economically practical for only new processes, though it has been demonstrated to be potentially useful for existing facilities.
- The *lack of an inherent safety infrastructure*, or a framework for evaluating IS systems, including technologies supporting IS approaches, and methodologies that permit IS to be integrated into technical, economic, safety and security design considerations. This includes the lack of consensus metrics for IS.
- The *lack of specific guidance* on how to conduct an inherent safety study, particularly for existing facilities and processes.
- A *lack of understanding* of IS principles and how to practically apply them to both new and existing facilities.

Therein lies the dilemma. Policy makers may view implementing IS as a relatively simple and effective way of minimizing or eliminating the hazards or consequences from process incidents, but in practice companies encounter obstacles since these requirements must be integrated with other design and operational considerations. Despite the attention devoted to it in the political arena, IS remains more of a philosophy than a codified process with a well-established and understood framework for evaluation and implementation. Both industry and regulators lack tools and measures to compare the inherent safety of different options or to determine what is "feasible." Therefore, policy debates over how best to encourage IS continue to be frustrating for all concerned.

14.2 EXPERIENCE WITH INHERENT SAFETY PROVISIONS IN UNITED STATES REGULATIONS

Unlike other process safety issues, IS is not easily regulated. For example, when the United States EPA promulgated its Risk Management Program (RMP) rule in 1996, some commenters recommended the Agency require facilities to conduct "technology options analyses" to identify inherently safer approaches. The US EPA declined to do so, stating that:

> *"PHA teams regularly suggest viable, effective (and inherently safer) alternatives for risk reduction, which may include features such as inventory reduction, material substitution, and process control changes. These changes are made as opportunities arise, without regulation or adoption of completely new and unproven process technologies. EPA does not believe that a requirement that sources conduct searches or analyses of alternative processing technologies for new or existing processes will produce additional benefits beyond those accruing to the rule already."*

In 2017, a final revised RMP Rule was published in the Federal Register (Ref 14.8 Revised RMP Rule). This represented the review ordered by then-President Obama's 2013 Executive Order (Ref 14.17 Executive Order) to the federal agencies responsible for regulating the safety and security of the chemical industry in the wake of the ammonium nitrate fire and explosion in West, TX in 2013. One of the revisions in the final revised RMP Rule was a provision to perform a Safer Technology & Alternatives Analysis (STAA) as part of the PHAs of RMP-covered processes. The revised final RMP Rule was delayed several times following its publication, but in 2018 the U.S. Court of Appeals for the District of Columbia Circuit vacated the delay and ordered the final rule be implemented.

US EPA may choose to undertake rulemaking to rescind the final revised RMP Rule in the future, but as of this writing the STAA provisions of the revised RMP Rule are in effect in the United States. The US EPA has set the compliance date for performing the STAAs to 2021. The statutory language of the law enabling the Chemical Facility Anti-Terrorism Standard (Ref 14.7 DHS) precludes Department of Homeland Security (DHS) from requiring implementation of specific security measures such as IS, yet the Department recognizes that "facilities are certainly free to consider IST options, and their use may reduce risk and regulatory burdens." (Ref 14.7 DHS). This approach is consistent with expert testimony given before the Senate Environment and Public Works Committee in 2006, as well as other references on the topic (Ref 14.10 Mannan).

As noted earlier, in addition to the federal US EPA RMP requirements, two jurisdictions in the United States—Contra Costa County, CA, and the State of New Jersey—adopted IS regulations in the previous decade. Recently, California has adopted revised process safety regulations for oil refineries in the state that include a strong IS component. This section explores the IS requirements and lessons learned from these three jurisdictions. The following sections describe the IS portions of these three jurisdictions.

14.2.1 Inherently Safer Regulatory Requirements – Contra Costa County, California, USA

This section focuses on a regulation for inherently safer studies in Contra Costa County, California, USA, providing a regulatory framework that can help companies evaluate there IS options. The first discussion covers the IS requirements of the Contra Costa County (CCC) Industrial Safety Ordinance (ISO) / Richmond Industrial Safety Ordinance (RISO), covers the specific regulations defining "feasible" IS studies, and then concludes with major accident/incident reduction results after the guidance and regulations were implemented.

Contra Costa County (CCC) forms the eastern boundary of the San Francisco and San Pablo Bay northeast of Oakland and San Francisco. The county is highly populated, with approximately 1.14 million residents as of 2016. Due to its extensive waterfront, the county has long been home to chemical/processing industry facilities, including oil refineries and chemical plants.

Following a series of serious industrial accidents, the County enacted the Industrial Safety Ordinance (ISO) in 1999 and has revised it several times since then, most recently in 2014 (Ref 14.4 CCC ISO). Designed to be "the most stringent in the United States, if not the world," (Ref 14.6 CCHS 2017) the ordinance expands on the California Accidental Release Prevention (CalARP) Program (Ref 14.1 CalARP) for petroleum refineries and chemical plants in the County. The City of Richmond, CA, located within Contra Costa County, adopted its own ordinance in 2001 (Ref 14.3 RISO). The Richmond Industrial Safety Ordinance (RISO) mirrors the county ordinance.

Inherent Safety Requirement of the CCC ISO/RISO. A facility covered by the CCC ISO or RISO is required to conduct an inherently safer systems analysis (ISSA) for each covered process as follows (Ref 14.4 CCC ISO):

- The facility conducts an ISSA on existing covered processes every five years.
- The facility conducts an ISSA in the development and analysis of recommended action items identified in a PHA.
- Whenever a major change is proposed at a facility that could reasonably result in a major chemical accident or release, the facility conducts an ISSA as part of the required management of change review.
- If an incident occurs and the incident investigation report or its associated root cause analysis recommends a major change that could reasonably result in a major chemical accident or release, the facility commences and completes an ISSA of the recommended major change as soon as administratively practicable after completion of the incident investigation or root cause analysis report.
- The facility conducts an ISSA during the design of new processes, process units and facilities. Immediately upon completion of the ISSA report advise CCHS of the availability of the ISSA report.
- The facility prepares a written report documenting each ISSA including identification and description of the inherently safer system(s) analyzed in the ISSA, description of the methodology used to analyze the inherently safer systems(s), the conclusions

of the analysis, the rationale for the conclusions, and the action plan with timeline to implement the inherently safer system(s) recommended in the ISSA.

- The facility selects and implement each inherently safer system identified in an ISSA report to the greatest extent feasible and as soon as administratively practicable. If a facility concludes that implementation of an inherently safer system is not feasible, the facility documents the basis for this conclusion in meaningful detail. The documentation includes sufficient evidence to demonstrate to the department's satisfaction that implementing the inherently safer system is not feasible and the reasons for this conclusion. *A claim that implementation of an inherently safer system is not feasible cannot be based solely on evidence of reduced profits or increased costs.* (emphasis added)

Inherently safer systems means inherently safer design strategies as discussed in the latest edition this book and means feasible alternative equipment, processes, materials, lay-outs, and procedures meant to eliminate, minimize, or reduce the risk of a major chemical accident or release by modifying a process rather than adding external layers of protection. (Ref 14.4 CCC ISO).

To help resolve any confusion regarding IS implementation, CCHS worked with the covered facilities to expand the original ISO program guidance on these requirements. The resulting IS guidance document, published by the CCHS in 2002 and revised in 2011 (Ref 14.5 CCHS), provides guidance for both new processes as well as existing processes. For new processes, the guidance identifies the following different project phases when reviews should be performed to confirm if IS is applicable: (Ref 14.5 CCHS)

- during the chemistry forming (synthesis) phase for product/process research and development to focus on the chemistry and process;
- during the facilities design scoping and development prior to completion of the design basis to focus on equipment and configuration; and
- during the basic design phase of the project.

The guidance contains more detailed questions and review items to apply during each of these project phases. The guidance does recognize that IS may not be applicable for all facilities and project phases in the same way or in the same sequence. IS should be examined as early as possible to minimize the impact on the project.

For existing processes, the guidance notes that the consideration of IS is limited to scenarios where a Major Chemical Accident or Release (MCAR) could reasonably occur. CCHS defines a MCAR as an incident that meets the definition of a Level 3 or Level 2 incident in the community warning system incident level classification system defined in the hazardous materials incident notification policy, as determined by the department; or results in the release of a regulated substance and meets one or more of the following criteria: (1) Results in one or more fatalities; (2) Results in at least twenty-four hours of hospital treatment of each of at least three persons; (3) Causes on- and/or off-site property damage (including clean-up and restoration activities) initially estimated at $500,000 or more. On-site estimates are performed by the stationary source. Off-site estimates are performed by appropriate agencies and compiled by the department; or (4) Results in a vapor cloud of flammables and/or combustibles that is more than 5,000 pounds (Ref 14.4 CCC ISO). Facility PHAs can be used to identify which scenarios have consequences (without crediting any active safeguards) that rise to the MCAR level. In most chemical/processing industry PHAs that are part of formal PSM programs, this is documented in PHA worksheets by describing the consequences so that offsite and onsite impacts that result from major releases are clearly identified. This is also usually supplemented by a qualitative or quantitative consequence or severity ranking that clearly indicates the severe consequences that are the hallmark of a MCAR.

Facilities can perform one of the following methods to ensure that ISS are considered and documented for the covered processes:

- An independent ISS analysis that is done in addition to a PHA. Checklists or guideword analysis contained in the guidance document that incorporates ISS can be used to accomplish this analysis.

- An ISS analysis that is incorporated into the existing PHA review process. This would require that the PHA for each process be re-done in its entirety to include an initial ISS analysis. Revalidated PHAs that examine only portions of the process may not be sufficient to satisfy the initial ISS review if the whole process is not evaluated. Checklists or guideword analysis contained in the guidance document that incorporates ISS can be used to accomplish this analysis.

Whichever type of ISS analysis is implemented by the facility the CCHS guidance specifies that the analysis be conducted and documented in the following manner:

- The facility will document the qualifications of the team facilitator/leader and team makeup, including positions, names, and any relevant experience or training.
- The facility will document the ISSs considered as well as those implemented.
- If the facility chooses to do an independent ISS analysis, the facility should document the method used for the analysis, what ISS were considered, and the results of each consideration. If the checklist for ISS was used, for items that were not considered, document why those items were not considered, i.e., not applicable or were already considered in previous consideration.
- The facility will document for the ISS considered and not implemented, the grounds that were used to make the feasibility determination. See CCHS's definition of feasibility below.
- The documentation for incorporating the guidewords for ISS into a HAZOP should be consistent with the documentation used during any HAZOP Study.
- For any other ISS analysis, the facility should document the ISS considered, the ISS implemented, and the ISS not implemented.
- The ISS analyses should be revalidated at least once every five years. The revalidation should incorporate improvements made in method since the last review was conducted or select a new method to perform the ISS analyses; ISS review(s) for all changes that have been made since the last ISS analysis; review of all major chemical accidents or releases or potential major chemical

> accidents or releases that occurred at the process under review; and review for any new and existing technologies not previously reviewed that can be incorporated that will make the process under review inherently safer. (Ref 14.5 CCHS)

CCHS does not specify a particular format for an ISSA report in either the ordinance or the guidance document. The guidance document does include IS checklists and guidewords to aid in the performance of the ISSA.

CCHS ISO/RISO Definition of "Feasible". CCHS defines "feasible" as "capable of being accomplished in a successful manner within a reasonable period of time, taking into account economic, environmental, legal, social, and technological factors." However, a claim of financial infeasibility "shall not be based solely on evidence of reduced profits or increased costs, but rather shall include evidence that the financial impacts would be sufficiently severe to render the inherently safer system as impractical." If a facility concludes that an inherently safer system is not feasible, the basis for this conclusion is documented in meaningful detail. This documentation includes (1) sufficient evidence to demonstrate to the county's satisfaction that implementing this inherently safer system is impractical, and (2) the reasons for this conclusion. The guidance provides additional clarity on what CCHS will and will not accept as a justification of infeasibility. Many of the criteria are similar to those developed by the US EPA and US OSHA. CCHS will provide an affirmative determination that the documentation, calculations and justifications for not implementing an ISS are acceptable. (Ref 14.5 CCHS)

Results. CCHS reported in 2017 that implementation of the ISO has improved and, in most cases, is being done as required by the ordinance. Since ISO covered facilities must annually update the information on their accident history in their Safety Plans and include how they have used IS processes and solutions within the last year, the county has significant information on how IS has been implemented in Contra Costa County as a result of the ISO. A summary of the inherently safer systems and risk reduction measures implemented by ISO-covered facilities since 2002 appears in Table 14.1.

In the 2007 ISO Annual Report, CCHS reported that the number and severity of Major Chemical Accidents or Releases (MCARs) have been decreasing since the implementation of the ISO. However, the small number of MCARs (fewer than a dozen total incidents in any given year since 1999) makes it difficult to demonstrate a linear trend, or to establish a direct causal relationship between the ISO and/or implementation of ISS and the number of incidences. Figures 14.1-14.3, taken from the 2017 ISO Annual Report (Ref 14.6 CCHS 2017), display, respectively, the number and severity of MCARs that have occurred in the county since 1999–incidents at the eight ISO/RISO and CalARP covered facilities,

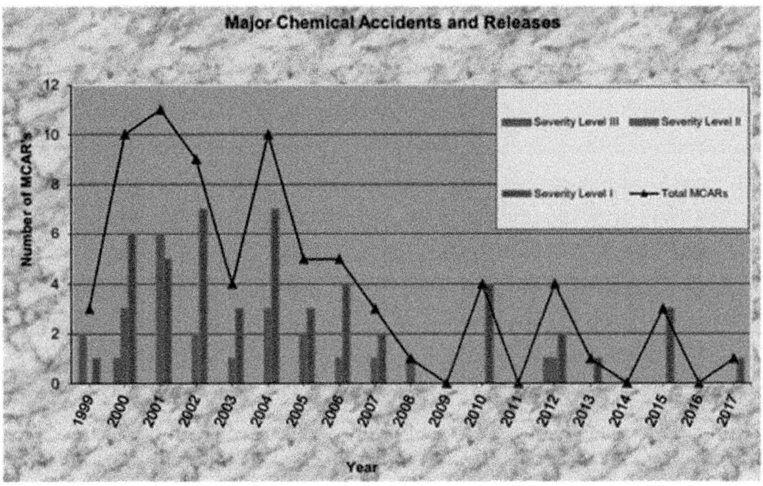

Figure 14.1: Major Chemical Accidents and Releases (MCAR)

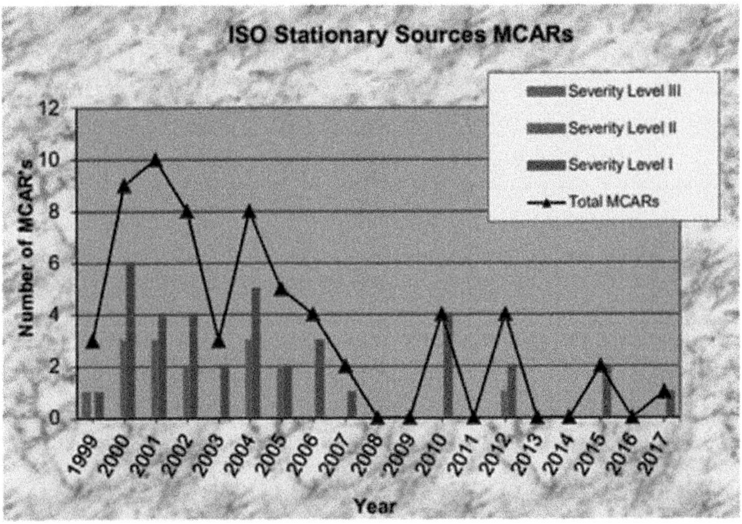

Figure 14.2: ISO Stationary Sources MCARs

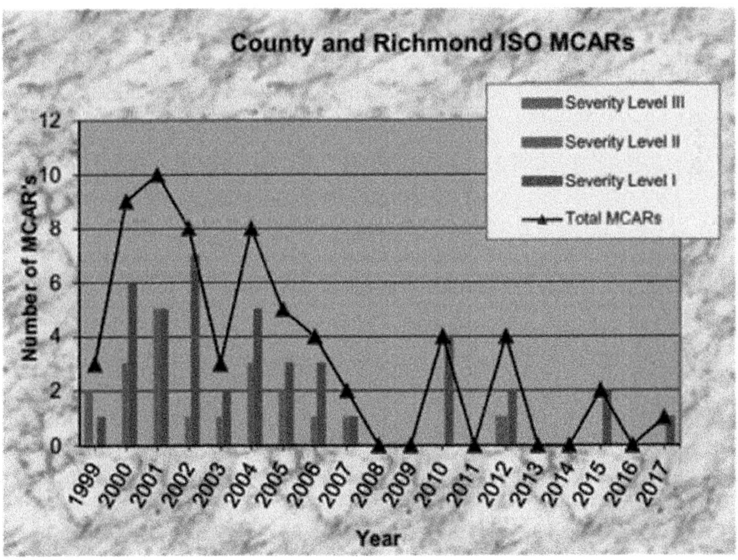

Figure 14.3: County and Richmond ISO MCARs

Table 14.1 Summary of Contra Costa County IST Measures Implemented from 2002 - Present

		2002-2006	2007-2012	2013-2018
Inherent	Minimize	• Reduction of inventory (36)	Reduction of inventory (9)	Reduction of chemical inventory from elimination of process equipment and/or piping or deinventory (14)
		• Eliminated a unit and regulated material (1)	Replaced old scrubber with reduced inventory (1)	Reduction of chemical inventory by combining or removing equipment from the process (13)
		• Incorporated existing equipment into a new unit and reduced piping (1)	Reduction of inventory by removing piping/equipment from the process (19)	Eliminated hazardous materials storage equipment or off-site transportation (6)
		• Eliminated hazard (2)	Remove process storage (15)	Eliminated hazardous material release points from equipment

				modification or removal (8)
		• Process re-design to eliminate need for a second tower (1)		Reduction of inventory by removing dead-leg piping (3)
		• Reduced the potential inventory		
		• Eliminated tankage (thus reducing inventory)		
		• Eliminated stored aqueous ammonia		
		• Removed excess piping		
		• Installation of a tubular reactor that reduced the inventory of ammonia		
		• Reduced inventory by eliminating tanks or reducing the		

		amount of piping		
		• Eliminated a process furnace		
	Substitute	Process change using less hazardous chemical (1)	Use a less hazardous chemical (3)	Changed chemical and equipment design to reduce potential of a hazard (4)
	Moderate	• Substituted a less hazardous chemical for a more hazardous chemical	Elimination of atmospheric PSVs (1)	Reduced the potential of a hazard by changing operating conditions (3)
		• Switched from anhydrous dimethyl amine to 60% aqueous dimethyl amine	Reduced potential of exposure by eliminating equipment/leak source (1)	Reduced the potential of a hazard by changing chemical (1)
		• Substituted a chemical with a less hazardous form of the chemical		
		• Eliminated a process furnace		

		• Decreased concentration (1)		
		• Less hazard process conditions or form (7)		
	Simplify	• Eliminated tankage that could cause a hazard	Equipment design change to eliminate dead leg potential (2)	Reduced the number of chemicals by consolidation of chemical usage (1)
		• Changed the process conditions to less hazardous process conditions	Removed out of service or unnecessary tasks, equipment or piping (11)	Simplified unit design and chemical inventory by changing/re-routing equipment (10)
		• Removed excess piping	Elimination of glass gauge to eliminate exposure to hazard (1)	Eliminated exposure potential by changing equipment design (2)
		• Eliminating a process furnace	Removed Sample stations at process unit (1)	

		• Removed out of service equipment or dead leg (8)	Eliminated equipment (3)	
		• Simplified process or eliminate equipment (13)		
		• Revised equipment design features (5)		
		• Eliminated cross contamination connections (3)		
Passive	Minimize	Revised equipment design features (1)	Revised equipment metallurgy design features (42)	Reduced potential of the hazardous condition by using smaller amount of chemical (5)
	Substitute		Replaced tubing and equipment with alloy not susceptible to chloride cracking (2)	Reduced potential of a hazard by changing the substance (21)
			Reduced exposure	

			potential with change of energy source, design feature (2)	
	Moderate	Reduced the potential of a hazard by eliminating the hazard or moving to an alternate location (2)	Reduced the potential of a hazard by moving to an alternate location (12)	Upgraded equipment metallurgy or design to reduce potential of a hazard (117)
			Reduced the impact of the hazard by adding sensors (1)	Reduced the potential of a hazard by using a less corrosive chemical (1)
			Revised equipment metallurgy, components, controls features or personnel placement (39)	Reduced hazardous conditions of impact of exposure by equipment design or layout features (28)
			Reduced hazardous conditions by equipment design features (7)	Reduced the potential of a hazard by relocating equipment (2)

	Simplify	• Simplified the process or piping (4)	Reduction of inventory by elimination of dead leg (7)	Change equipment configuration or design to reduce potential of a hazard (9)
		• Reduced potential of a hazard by design features, heating media and equipment location (25)	Reduced potential of a hazard or the frequency by changing design features, temperature and/or pressure design rating of equipment and/or installation of equipment (43)	Reduced frequency of exposure by changing equipment design (3)
		• Improved overpressure or temperature design rating of equipment (19)	Upgrade tubing and equipment metallurgy to reduce potential of a hazard or the frequency (3)	Reduced potential of the hazardous condition by simplified design (1)
		• Eliminated hazard to downstream unit without active device by removing		Simplified unit design and chemical inventory by changing/re-routing equipment (2)

		contaminants from system (1)		
		• Reduced potential overpressure hazard by equipment elimination (8)		
		• Reduced the potential of a hazard by eliminating liquid accumulation points or vibration stress (5)		
		• Reduced the frequency of a hazard (5)		
		• Hard piping for hydrofluoric acid transfer instead of using a mobile trailer		
		• Installed vessels capable of withstanding the highest pressure scenario		

		• Piping system was modified to naturally restrict flow without use of active controls		

Active	Moderate		Reduced potential of exposure by adding alarm limits (2)	Reduced the potential of a hazard by adding equipment, or controls (16)
			Reduced potential of exposure by adding controls (3)	
Procedural	Simplify		Reduced potential of error by standardizing into procedure (2)	Reduced potential of error by standardizing into a procedure (10)
	Moderate			Developed procedure to reduce potential for catastrophic releases (1)

Lessons Learned from the ISO.

Applicability: Experience under the ISO confirmed that first- and second-order inherent safety is most readily applicable to new and modified facilities and processes. Additional layers of protection have been most commonly implemented in existing processes, such as those identified through PHAs. However, broad application of IS across all four strategies of process risk management is more practical and may result in novel risk reduction ideas. The mandate to consider the full spectrum of IS options does seem to have prompted new ways of looking at reducing hazards and risks and appears to have contributed to overall risk reduction at these facilities.

Guidance: Industry needs clear guidance on how and when to apply IS measures. The eight facilities subject to the ISO initially found it difficult to apply ISS to existing processes. The 2002 guidance, which was updated most recently in 2011, provided greater and needed detail on how and when to apply ISS and significantly improved the regulator's and facility's understanding of what was expected under the ordinance. Before the guidance was issued, the majority of ISO submissions involving PHA action items did not actually involve application of IS, and the distinction between IS and risk reduction in general was often lost. However, the guidance did not resolve all implementation questions. For example, CCHS and facilities may still disagree about what is reasonable and what is feasible. Nevertheless, after the guidance was published, many more IS opportunities were found and applied. Inherent Safety is best applied by those knowledgeable of, and properly trained in, the process. Continued guidance and additional training are needed for PHA or IS review teams to know how to do this effectively.

Expectations: Both the public and regulators may have higher risk reduction expectations of IS approaches than what is ultimately found to be feasible in a given situation. The public and regulators expect to see a realistic effort to apply first- and second-order inherent safety measures to achieve major risk reduction. However, only considering first- or second-order IS may be at odds with other practical and cost-effective risk reduction options, especially for existing processes. This difference between public and industry perceptions of risk - and a misunderstanding or lack of appreciation of the impacts of some IS modifications - may create a barrier to both common expectations, and to an accepted definition of "feasibility" between the public and industry. Any move in an inherently safer direction is likely to be a positive risk reduction move and should be encouraged.

14.2.2 New Jersey Toxic Catastrophe Prevention Act (TCPA) and Prescriptive Order for Chemical Plant Security

This section focuses on a regulation for inherently safer studies in the state of New Jersey (NJ), USA, providing another regulatory framework that can help companies evaluate there IS options. The first discussion covers the IS requirements of the NJ Toxic Catastrophe Prevention Act (TCPA), covers its Prescriptive Order for Chemical Plant Security, covers its specific regulations defining "feasible" IS studies, discusses results after implementing the Prescriptive Order, and some lessons learned from this approach.

New Jersey is the United State's most densely populated state. It has a large number of chemical and petroleum facilities, and is also a locus of pharmaceutical, biotechnology, and other life-science industries. These and other industrial plants that produce or use hazardous chemicals are clustered in the Delaware and Passaic River valleys, and around the well-developed transportation corridor linking the New York and Philadelphia metropolitan regions.

The combination of critical chemical infrastructure in close proximity to high population centers has provided an incentive for the state to enact particularly stringent programs to reduce the risk of release - both accidental and intentional - from such facilities. The NJ state Toxic Catastrophe Prevention Act (TCPA) regulation (Ref 14.11 TCPA) was adopted in 1987 in the wake of the Bhopal incident to protect the public from catastrophic accidental releases of extraordinarily hazardous chemicals, and was the United State's first holistic, multi-element process safety regulatory program. The success of the TCPA program influenced the enactment of the two titles in the Clean Air Act of 1990 that authorized US OSHA's PSM Standard and EPA's RMP Rule.

NJ Toxic Catastrophe Prevention Act. New Jersey enacted the Toxic Catastrophe Prevention Act legislation in 1985 (Ref 14.16 NJSA) and adopted regulations to implement the law in 1987 (Ref 14.11 TCPA). The TCPA regulations are revised every five years by statute (the law "sunsets" and must be re-authorized), and these revisions have added several provisions, including IS. The purpose of the TCPA is to reduce the risk of "potentially catastrophic discharges of hazardous substances in the environment." The legislature concluded that the ability to respond and mitigate a release was critical, but insufficient, and that

> *"...the single most effective effort to be made is toward prevention of those environmental accidents by anticipating the circumstances that could result in their occurrence and taking those precautionary and preemptive actions required." (emphasis added)*

While not using the term "inherent safety," the statute did require that extraordinarily hazardous substance risk reduction work plans document:

> *"any alternative processes, procedures, or equipment which might reduce the risk of a release of an extraordinarily hazardous substance while yielding the same or commensurate results, and the specific reasons they are not employed."*

The New Jersey state legislature envisioned that facility owners/operators would seriously consider changes in process equipment, and even chemical substitutions, to reduce the risk of accidental releases. The original TCPA regulations also contained an IS-like provision that required owners/operators to conduct "state-of-the-art" (SOA) reviews of existing equipment when PHAs and risk assessments required by the regulations were revalidated every five years. The SOA reviews were required if PHA/risk assessment hazard scenarios caused offsite consequences (i.e., exceeded the toxic limits of concern at the site boundary) and the frequency of the release scenario exceeded a 1 X 10-4 per year.

The TCPA regulations were significantly revised in 2003 to include an explicit definition of, and a requirement to consider, inherently safer technology. Owners/operators of newly designed and constructed covered processes are required to evaluate inherently safer technologies, and to document recommendations from these evaluations, regardless of whether they were incorporated in the new process. The regulations adopted the definition of "inherently safer technology" from the first edition of this book.

> *"Inherently safer technology" means the principles or techniques incorporated in a newly designed and constructed covered process to minimize or eliminate the potential for an Extraordinarily Hazardous Substance (EHS) accident that includes, but is not limited to, the following:*
>
> 1. *reducing the amount of EHS material that potentially may be released;*
> 2. *substituting less hazardous materials;*
> 3. *using EHSs in the least hazardous process conditions or form; and*
> 4. *designing equipment and processes to minimize the potential for equipment failure and human error.*

In explaining the 2003 regulations, the New Jersey Department of Environmental Protection (NJDEP) called for consideration of the full range of inherent safety options, recognizing

> *"that every (inherently safer technology) technique will not be suitable for every process and will rely on the owner or operator's evaluation of which techniques, if any, will be appropriate for a particular newly designed and constructed process."*

NJDEP also limited application of inherent safety as follows:

> *"newly designed and constructed processes," concluding that "the most reasonable and effective time for an owner or operator to evaluate inherently safer technology is during the design phase of a process, before it has been built."*

These policy positions – describing "inherently safer technology" in terms of principles or techniques, retaining the full range of options, and applying only during new design or construction – are critical. They demonstrate that, as late as 2003, New Jersey recognized inherent safety as a philosophy to be incorporated in the design process, rather than an endpoint to be reached.

In May 2008, NJDEP proposed amendments to the TCPA rules that expanded the program in two ways: (1) *all* TCPA facilities—not just those with the potential for offsite consequences - must conduct IST reviews of any process with an EHS that meets or exceeds the regulatory threshold; and (2) IST reviews must include both *existing* and new processes, rather than just newly designed and constructed covered processes. The new rule also revised the TCPA application to "releases" not just "accidents." The new rule continued the trend towards a more comprehensive application of inherent safety in New Jersey. The 2008 amendments also deleted the earlier requirement for state-of-the-art reviews in lieu of the new, more comprehensive requirements for IST reviews.

Prescriptive Order for Chemical Plant Security. The New Jersey Domestic Security Preparedness Task Force was established by the 2001 New Jersey Domestic Security Preparedness Act as a cabinet-level body responsible for setting state homeland security and domestic preparedness policy. The Task Force was part of the newly-formed Office of Counter-Terrorism (now the NJ Office of Homeland Security and Preparedness). The Task Force subsequently adopted security standards, guidelines and protocols for 16 critical infrastructure sectors, including the chemical sector.

In 2005, the State of New Jersey issued a Prescriptive Order for chemical plants, requiring 157 facilities to comply with the "Chemical Group Security Assessment and Best Practices Report" previously adopted by the state's Domestic Security Preparedness Task Force (Ref 14.15 NJPO). Specifically, facilities were required to conduct security vulnerability assessments and to develop a security prevention, preparedness and response plan. Chemical facilities subject to TCPA were also explicitly required to

> *"...conduct a review of the practicability and the potential for adopting inherently safer technology (IST) as part of the assessment and plan," and to submit a report on the IST review to the state."*

The order adopted the TCPA definition of IST, but went further than the TCPA by requiring that

> *"...such review shall include analysis of whether adoption of IST alternatives is practicable, and the basis for any determination that implementation of IST is impractical."*

NJDEP published implementation guidance for performing the IST reviews in 2006 and this guidance has been revised several times, most recently in 2017. (Ref 14.13 NJ IST) The guidance explained that,

> *"...an analysis of IST can be done in all phases of a plant's life, such as during the conceptual phase, design phase, or operational phase. However, an operating plant may have a limited option of IST application, considering the practicability and cost-effectiveness. For an existing facility, an application of multiple levels of engineered safeguards, coupled with practicable and cost-effective technological or design changes, may accomplish the same degree of inherent safety in a process plant." (Ref 14.13 NJ IST). The guidance went on to state that*
>
> *"IST may be applied regardless of release likelihood to reduce the hazard of an EHS."*

The guidance called for the following questions to be answered for each of the four criteria in the IST definition (minimize, substitute, moderate, simplify):

- Are there any available IST alternatives over the current process?
- Is adoption of IST alternatives practicable?
- What is the basis for any determination that implementation of IST is impractical?
- What past IST and risk reduction measures have already been incorporated into the current process?

The guidance also contained provisions for the qualifications of the persons performing the IST reviews, how they were to be documented, and a definition of feasibility. The guidance specified that the technology should be available at reasonable cost commensurate with the anticipated reduction or elimination of the hazard; it was generally not expected that research into new technologies would be conducted or that a major process modification or replacement be instituted to implement an inherently safer alternative. (Ref 14.13 NJ IST)

New Jersey TCPA Definition of "Feasible". The current (2017) version of the TCPA guidance contains more detailed criteria for determining the feasibility or infeasibility of an IS measure. These examples are not intended to be all-inclusive. Also, a clarification is included whether a qualitative and/or quantitative justification is expected. To date this guidance on the feasibility of IS options is the most detailed that has been published by a regulator or any organization.

Environmental and Public Health and Safety Feasibility

- There would be a significant negative environmental impact (Consideration of water resources, water pollution, air pollution, solid and hazardous wastes, noise, etc.)
- The IST could decrease the hazard but would increase the overall risk.
- The risk would be shifted to another location where the risk would be the same or higher (If infeasibility is claimed because of risk, document the difference in frequency and/or consequences with and without the IST).
- All of the above require both qualitative and quantitative justification.

Legal Feasibility

- The IST could result in a conflict with existing federal, state, or local laws.
- The IST could violate a license agreement and the license agreement cannot be modified and must remain in effect
- Qualitative justification is required for both.

Technological Feasibility

- Is in conflict with Recognized and Generally Accepted Good Engineering Practices (RAGAGEP)
- Product quality specifications cannot be met (qualitative and quantitative).
- Availability of materials (qualitative and quantitative).
- Space restrictions (qualitative and quantitative).
- Impact on production rate (qualitative and quantitative).
- Commercially available or not (qualitative).

Economic Feasibility

If a given IST is claimed to be infeasible due to economic reasons the infeasibility justification should address the following elements:

- Life Cycle Analysis
- Capital investment, including design and implementation

- Net operating costs
- Change in the cost of materials including transportation and handling related costs
- Change in energy consumption
- Change in human costs such as number of operators, training
- Any other direct manufacturing costs
- Net regulatory compliance cost, change in fees
- Demolition and future cleanup and disposal cost
- All of the above criteria require a quantitative justification (cost/benefit analysis)

Generally, an IS measure is feasible if it has been successfully applied to similar processes or similar situations unless there are unique circumstances at the facility. The justifications should highlight those unique circumstances and how they relate to the feasibility factors. (Ref 14.13 NJ IST)

It is important to point out that, while the Prescriptive Order and its IST review requirement were driven by security concerns and the potential for intentional releases, efforts to comply with the Prescriptive Order also served to address the potential for accidental releases. In other words, the required review would address the full range of IST strategies, and not only substitution, minimization, and moderation—the most effective strategies to reduce security-related risk.

Results of Implementing IST Under the Prescriptive Order. According to NJDEP, of the 157 facilities subject to the Prescriptive Order, more than 98% complied within the 120-day deadline established in the Order. Of those, 32% provided a schedule to implement additional IST or other risk reduction measures, and 19% identified additional IST or risk reduction measures. The remaining 49% of the facilities had no additional recommendations and 80% of the facilities concluded that at least some of the IST or risk reduction measures identified during the IST evaluation were infeasible for their operations. (Ref 14.12 Sondermeyer)

Based on the results of the IST review program required under the Prescriptive Order, New Jersey believes that evaluating inherently safer technology is not overly burdensome on industry and is an effective tool for critically evaluating the risk reduction opportunities available at a specific facility. Furthermore, NJDEP suggests that IS analysis is simply good business practice for any facility storing or utilizing extraordinarily hazardous materials from an economic, worker safety and regulatory compliance standpoint. (Ref 14.12 Sondermeyer). Experience with IS under the Prescriptive Order served as justification for expanding the IS program for all TCPA facilities.

In 2010 NJDEP published a summary of the IS reviews performed by TCPA registrants in 2008 and 2009 after the requirement for IS reviews was added to the TCPA beyond the Prescriptive Order IS reviews. The results of these IS reviews are shown in Table 14.2. (Ref 14.14 NJ IS Summary)

Learnings from New Jersey. The State of New Jersey has viewed IS as an option in their security, as well as their process safety regulatory arsenal, believing that a facility with less hazardous chemicals will be less attractive to terrorists and, if attacked, have a reduced probability of a serious accidental release and/or reduced consequences.

One way in which the TCPA has driven the concept of inherent safety has been by inspiring companies to seek ways to eliminate or reduce the amount of EHSs (i.e., the covered chemicals) handled to a level below the threshold quantity to which the regulations apply. In the almost 30 years since the TCPA was enacted, the number of TCPA-regulated facilities, as well as the quantity of EHSs registered in the state, have dropped significantly. This has had the benefit of reducing risk as well as avoiding regulatory requirements. It should be noted, however, that many TCPA covered facilities voluntarily reduced their EHS inventories before New Jersey issued the Prescriptive Order or adopted the IS provisions in the TCPA.

Table 14.2 Summary of IST Measures Implemented or Scheduled by Sector

Sector	# of Facilities Submitted an IST Report	Total # of IST Measures Implemented or Scheduled	# of Facilities Reporting One or More Measures to be Implemented	Percentage of Facilities Implementing IST Measures	Maximum # of ISTs to be Implemented by a Facility	# of Facilities Eliminated EHS Use by Imple-menting IST
Chemical	41	77	18	44	11	0
Water/ wastewater	13	15	7	54	4	2
Refinery	4	10	2	50	3	0
Food	14	35	9	64	7	0
Power	6	1	1	17	1	0
Other	7	9	4	57	4	0
Total	85	143	41	48	11	2

14.2.3 Inherently Safer Systems Requirements – California Accidental Release Prevention (CalARP) Regulations

This section focuses on a regulation for inherently safer studies in in the state of California, USA, providing another regulatory framework that can help companies evaluate there IS options. As a result of the fire in one of the crude units at the Chevron refinery in Richmond, CA in 2012, the State of California revised its two process safety regulations at the state level: the California Accidental Release Prevention (CalARP) regulations (Ref 14.1 CalARP), and CalOSHA's PSM regulation (Ref 14.2 CalOSHA). These revised regulations became effective in 2017 and currently apply only to oil refineries in the State of California.

These revisions represent a significant modification to process safety regulatory requirements in California and include a requirement to include IS as part of a Hierarchy of Controls Analysis (HCA) in the following manner:

- Identify, analyze, and document all inherent safety measures and safeguards (or where appropriate, combinations of measures and safeguards) in an iterative manner to reduce each hazard to the greatest extent feasible.
- Identify, analyze, and document relevant, publicly available information on inherent safety measures and safeguards. This information includes inherent safety measures and safeguards that have been: (A) achieved in practice by for the petroleum

refining industry and related industrial sectors; or, (B) required or recommended for the petroleum refining industry, and related industrial sectors, by a federal or state agency, or local California agency, in a regulation or report.

For each process safety hazard identified in the HCA, the team develops written recommendations to eliminate hazards to the greatest extent feasible using first order inherent safety measures. The team develops written recommendations to reduce any remaining hazards to the greatest extent feasible using second order inherent safety measures. If necessary, the team also develops written recommendations to address any remaining risks in the following sequence and priority order:

- Effectively reduce remaining risks using passive safeguards;
- Effectively reduce remaining risks using active safeguards;
- Effectively reduce remaining risks using procedural safeguards.

The HCA report identifies the IS measures identified in the study, as well as the rationale for the IS measures and safeguards recommended by the HCA team for each process safety hazard. Additionally, the owner/operator may change a team recommendation if the owner/operator can demonstrate in writing that an alternative inherent safety measure would provide an equivalent or higher order of inherent safety, or, for a safeguard recommendation, an alternative safeguard would provide an equally or more effective level of protection. The HCAs are due to be completed in California refineries during the 2018-2020 timeframe. (Ref 14.1 CalARP).

These revisions to the CalARP and CalOSHA PSM regulations go beyond previous IS regulations in that they specifically require that first order and second order IS measures be implemented. Current process safety regulations in Contra Costa County and New Jersey require IS reviews and justification of the infeasibility of IS measures identified and not implemented. *The HCA requirements do not include a review of the feasibility of the available or applicable inherent safety measures but requires that inherent safety measures must be used to "reduce the hazard to the greatest extent feasible."* This is an important change in how inherent safety has been regulated thus far.

14.2.4 Safer Technology & Alternatives Analysis – Revised US EPA Risk Management Program (RMP) Rule

This final section on specific United States regulations with IS requirements focuses on the 2018 the U.S. Court of Appeals for the DC Circuit decision to vacate the delay of the final revised RMP Rule published in the Federal Register in 2017. The revised RMP Rule contains a provision to perform a Safer Technology & Alternatives Analysis (STAA) as part of performing PHAs on RMP-covered processes.

US EPA modified the PHA provisions in the RMP Rule by adding a requirement for certain industry sectors to conduct a STAA and to evaluate the practicability of any inherently safer technology (IST) identified. The practicability study will determine the costs and assess the reasonableness of implementing technology alternatives. US EPA limited the applicability of this requirement to owners or operators of facilities with RMP Program 3 regulated processes in North American Industrial Classification System (NAICS) codes 322 (paper manufacturing), 324 (petroleum and coal products manufacturing), and 325 (chemical manufacturing). In the proposed rulemaking, US EPA specified that the STAA would consider, in the following order of preference:

- IST or inherently safer design (ISD),
- Passive measures,
- Active measures, and
- Procedural measures.

US EPA further indicated that the owner or operator would be able to evaluate a combination of these risk management measures to reduce risk at the process. US EPA did not mandate the adoption of any IST found to be practicable in part because we recognize that a passive measure or other approach on the STAA hierarchy may also be effective at risk reduction and left the adoption of particular accident prevention approaches to owners' and operators' reasonable judgment.

US EPA also added several definitions that relate to an STAA. US EPA defined active measures to mean risk management measures or engineering controls that rely on mechanical, or other energy input to detect and respond to process deviations. Some examples of active measures included alarms, safety instrumented systems, and detection hardware (such as hydrocarbon sensors). "Passive measures" means risk management measures that use design features that reduce the hazard without human, mechanical, or other energy input, e.g., pressure vessel designs, dikes, berms, and blast walls. US EPA defined procedural measures to mean risk management measures such as policies, operating procedures, training, administrative controls, and emergency response actions to prevent or minimize incidents.

US EPA substituted the term "practicability" for "feasible" in the final rule and defined the new term to mean capable of being successfully accomplished within a reasonable time, accounting for economic, environmental, legal, social, and technological factors. Note that the final definition is the same that US EPA used in the proposed rule for "feasible." US EPA did not specifically define "economic, environmental, legal, social, and technological factors" but did refer to New Jersey's definition of these terms in their IST guidance document as examples of how these terms can be defined with respect to IS. However, the use of New Jersey's definition in the implementation of STAA in the RMP Rule is not mandatory. US EPA believes that the definition of "practicability" in the final rule provides sufficient flexibility for the owner or operator to determine whether an IST or ISD considered could be successfully accomplished, and that the same tools and methods that facilities currently use for their PHA can be used to identify and measure hazards and risks of any safer alternative options. US EPA further clarified in the definition that environmental factors would include consideration of potential transferred risks for new risk reduction measures.

For IST or ISD, the proposed definition meant risk management measures that:

- Minimize the use of regulated substances.
- Substitute less hazardous substances.
- Moderate the use of regulated substances.
- Simplify covered processes in order to make accidental releases less likely or the impacts of such releases less severe.

14.3 ISSUES IN REGULATING INHERENT SAFETY

Industry and policymakers have a major opportunity and a major challenge ahead to develop effective programs—whether voluntary industry initiatives or government regulations—to encourage the broad adoption of IS. First, both industry and policymakers need a fuller and more consistent understanding of IS - what it is and how it can be applied. Secondly, new analytical tools for conducting IS reviews and measuring progress, as well as decision-making criteria, will be needed (see Chapter 12).

While it is generally accepted that inherently safer technology has the potential to reduce both accidental and intentional process safety hazards, the implementation, and especially regulation, of IST is not generally straightforward. As the Contra Costa County and New Jersey experience has demonstrated, there is limited feasibility for implementing IS in existing facilities due to costs, as well as other tradeoffs, such as risk transfer. Industry recognizes that IST is not a "silver bullet" to eliminate security and accidental risks, including the recognition that tradeoffs are often involved which can simply change (or increase) the risk and not reduce it. One such example is the use of smaller shipping containers, which may reduce the consequences of a container leak, but will generally increase the risk in the transportation sector offsite due to more frequent shipments and more frequent human operations, i.e., connecting and disconnecting the containers from the process. Cost is also usually a factor, and the application and management of process safeguards (active, passive, and procedural) often provides an adequate and more cost-effective level of protection compared with alternative and sometimes radical process re-designs.

IS concepts are worthwhile as engineering design guides and as a general philosophy for designing, constructing, maintaining, and operating chemical process facilities. The ongoing concern of industry is that IS alternatives, once identified, will be mandated regardless of feasibility, potentially causing economic or even safety consequences to the facility or to the transportation system. As with other process safety/risk management regulations, the way in which an IS option becomes a requirement and is enforced is the key to the success of such an initiative. Based on the experience in New Jersey, however, it is unlikely that requirements for conducting IS reviews and implementing such technology "where practicable" will necessarily result in large-scale risk reduction against security-related risks.

14.3.1 Consistent Understanding of Inherent Safety

Misunderstandings or misperceptions about IS tend to localize around four concepts – *goals, applicability, scope* and *economic feasibility*.

The goal of both a safety and security program should be to reduce risk. Inherent safety is an approach to reducing and managing risks; it is not an end in itself. IS policies and regulations will be most successful when they clearly state a risk reduction and management goal with a recognition that some risks are inherent to the production of some critical goods and services, and that such risks can be managed within acceptable ranges.

IS may be applicable to existing as well as new facilities and processes. There may be a perception that IS is relevant only for new facilities and that there are no feasible opportunities once the process is operational. While it is true that the potential for major improvements may be greatest during process development, this book has demonstrated that facilities have reduced or even eliminated hazards or have managed change to avoid new hazards by applying IS methods throughout the facility life cycle. The majority of the applications for IS are with the installed industrial base, whereas the feasibility of applying IS to the fullest diminishes as the facility is built. This leaves many companies where new processes (and particularly new technologies) are rarely implemented with fewer occasions to practice the methods.

The scope of IS is not limited to hazard reduction., IS concepts are applicable to the layers of protection surrounding the remaining hazards. A narrow view argues that IS only applies to major changes in the degree of hazard, while a broader viewpoint finds any changes that increase safety through the application of IS principles to be an advantage. This includes the programmatic aspects of PSM programs (see Chapter 11).

Changes to facilities and processes must be economically feasible. Costs are a primary concern when considering modifications to existing facilities. Inherent and passive approaches are strategic, usually must be implemented early in the facility development, and have broad and wide-ranging impacts on the process design. Tactical approaches, including the active and procedural categories of the hierarchy of controls, can be implemented late in the design process and are characterized by repetition and high costs associated with maintenance (Ref 14.9 Hendershot). Costs have to be evaluated on a holistic basis, and it has been proven that the life cycle costs and risks may be reduced after implementing IS measures. Both the Contra Costa County and New Jersey programs explicitly include a consideration of economic feasibility as part of an IS evaluation.

14.3.2 Needed Tools

Chapter 12 discusses a number of tools and approaches for evaluating different and potentially conflicting IS options and measures. Additional tools and approaches will be needed to support IS regulatory programs including those to support systematic IS reviews, and economic evaluation and performance evaluations.

Systematic review methods. Methods for systematic IS reviews are needed to provide consistent and comparable IS evaluations within and across facilities. Such an approach provides greater assurance that dissimilar facilities will conduct reasonably equivalent IS evaluations, and that the results of those evaluations will be understandable and acceptable to regulators and the public.

Economic valuation methods. Since economic feasibility is a critical factor in identifying viable IS opportunities, better methods for estimating this variable would benefit both regulatory and voluntary IS programs. Methods to estimate the value of IS and to quantitatively assess whether a given process is "as inherently safe as is practicable" are currently unavailable or unproven. Though examples are increasingly available in the literature, case studies documenting the economic benefit of IS projects do not cover a wide array of industrial situations.

Evaluation of performance. A constraint to regulation is the lack of consensus on appropriate IS metrics. Assuming that the regulation is performance-based, rather than specification-based, metrics should be developed to help provide consistency. It is difficult to measure the effectiveness of IS regulations simply by inferences from the rate or severity of incidents.

> *"Regulation to improve inherent safety faces several difficulties. There is not presently a way to measure inherent safety. Process plant complexity essentially prevents any prescriptive rules that would be widely applicable. It would seem that legislation could explicitly require facilities to evaluate inherently safer design options as part of their process hazard analysis. But inherent safety would be almost impossible to enforce beyond evaluation because there are unavoidable technical and economic issues"* (Ref 14.10 Mannan).

This is particularly problematic when an IS opportunity appears to conflict with another safety, health or environmental policy objectives. IS performance metrics could help evaluate such risk tradeoffs.

14.4 SUMMARY

Seeking greater use of IS strategies and approaches has the potential to improve the safety, security and even economic performance of a process. However, inherent safety should not be seen as an end in itself, or even the mandatory strategy to reduce risk. Rather, it should be seen as one strategy to be employed to reach a risk reduction target.

Contra Costa County, and the states of California and New Jersey require consideration of IS as part of their chemical security (New Jersey) and process safety (California, New Jersey, and Contra Costa County) programs. Facilities have complied (or will comply) with the IS requirements in these jurisdictions, but in the absence of consensus IS metrics, it is difficult, if not impossible, to evaluate the contribution these requirements have made to safety and security. By March 2021 RMP-covered facilities in certain NAICS codes will also have complied with the revised RMP Rule that requires performing a STAA and considering the practicability of IST modifications that were identified during the STAA

Policymakers have, and continue to, consider regulation as a means to accelerate the application of inherent safety in some of the higher risk processing industries. Both industry and policymakers will benefit from a deeper understanding of inherent safety – goals, applicability, scope, and economic feasibility. Most important is the development of new and improved tools and metrics for evaluating the contribution of IS to risk reduction and economic assessments.

14.5 REFERENCES

14.1 California Code of Regulations, California Office of Emergency Services, *Accidental Release Prevention (CalARP) Regulations*, Title 19, Division 2, Chapter 4.5.

14.2 California Code of Regulations, *Process Safety Management for Petroleum Refineries*, Title 8, Section 5189.

14.3 City of Richmond, CA, *Richmond Industrial Safety Ordinance (RISO),* Municipal Code, Chapter 6.43.

14.4 Contra Costa County (CCC), California, *Industrial Safety Ordinance (ISO),* County Ordinance Chapter 450-8.

14.5 Contra Costa Health Services (CCHS), California, *Industrial Safety Ordinance Guidance Document,* Section D, June 2011.

14.6 Contra Costa County, California, *Industrial Safety Ordinance Annual Performance Review & Evaluation Report*, Contra Costa Community Health Hazardous Materials Programs, 2017.

14.7 Department of Homeland Security, Interim final rule implementing the Chemical Facility Anti-Terrorism Standard, April 9, 2007, 72 Fed. Reg. 17718.

14.8 Environmental Protection Agency, Final Risk Management Rule, 40 CFR 68, January 13, 2017, 82 Fed. Reg. 4594.

14.9 Hendershot, D.C., Safety through design in the chemical process industry: Inherently safer process design Presented at the Benchmarks for World Class Safety Through Design Symposium, Institute for Safety Through Design, National Safety Council, 1997.

14.10 Mannan, S. White Paper-Challenges in Implementing Inherent Safety Principles in New and Existing Chemical Processes, Texas A&M University, 2002.

14.11 New Jersey Administrative Code, *Toxic Catastrophe Prevention Act (TCPA) Regulations*, N.J.A.C. 7:31.

14.12 Sondermeyer, G. Testimony of NJDEP Director of Operations Gary Sondermeyer, before the US House of Representatives Committee on Homeland Security, December 12, 2007.

14.13 State of New Jersey Department of Environmental Protection, Bureau of Release Prevention, *Guidance on Inherently Safer Technology*, September 25, 2017.

14.14 State of New Jersey Department of Environmental Protection, Bureau of Release Prevention, *Inherently Safer Technology (IST) Implementation Summary,* January 10, 2010.

14.15 State of New Jersey, Domestic Security Preparedness Task Force, Domestic Security Preparedness Best Practices Standards at TCPA/DPCC Chemical Sector Facilities (New Jersey Prescriptive Order - NJPO), November 21, 2005.

14.16 State of New Jersey, *Toxic Catastrophe Prevention Act (TCPA)* NJSA 13:1K-19 et. seq.

14.17 White House, Executive Order 13650: Improving Chemical Facility Safety and Security, August 1, 2013.

15. Worked Examples and Case Studies

15.1 INTRODUCTION

This chapter illustrates the application of IS principles and concepts in both idealized and actual situations. It also includes a post hoc consideration of IS opportunities as applied to the Bhopal tragedy, dramatically illustrating the potential benefits to both the facility and the surrounding community from identifying and implementing IS opportunities.

15.2 APPLICATION OF AN INHERENT SAFETY STRATEGIC APPROACH TO A PROCESS

As discussed in Chapter 8, inherent safety (IS) concepts can be considered throughout the life cycle of a process. The following example illustrates the concepts described in Chapter 2 (see Figure 2.3), as applied over the life cycle of a process.

Reactive Chemicals, Inc., a fictional coatings industry supplier, is planning to install a new polymerization unit to produce Intermediate C and Final Product Z. The final product goes into various coatings industry applications. Industry expectations are for lower solvent formulations of this type polymer. The following illustrates the processes involved:

Intermediate production: A + B = C

In the intermediate reaction, raw material A is reacted with raw material B to produce intermediate C. Current production is in a batch reactor with all materials, including the catalyst, in the initial charge.

Raw material A is flammable (flash point <100°F), toxic, and supplied and stored in bulk. Raw material B is a reactive monomer that is corrosive (to human tissue) and combustible (flash point >100°F) and is typically inhibited with hydroquinone (HQ) or methoxyhydroquinone (MEHQ). Like Raw Material A, it is supplied and stored in bulk. The catalyst used for the intermediate production is boron trifluoride (BF3), a toxic gas supplied in cylinders. The reaction is carried out at 200 °F and at slight positive pressure.

After production of C, the intermediate is batch distilled under full vacuum and the purified C is collected, re-inhibited with MEHQ, and stored at ambient conditions under air. C is a reactive monomer with a flash point above 200 °F. Distillation bottoms are drummed for disposal as a reactive waste.

Final Product Production: C + D = Z

Intermediate C is polymerized with raw material D in a batch reaction producing final product Z, which is diluted in solvent E. Current production is in a batch reactor with all materials including initiator and solvent in the initial charge. The reaction is conducted at atmospheric pressure, and cooling is achieved by solvent reflux and supplemented by a reactor jacket. Available plant cooling water is used.

Raw material D is a reactive, corrosive (to human tissue) monomer, and Solvent E is flammable and considered toxic. Both materials are supplied and stored in bulk. The initiator is a peroxide type which requires refrigerated storage.

The final product Z is a polymer which, by itself, is non-toxic and nonreactive. However, in the current solvent, the product is flammable and toxic (see Figure 15.1).

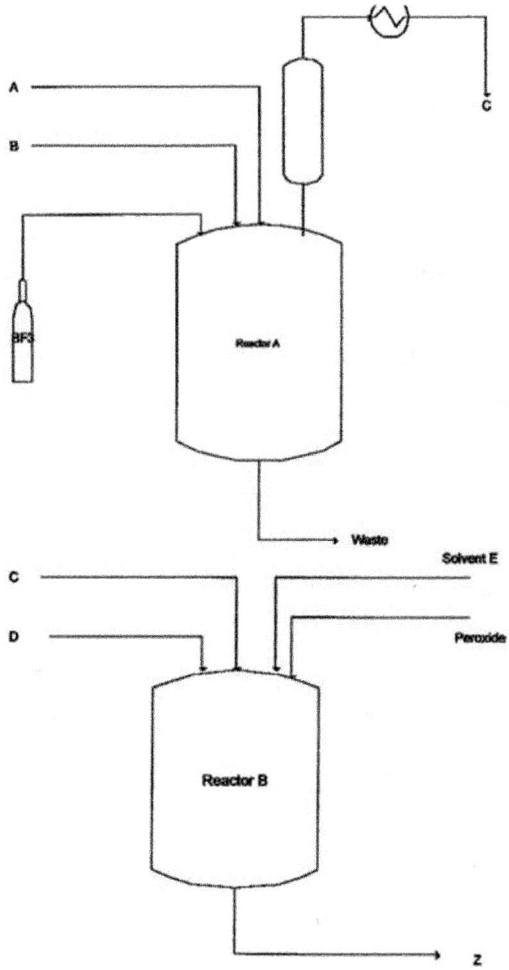

Figure 15.1. Initial Process

The worksheet that follows demonstrates how different orders of inherent safety could be incorporated, but also how inherent safety concepts could be incorporated through the life cycle of the process, as illustrated in Figure 15.2.

Reactor Life Cycle Stages Examples							
	Life Cycle Phase						
Examples of Inherent Safety Order	**Research and Development**	**Conceptual Design Phase**	**Detailed Engineering**	**Construction and Startup**	**Routine Operations**	**Plant and Process Modification**	**Decommissioning**
First order Inherent Safety Examples	Flammable polymerization solvent replaced with non-toxic low VOC solvent and water	Wiped Film Evaporater chosen to replace batch distillation of intermediate C	Elimination of flammable solvents allowed expansion of electrically unclassified areas during design	Use water/ surfactant cleaning methods to eliminate flammable solvent cleanings during startup	Use water/surfactant cleaning methods to eliminate flammable solvent cleanings during product changeovers, Waste can now go through site waste treatment process	Redesign polymerization catalyst/inhibitor package to eliminate hazardous peroxide use	Complete elimination of hazardous materials (including residuals) from the process using water and surfactants for cleaning.
Second order Inherent Safety Examples	• Initial reaction inhibitor package replaced with phenothiazine (PTZ) which allows reaction to be run under nitrogen (reducing flammability	Decision made to add PTZ inhibitor to monomer B in storage tank (upon receipt) reducing likelihood of runaway in storage	• Batch size calculated to utilize full container of BF3 reducing likelihood of mischarge and inadvertent release of BF3	Minimum inventory in process during startup to minimize any effects	Improve inventory management system of BF3 catalyst to minimize plant inventory (i.e., just-in-time raw material deliveries)	• Replace centrifugal pumps with mag drive pumps to reduce leak potentials and polymerization initiation potential	Identification of technology to immobilize potentially hazardous materials generated during demolition and decommissioning to minimize long term safety, health

Reactor Life Cycle Stages Examples							
	Life Cycle Phase						
Examples of Inherent Safety Order	**Research and Development**	**Conceptual Design Phase**	**Detailed Engineering**	**Construction and Startup**	**Routine Operations**	**Plant and Process Modification**	**Decommissioning**
Second order Inherent Safety Examples	concerns) and reduces likelihood of runaway reaction • Second reaction changed to semi-batch which limits runaway reaction potential		• Monomer storage designed with indirect heating (tank inside a heated room) to prevent overheating as a runaway scenario			• Replace filter press with bag filtration to reduce waste disposal.	and environmental hazards from the demolition waste.
Layers of Protection Examples			• Polymer-ization vessel designed to contain runaway (P & T consistent with adiabatic heat rise) for worst case scenario	Incorporation of human factors features (i.e., accessibility, maintainability, readability)	• Sampling procedures and methods changed to reduce exposures during in-process and product sampling	• Leak detection system installed in BF3 area with mitigation system • Incorporation of human	Mechanical isolation of process to minimize potential for hazardous material release

Reactor Life Cycle Stages Examples							
	Life Cycle Phase						
Examples of Inherent Safety Order	**Research and Development**	**Conceptual Design Phase**	**Detailed Engineering**	**Construction and Startup**	**Routine Operations**	**Plant and Process Modification**	**Decommissioning**
Layers of Protection Examples			• DCS and SIS systems , secondary containment for all vessels, pressure relief and mitigation systems • Human factors analysis completed during detailed design to facilitate accessibility, maintainability and readability.		• BF3 charging system optimized to reduce potential for exposures during cylinder connection/disconnection activities.	factors features (i.e., accessibility, maintain-ability, readability) • Provision of additional interlocks, alarms, control systems, secondary containment, pressure relief and mitigation systems subsequent to PHA revalidations.	

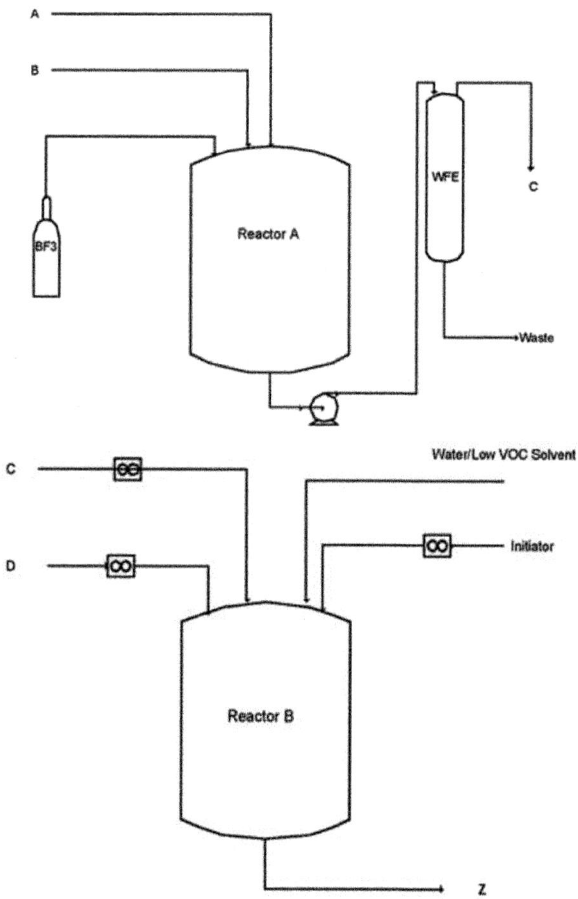

Figure 15.2 Modified process

15.3 CASE STUDIES FROM CARRITHERS

The case studies and examples in the following section are taken from a presentation entitled, "It's Never Too Late for Inherent Safety," by G.W. Carrithers, A.M. Dowell, and D.C. Hendershot (Ref 15.2 Carrithers).

15.3.1 An Exothermic Batch Reaction

An existing semi-batch process is used to carry out an exothermic reaction:

$$A + B \xrightarrow[\textit{Solvent S}]{\textit{Catalyst C}} D + HEAT$$

In addition to the highly exothermic nature of this reaction, there is an additional hazard. If Reactant B is significantly overcharged (double charge or more), a side reaction can occur that generates thermally unstable by-products, and a runaway reaction can result. Similarly, if the batch temperature gets too high, there is a potential for a thermal runaway due to undesired side reactions.

A simplified version of the process equipment is shown in Figure 15.3. Reactant A is dissolved in Solvent S in Weigh Tank A, and the solution is then fed to the reactor. Catalyst C is added to the reactor, and cooling is started to the reactor cooling coils. Reactant B, the limiting reagent, is then gradually fed to the reactor by gravity addition from Weigh Tank B, which has been pre-charged with the proper amount of reactant. The feed rate of Reactant B is controlled by the batch temperature. The process has a safety instrumented system (SIS) that takes it to a safe state when abnormal operation is detected. The SIS takes action if the reactor agitator fails, or cooling fails, or high reactor pressure or temperature is detected. All of the Safety Instrumented Functions (SIFs) stop the feed of the limiting reagent, Reactant B, by closing the feed flow control valve and a dedicated block valve.

As a result of the findings from a PHA, the system was modified as shown in Figure 15.4. The modified system includes several inherent safety features, which were implemented in this existing plant with a relatively small investment.

1. The Reactant B feed tank has been moved to the floor below the reactor level, and the feed is now controlled by a metering pump. In case of high temperature, the metering pump is turned off and a block valve is closed by the SIS on reactor high temperature. It is less likely that there will be a significant leak through the

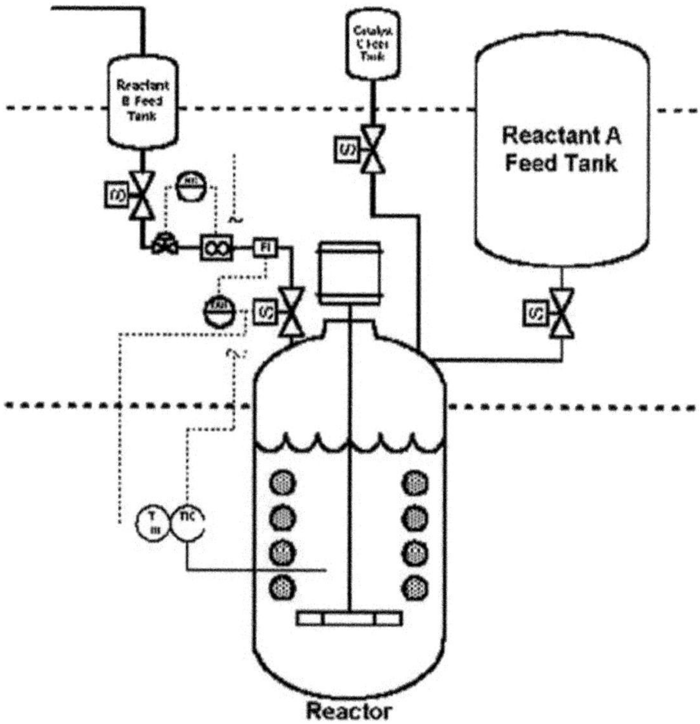

Figure 15.3 Original batch reaction system

2. metering pump, as compared to the original gravity feed where the driving force for the Reactant B flow is always present.
3. The maximum flow rate of the metering pump is not capable of generating more heat from reaction than the reactor cooling capacity can manage. Therefore, it is not possible to overheat the reactor by feeding Reactant B at a rate that exceeds the reactor's capability to remove heat.
4. The maximum capacity of the Reactant B feed tank has been reduced to exactly one batch charge. In this case, the same Reactant B feed tank was re-used, but it was relocated to the lower floor. To reduce its maximum capacity, an overflow was added to the side of the tank at the desired level, with the overflow piped back to the Reactant B storage tank.

5. In the revised system, Reactant B is charged to the feed tank through a three-way valve to the bottom of the feed tank. The three-way valve allows flow either from the storage tank to the Reactant B feed tank, or from the feed tank to the reactor. It is not possible to pump Reactant B directly from the storage tank to the reactor. This system makes it much more difficult to overcharge Reactant B.

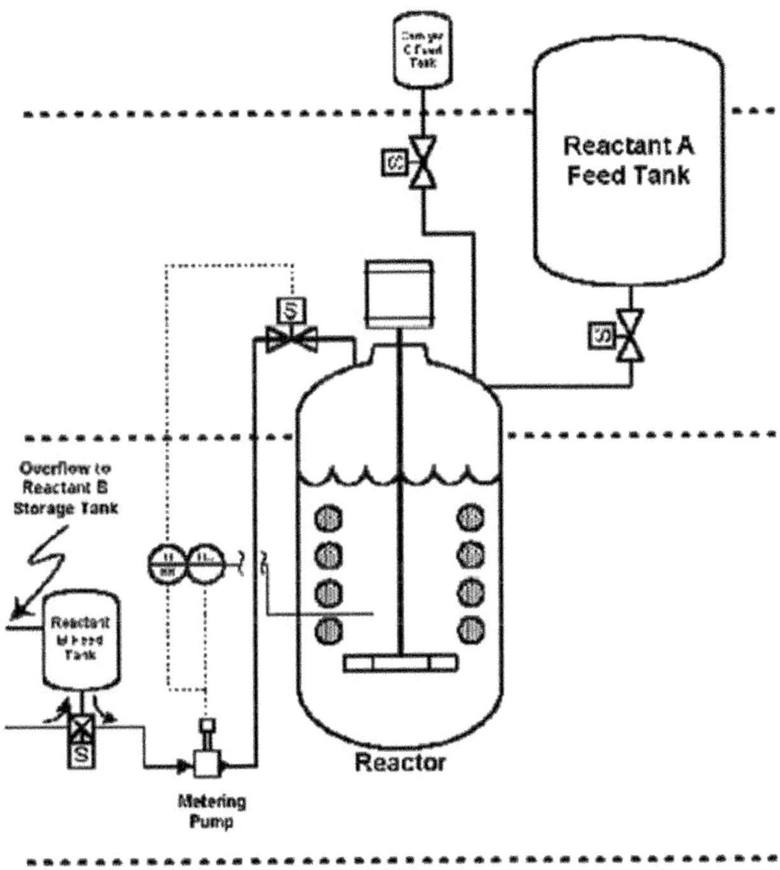

Figure 15.4: Modified, inherently safer batch reactor system

These modifications represent significant improvements to the inherent safety of the existing plant. Following the PHA, a quantitative risk analysis (QRA) of this reactor system was completed. The QRA showed a total risk reduction of about a factor of 1000 when compared to the original system.

15.3.2 Refrigeration of Monomethylamine

Refrigeration can be an effective way of minimizing the effects of a release of a material with a low boiling point (Ref 15.3 CCPS). Refrigeration reduces the hazard potential by:

- Reducing the storage pressure, preferably to near ambient, thus reducing the release rate in case of a leak in the system.
- Eliminating or greatly reducing the initial flash of material in case of a leak, because the material is not stored as a superheated liquid above its ambient pressure boiling point.
- Eliminating or reducing the atomization of liquid which does not flash. The very small droplets will not "rain out," but instead will travel downwind in the vapor cloud, where they will eventually evaporate due to heat absorbed from the surrounding atmosphere.

Monomethylamine (H2NCH3), a flammable gas with a strong, ammonia-like odor, was stored in a pressure vessel at ambient temperature. The vapor pressure of monomethylamine is about 57 psia (3.9 atm. absolute) at 20 °C. Refrigeration of the monomethylamine was identified during a hazard analysis as an option for reducing the hazard zone in case of a leak from the system. The potential impact of this change was evaluated using PHAST®. (Ref 15.5 DNV). Table 15.1 shows the results of the consequence modeling study for one example release scenario—the rupture of a 2-inch (5.1 cm) line between the liquid monomethylamine storage tank and the transfer pump to the manufacturing process. The ERPG-3 (Ref 15.1 AIHA) concentration of 500 ppm was selected as a representative concentration to illustrate the reduction in the hazard distance which could be attained by refrigerating the monomethylamine.

Refrigeration of monomethylamine has a significant impact on the hazard zones. The modified system is also inherently safer because the storage tank is capable of containing the monomethylamine at ambient temperature. It does not rely exclusively on the proper operation of the refrigeration system to ensure that the vapor pressure does not exceed the design pressure of the storage tank.

Table 15.1: Effect of Refrigeration on Distance to ERPG-3 Concentration for a 2-inch (5.1 cm) Monomethylamine Pipe Rupture

Monomethylamine Storage Temperature (°C)	Distance to ERPG-3 (500 ppm) Concentration, miles (km)
10	1.2 (1.9)
3	0.7 (1.1)
-6	0.4 (0.6)

15.3.3 Elimination of a Chlorine Water Treatment System

Chlorine is often used for water treatment in a chemical manufacturing facility. Because the chlorine handling facilities are often in the utilities area, they may not be recognized as presenting a significant potential hazard.

One ton cylinders of chlorine were used for disinfecting water at a chemical manufacturing facility, with liquid chlorine piped through ½-inch (1.3 cm) pipe to an injection nozzle, where it was mixed with the water being treated. Modeling of the consequences of a rupture of the ½-inch (1.3 cm) liquid chlorine pipe indicated that the distance to the ERPG-3 chlorine concentration of 20 ppm (Ref 15.1 AIHA) could be as far as about 1 mile (1.6 km), depending on weather conditions. An alternative chlorination system was considered that uses a chlorine vaporizer immediately adjacent to the cylinders. In this system, the long outdoor transfer line from the chlorine cylinder storage feed area to the water treatment area contained chlorine as a gas, instead of as a liquid, significantly reducing the inventory in the transfer pipe. This alternative reduced the distance to the ERPG-3 concentration of 20 ppm chlorine to about 0.5 miles (0.8 km) for the release scenario of rupture of the chlorine transfer line.

For this installation, alternative water disinfecting systems were also investigated. It was found to be feasible to use sodium hypochlorite treatment, and this essentially eliminated the hazard of a chlorine vapor cloud entirely.

15.3.4 Reduction of Chlorine Transfer Line Size

A manufacturing process included a chlorination process using liquid chlorine. The original facility used a 2-inch (5.1 cm) transfer line from the chlorine storage facility to the manufacturing building. A hazards review questioned the line size, and it was determined that it could be reduced to 1 inch (2.5 cm) without impacting the manufacturing process. Table 15.2 shows the impact of this reduction on the hazard zone resulting from the potential failure of the transfer pipe. Several typical weather conditions are considered. For purposes of this example, the hazard zone was defined as the distance to the ERPG-3 concentration for chlorine, 20 ppm. For all weather conditions, the distance to the ERPG-3 concentration was reduced by a factor of two to three.

15.3.5 Substitution of Aqueous Ammonia For Anhydrous Ammonia

Dilution of a hazardous material can be an important strategy for improving the inherent safety of a chemical process or storage facility. It can reduce the storage or vapor pressure of a hazardous material and reduce the atmospheric concentration of hazardous vapor from a spill.

Approximately 500,000 pounds (227,000 kg) of anhydrous ammonia were stored in a large pressurized storage tank. The tank was rated for 75 psig (90 psia, 6.2 atm. absolute) working pressure, and it had a pressure relief valve set for 72 psig (87 psia, 6.0 atm. absolute). This value compares to the vapor pressure of ammonia of about 93 psig (108 psia, 6.4 atm. absolute) at 60 °F (16 °C). The ammonia was kept under refrigeration to maintain the storage tank pressure below the tank pressure rating and relief valve set point. Occasionally, the refrigeration system would fail, and the storage tank pressure would slowly increase

Table 15.2: Effect of Reduction of Line Size on Hazard Zone from Potential Failure of a Chlorine Transfer Line

Chlorine Transfer Line Diameter	Distance in kilometers (miles) to ERPG-3 Concentration for Chlorine (20 ppm) (Accident Scenario - Line Rupture)		
Weather Conditions	D Stability 3.4 mph (5.5 km/h) Wind Speed	D Stability 11.2 mph (18.0 km/h) Wind Speed	F Stability 3.4 mph (5.5 km/h) Wind Speed
2 inch (5.1 cm)	3.4 (5.5)	1.3 (2.1)	4.2 (6.8)
1 inch (2.5 inch)	1.2 (1.9)	0.4 (0.65)	1.7 (2.7)

as the ammonia temperature rose. There was a potential to open the relief valve if the refrigeration system could not be returned to service before the relief valve set point was reached.

The system was reviewed, and the team determined that 28% aqueous ammonia could be substituted for anhydrous ammonia in all of the processes and products where it was currently used with minimal impact—including economic considerations—on these manufacturing processes.

To illustrate the benefit of substituting aqueous for anhydrous ammonia, the consequences of two release scenarios were modeled using PHAST®:

- Release of the entire contents of the ammonia storage tank over a 10 minute period. This scenario is equivalent to the "worst-case" scenario and is a requirement of the US EPA Risk Management Program (RMP) regulations (Ref 15.14 US EPA).

- Failure of a 2-inch (5.1 cm) diameter ammonia transfer pipe between the storage tank and the ammonia transfer pump. This scenario might be considered an appropriate "alternative release scenario" as required by the EPA Risk Management Program regulations (Ref 15.14 US EPA).

The results of the analysis are summarized in Figure 15.5 for both potential release scenarios. The use of aqueous ammonia greatly reduces the downwind ammonia concentration in the resulting vapor cloud.

The aqueous ammonia system also does not require a refrigeration system, which greatly simplifies the storage facility. The storage tank is now capable of containing the ammonia under all ambient temperature conditions, eliminating the reliance on proper operation of the refrigeration system to maintain a storage tank pressure less than the relief valve set pressure.

This case study also illustrates the importance of periodic review of all facilities that includes a search for inherently safer design options. There had, at one time, been a requirement for anhydrous ammonia at this site. It was logical, and safer, to use a single ammonia storage facility for anhydrous ammonia, rather than two separate systems. However, the process that required anhydrous ammonia had been shut down, and it was now possible to convert the other processes to the use of aqueous ammonia, resulting in an inherently safer system.

15.3.6 Limitation of Magnitude of Deviations for Aqueous Ammonia

The hazard of a substance can be reduced by limiting the possible magnitude of process deviations. By designing equipment to be more tolerant of error (or more robust), the frequency of a release can be reduced (Ref 15.3 CCPS).

For the reasons listed in the previous example, an aqueous ammonia system was installed to supply ammonia for reaction in a process that could tolerate the water solution. The system was initially installed using an available surplus tank of 10 psig (25 psia, 1.7 atm absolute) design pressure, and fitted with a 10 psig (25 psia, 1.7 atm absolute) relief valve. The system injected anhydrous ammonia in ratio with water into an eductor in a circulating heat exchanger loop (Figure 11.6). For the purpose of this example, the desired concentration was about 30 wt.%, and the normal operating temperature was about 100 °F (38 °C) or less.

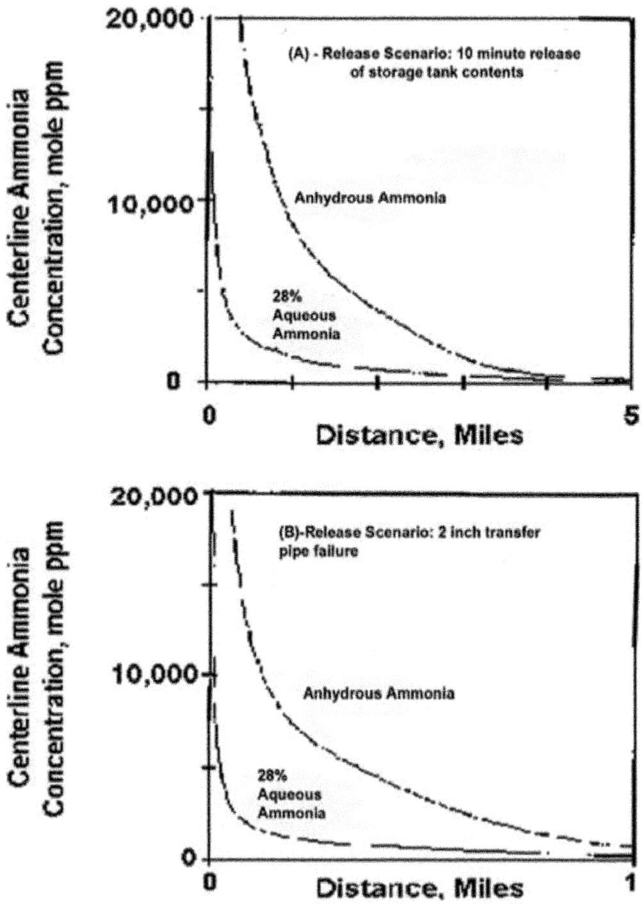

Figure 15.5: Comparison of centerline vapor cloud concentration

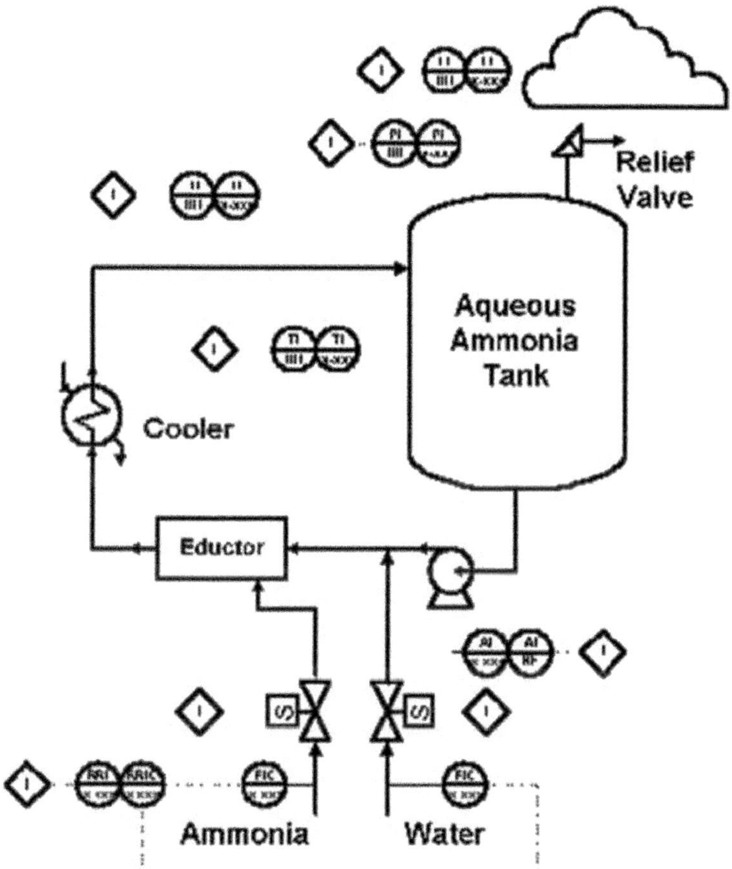

Figure 15.6: Aqueous Ammonia: Limitations of Magnitude of Deviations

Table 15.3 gives the total vapor pressure of aqueous ammonia for conditions near the design point. The tank design pressure will be exceeded, and the relief valve will open if the temperature rises, or if the concentration increases, even by small increments, as shown by the shaded area and the black border on the right side.

Table 15.3: Total vapor pressure of aqueous ammonia (Ref 15.12 Perry)

Temperature	Wt. % NH3 in solution			
°F	19%	24%	29%	34%
90	7.41 psia	11.4 psia	17.2 psia	25.5 psia
100	9.3 psia	14.2 psia	21.3 psia	31.2 psia
110	11.6 psia	17.6 psia	26.1 psia	37.8 psia
120	14.4 psia	21.5 psia	31.7 psia	45.6 psia
130	17.7 psia	26.2 psia	38.2 psia	54.6 psia
140	21.5 psia	31.5 psia	45.7 psia	65.8 psia
			↑ 10 psig Tank Threshold of Release	↑ 50 psig Tank Threshold of Release

Initiating causes for a pressure surge leading to a release include:

- loss of flow in the circulation loop
- upset in the ratio control
- loss of coolant to the heat exchanger
- high temperature of coolant in the heat exchanger
- upsets in the ratio control

During plant operation, several pressure surges from these causes opened the relief valve. The resulting cloud required evacuation of adjacent process units. The emergency response was to use a fire monitor to knock down the cloud while the unit corrected the control or heat transfer problem and waited for the tank pressure to come down. To prevent releases, an SIS was implemented that isolated the make-up ammonia when there was high temperature, high pressure, improper feed ratio, high level, or loss of the circulating pump. The SIS had to be designed to have a fast response time, since temperature and pressure rose quickly during an upset.

A new tank of 50 psig (65 psia, 4.4 atm absolute) design pressure was installed to make the plant less sensitive to upset. The concentration would have to reach about 34% at a temperature of 140 °F (60 °C) to cause a release, shown by the heavy black border in the lower right corner of Table 15.3. The new tank essentially eliminated releases from the relief valve. The SIS did not have to act as quickly, and, perhaps, some shutdown initiators could be eliminated.

This example illustrates the fact that the cheapest equipment—even if it is free—may not always be the most cost-effective option, especially when the economics of the consequences of releases and the cost of SISs are considered.

15.3.7 A Vessel Entry Example

Nitrogen is often piped to process vessels for process reasons, as well as to inert equipment handling flammable materials. When it is necessary to enter such vessels for inspection or maintenance operations, one of the concerns is ensuring that the nitrogen is properly disconnected or otherwise isolated from the vessel. Figure 15.7 schematically shows an inherently safer approach for being sure that the nitrogen has been disconnected *prior* to vessel entry. All nitrogen lines to the vessel are fed through a single pipe, which passes through a hose or removable section of pipe directly across the vessel manway. It is not possible to open the vessel manway without first removing the nitrogen hose or removable pipe section. Figure 15.8 shows how this has been implemented for two vessels.

Of course, this design does not ensure that the atmosphere inside the vessel is safe for entry, and all permit and vessel entry procedures are still needed. There are many other mechanisms that could lead to a hazardous atmosphere inside the vessel. For instance, it may contain hazardous materials, the vessel may not have been properly purged prior to entry, and other process pipe connections may not be as reliable. However, by making it very difficult to avoid disconnecting the nitrogen supply to the vessel when opening the manway, this design does provide an inherently more reliable mechanism for ensuring that one hazard to personnel involved in a vessel entry operation has been eliminated.

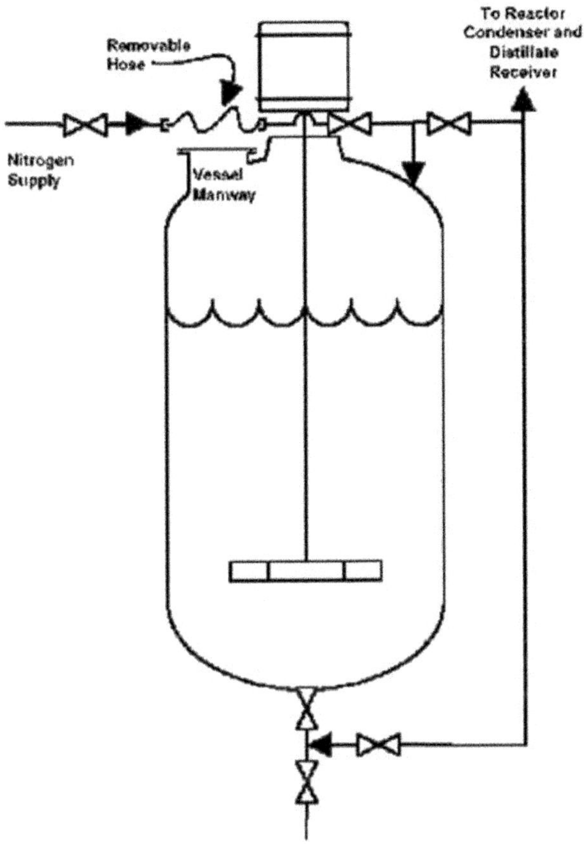

Figure 15.7: Nitrogen supply line routed over the vessel manway. The nitrogen must be disconnected to open the manway.

Figures 15.8a and 15.8b: Two examples of nitrogen supply lines piped to vessels across the manway. The manway cannot be opened without disconnecting the nitrogen.

15.4 PROCESS ROUTE SELECTION – EARLY R&D EXAMPLE

Inherent safety has been institutionalized into procedures that researchers must follow when developing a process chemistry and/or a process design. Researchers are required to review hazards and document a process hazards analysis (PHA) for each experimental set-up and/or significant change in that set-up. A checklist is a required part of that PHA effort and inherent safety (IS) questions are included in the instructions for completing the checklist.

The format (template) for technical reports on product, chemistry, and process development includes a section on IS, as does the report format for applying for permission to seek a patent. The principal instruction on the template for each is similar to the following:

> *"If the process or chemistry being addressed is aimed at the implementation of a new manufacturing process, or improvements/changes to an existing manufacturing process, then a discussion of the anticipated hazard/risk level at the commercial scale must be included. This discussion should be from an IS perspective and include consideration of the quantity of hazardous materials involved, and the severity of process conditions. The use of a standard "index" sheet—a form that dictates IS be considered—is required."*

The index sheet (Ref 15.9 Heikkila) is a chart that gives five levels of definition (from low to high) for toxicity, flammability, and reactivity (i.e., the material factor), for quantity (i.e., the quantity factor), and for reaction severity, pressure, temperature, potential for corrosion or erosion, dust content, operability, and experience (i.e., the process factor). The researcher is asked to assign a "level of severity" to each factor, and to sum them.

The higher the resulting number, the more hazardous the chemistry or process is, from an IS standpoint. Several alternative chemistries and/or processes must be proposed, and the "index" sheet used for them as well. If competing chemistries and/or processes exhibit a lower total "level of severity," the researcher is obligated to defend his choice. No chemistry that exhibits "severe" factors in all categories is accepted.

15.5 EXAMPLE OF AN INHERENTLY SAFER STUDY OF A STEAM PRODUCTION FACILITY

The following comes from a paper written and presented by Karen Study at a 2005 safety conference sponsored by the Mary Kay O'Connor Process Safety Center in College Station, Texas. (Ref 15.13 Study)

Choosing an inherently safer alternative may seem straightforward. However, sometimes what initially seems to be the most obvious IS alternative may not actually provide the best overall risk reduction. In this case study, an "inherently safer" alternative was selected and later discarded due to issues uncovered during the detailed design phase. The option ultimately chosen was inherently safer than both the original design and the "inherently safer" alternative.

15.5.1 Facility Description

The unit produces a large amount of steam using a multiple burner boiler with natural gas and a low BTU off-gas as its fuel sources. The boiler waste gas (flue gas) is sent to an elevated stack where it is discharged to the environment. This flue gas is mainly nitrogen and water, with oxygen and carbon dioxide. As with all boilers, there is also NOx present in the flue gas. A team was formed to assess different NOx reduction options. After evaluating several options to achieve the required NOx emission reduction targets, the design team chose to install a Selective Catalytic Reactor (SCR).

15.5.2 Initial Design Proposal (Liquid Anhydrous Ammonia)

To supply ammonia to the SCR, the design team chose to tap into an existing liquid anhydrous ammonia piping header that supplied a nearby processing unit. Piping was minimized as much as possible, to ~600 feet of 2 inch pipe. A vaporizer skid, which used steam to vaporize the liquid ammonia prior to injecting into the SCR, was to be installed near the boiler. See Figure 15.9 for a high-level overview of this option.

After the option was selected, the process safety group was consulted to provide input. Due to concerns regarding incrementally increasing risks associated with the current liquid anhydrous ammonia piping system, the safety group recommended using aqueous ammonia available from a nearby processing unit. This seemed to be a

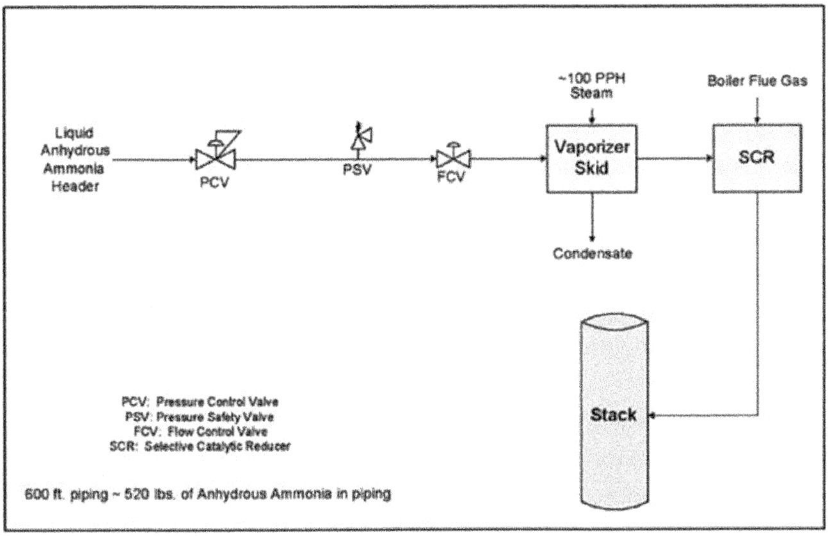

Figure 15.9: Initial ammonia supply proposal: Liquid anhydrous ammonia supply (Ref 15.13 Study)

straightforward alternative (see Figure 15.9 for anhydrous/aqueous ammonia comparison).

15.5.3 Aqueous Ammonia Design Proposal

An aqueous ammonia user in the plant could deliver it to the boiler facility via a connection downstream from the storage tank. Since this tank was much further from the boiler than the anhydrous ammonia header, the length of piping required was much greater. The tank also required new positive displacement pumps to supply the aqueous ammonia. And, a temporary supply alternative had to be built into the design since periodic shutdowns were required at the aqueous ammonia tank. To accommodate these periodic supply needs, additional connections and provisions were made for tank truck deliveries of aqueous ammonia. See Figure 15.10 for an overview of the aqueous supply option.

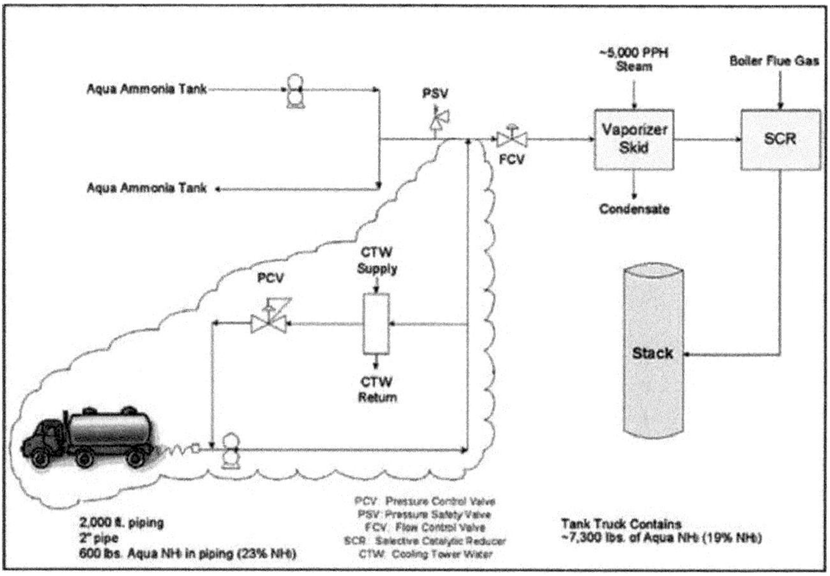

Figure 15.10: Aqueous Ammonia Supply Proposal

The design team conducted a Hazard and Operability analysis that raised several concerns regarding the tank truck delivery system and associated operations:

- the risk of spills and operator errors for the tank truck portion of the delivery system was higher for this option than an anhydrous ammonia system
- higher capital, operating and maintenance costs
- the reliability of the addition of pumps in the system.

So, the project was recycled back to the option selection phase.

15.5.4 *Final Round of Option Selection*

During the new round of option selection, the use of anhydrous ammonia vapor was evaluated. This option had not been fully evaluated previously based on early assumptions that:

- the available anhydrous ammonia vapor was insufficient for the SCR.
- aqueous ammonia was a safer alternative.
- the vapor stream would be more difficult to control than the liquid stream.

A nearby processing unit that vaporized anhydrous ammonia from the plant header prior to use was evaluated and, based on an evaluation of the downtime of their system, it was determined that anhydrous ammonia vapor could be supplied to the SCR at a sufficiently high on-stream rate, without a secondary supply system.

Flow control issues could be addressed by using redundant instruments that had a good operating history in similar service. To prevent condensation in the transfer piping, the ammonia pressure was dropped to 25 psig at the tie-in point to the header. This made the risks associated with the use of anhydrous ammonia vapor less than if high-pressure vapor was used. Redundant instrumentation for control valves and letdown regulators reduced downtime for the system. A small amount of low-pressure steam was added to the ammonia prior to injection into the SCR as a diluent to more evenly disperse the ammonia in the catalyst bed. The anhydrous ammonia vapor option is depicted in Figure 15.11. (Redundant instrumentation is not shown.)

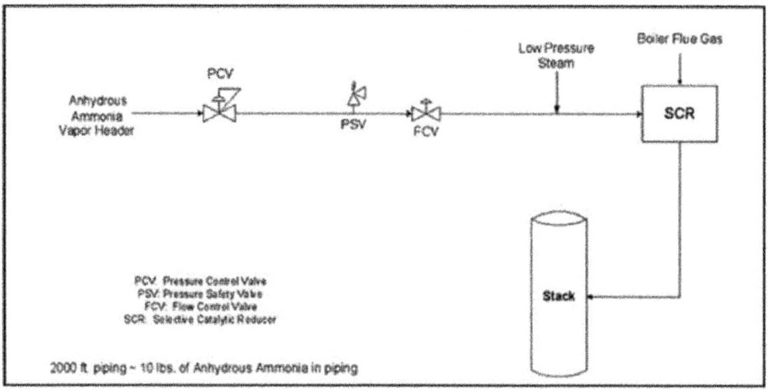

Figure 15.11: Anhydrous ammonia vapor option

15.5.5 Consequence Analysis

For comparison, the designers modeled a complete break in the piping supplying the SCR, resulting in an almost instantaneous release of the entire pipe contents. The consequence analysis for liquid anhydrous ammonia estimated a maximum distance of 1,170 feet to the IDLH concentration under the worst-case weather conditions (F atmospheric stability, 1.5 m/s wind speed). For the aqueous ammonia piping system using the same leak scenario and weather conditions, the impact distance for IDLH level results was 930 feet. The ammonia vapor scenario impact distance was 125 feet.

Another issue with the aqueous ammonia system is the temporary tank truck needed when the primary source is not available. Connecting and disconnecting hoses to the truck introduces new release scenarios involving smaller amounts of ammonia in close proximity to production operators. Plus, the large tank truck inventory greatly increases risks during a catastrophic failure scenario.

Modeling aside, one significant difference among all three options is simply the mass of material inside the transfer piping and associated equipment. The anhydrous ammonia vapor system contained an order of magnitude less ammonia than the other piping systems, and two orders of magnitude less than the aqueous ammonia tank truck. See Table 15.4 for a comparison of the options based on mass differences.

15.5.6 Conclusion and Action

The design team ultimately chose to use anhydrous ammonia vapor to supply the SCR. Receiving ammonia in the anhydrous vapor form was determined to be an economically viable option, and the safety analysis indicated that the vapor form of anhydrous ammonia was safer than design options incorporating either liquid anhydrous or aqueous ammonia.

In comparing both the anhydrous vapor and aqueous liquid options to the liquid anhydrous option, using the vapor decreases risks along the piping run without introducing new risks to the system. Based on modeling analysis performed for the piping run, the liquid aqueous

Table 15.4: Mass Comparison for Ammonia Transfer Options

Option	Piping Length (ft.)	Volume (ft3)	NH3 Mass (lbs.) *
Anhydrous Ammonia Liquid	600	14	520
Aqueous Ammonia (23 wt.% NH3)	2,000	47	600
Aqueous Ammonia Tank Truck (19 wt.% NH3)	N/A	652	7,300
Anhydrous Ammonia Vapor	2,000	47	10

**Density of ammonia and water at 80 °F; 37 lb./ft3 and 62 lb./ft3 respectively*

ammonia option did have a slightly lower risk compared to the liquid anhydrous ammonia system. However, the aqueous system introduced additional risks by the need to have a temporary supply station. Such a station would have increased risk to operating personnel due to hose handling while hooking up and disconnecting the tank truck. Not only did the anhydrous vapor option not require such a secondary supply, but the piping system contained an order of magnitude less ammonia than the other systems considered.

Judging both the aqueous and anhydrous vapor options from an economic standpoint, the vapor system also had lower capital costs, as well as lower projected operating costs, since very little steam was required prior to injection into the SCR. The vapor system had lower maintenance costs since no pumps were required and was predicted to be more reliable due to the simple design and absence of moving parts.

15.5.7 Conclusion

This project illustrates the principle that decision-making between inherently safer designs involves evaluating several different metrics, including volume of hazardous material, distances affected by a release, frequency of release (for example, during those connections for aqueous), risk (including consequence severity and frequency), and life cycle cost. The option selected depends upon the metrics used in the decision-making, and the weighting factors among those metrics. There is not a single metric that can be considered as "correct" for selecting one inherently safer design over another option. It always involves trade-offs.

15.6 CASE STUDY: BHOPAL

The following case study is adapted, with permission, from an article called "How to Make Inherent Safety Practice a Reality," written by Faisal I. Khan, and Paul R. Amyotte, and published in the February 2003 issue of The Canadian Journal of Chemical Engineering. (Volume 81, pp. 2-16)

Several thousand kilograms of a toxic gas, rich in methyl isocyanate (MIC), were released into the atmosphere from the Union Carbide India, Ltd. pesticide facility. MIC was the intermediate product of carbaryl (a pesticide) manufacturing and was stored in three horizontal stainless steel vessels. The MIC was usually stored under refrigerated conditions, but in the days before the incident, the refrigeration unit had been shut down. Water entered an MIC storage tank, creating an exothermic reaction which heated the MIC beyond its boiling point. The concrete mounds above the tank cracked, releasing MIC vapor through a relief valve. The scrubber and flare systems designed to remove and destroy the MIC were not in operation on the day of the incident. The Bhopal facility relied on engineered and procedural safeguards. Khan and Amyotte analyzed this incident to identify inherent safety considerations that might have been prevented or mitigated the tragedy.

15.6.1 Minimization

MIC was an intermediate in carbaryl production. The facility could have stored less, as evidenced by the fact that MIC stocks had been reduced by 75% within a year of the incident. Furthermore, by changing the release diameter—the parameter reflecting chemical inventory—from a standard 50-mm orifice to 30 mm, the CEI-hazard distance, calculated using Dow's Fire and Explosion Hazard Classification Guide (Ref 15.7 XX Dow 1994b) and Dow's Chemical Exposure Index Guide (Ref 15.6 Dow 1994a) would be reduced by 28%. If the MIC inventory (or release rate) had been minimized, the incident potential would have been reduced (2nd order IS measure).

15.6.2 Substitution

Carbaryl was manufactured by reacting phosgene and methylamine to produce MIC, which was then reacted with alpha-naphthol. An alternative process reacts the phosgene and methylamine in a different order to avoid the production of MIC. In this alternate process route, phosgene is reacted with alpha-naphthol, and then the intermediate—a less hazardous chloroformate—is reacted with methylamine (Figure 15.12). Application of the substitution principle at the process route selection stage could have played a role in averting the incident consequences (First order IS measure).

15.6.3 Moderation

Moderation of the storage conditions (temperature) would have been enabled by the refrigeration system had it been operating. Rather than being stored at 0°C or lower, as standard procedures required, the MIC was actually at ambient temperature — obviously much closer to its boiling point of 39.1 °C. With the contaminant presence leading to an exothermic reaction and elevated MIC temperatures and vapor generation, it is not clear how pressure might have been effectively moderated. However, a 90% reduction in the operating pressure involved would have resulted in a 60% decrease in the CEI-hazard distance.

15.6.4 Simplification

The Bhopal facility relied on end-of-pipe monitoring and control systems attached to the storage vessels, the reliability of which required they be

Methyl Isocyanate Route:

$$CH_3NH_2 + COCl_2 \rightarrow CH_3CNO + 2HCl$$

Methylamine *Phosgene* *Methyl Isocyanate* *Hydrochloric Acid*

$$CH_3CNO + C_{10}H_8O \rightarrow C_{12}H_{11}NO_2$$

Methyl Isocyanate *Napthol* *Carbaryl*

Non-Methyl Isocyanate Route:

$$C_{10}H_8O + COCl_2 \rightarrow C_{11}H_7ClO_2 + HCl$$

Napthol *Phosgene* *Napthol Chloroformate* *Hydrochloric Acid*

$$C_{11}H_7ClO_2 + CH_3NH_2 \rightarrow C_{12}H_{11}NO_2 + HCl$$

Napthol Chloroformate *Methylamine* *Carbaryl* *Hydrochloric Acid*

Figure 15.12. Alternative route for carbaryl production (Ref 15.4 Crowl)

in good working order. Reliance on such systems can create a dual problem: the safeguards may not be available when needed, and their existence may provide a false sense of security for process operators who may ignore initial warning signs, such as a pressure increase. Many of these issues are brought into focus when one analyzes a system through use of the simplification principle.

15.7 EXAMPLE: INHERENTLY SAFER PROCESS FOR PRODUCTION OF TRIALKYL PHOSPHATE ESTERS

The traditional route to manufacturing alkyl phosphate esters involves reacting phosphorus with chlorine to produce phosphorus trichloride, oxidizing the phosphorus trichloride to phosphorus oxychloride, then reacting phosphorus oxychloride with an excess of the desired alkanol, typically with concurrent vacuum removal of the generated hydrogen chloride gas, followed by washing/neutralization, and vacuum distillation of the final phosphate ester product. This process may be carried out in one facility, or more typically, the phosphorus oxychloride is produced by one company/facility and is then shipped to another company/facility for the phosphate ester production.

This process entails handing several highly hazardous substances – white phosphorus, chlorine, phosphorus trichloride, phosphorus oxychloride, and anhydrous hydrogen chloride gas. The phosphorus trichloride and phosphorus oxychloride intermediates are highly hazardous, water reactive and corrosive. The process uses chlorine for its chemical potential only – it is essentially discarded as dilute HCl (following water scrubbing) at the end of the reaction sequence. Even if this acid is collected and reused for neutralization, it ultimately ends up as salt water somewhere.

A more direct, inherently safer process for producing trialkyl phosphates has been developed by one company and demonstrated in the laboratory. In this process, phosphorus is burned in air, and the produced phosphorus pentoxide (P2O5) is "scrubbed" in a recirculating packed scrubber containing the alkanol of choice as the "scrubbing" media. By controlling the overall process pressure, alkanol excess, and recirculation rates, the water produced in the reaction can be removed sufficiently to avoid excessive partial ester formation and resultant yield loss. Commercially viable yields were reported for this "direct ester" process.

This new process eliminates the use of chlorine, with its inherent handling hazards, eliminates the formation of, and handling hazards associated with, phosphorus trichloride and phosphorus oxychloride, and consequently eliminates the generation, handling, and disposal of the HCl wastestream.

15.8 SUMMARIES IN BRIEF: EXAMPLES BY IS STRATEGY

Implementation of inherently safer design principles over the process life cycle has been increasingly practiced by the chemical processing industry. Khan and Amyotte published an extensive list of examples in their article, "How to Make Inherent Safety Practice a Reality" (Ref 15.10

Khan), abbreviated here as KA. Overton and King in their article, "Inherently Safer Technology: An Evolutionary Thing," describe inherent safety applications by The Dow Chemical Company (Ref 15.11 Overton), abbreviated here as OK. This section will present worked examples and case studies which represent the principles and concepts discussed in this book. They are illustrative of the many approaches that companies may take to achieve inherently safer processes.

15.8.1 Minimize

Use smaller quantities of hazardous substances.

- Although the reaction of nitric acid and glycerin is quite rapid, early industrial implementation of this chemistry used large batch reactors. Current technology has enabled efficient contacting of the reactants, making the actual reaction process continuous and rapid. These continuous reactors are orders of magnitude smaller than the older technology batch reactors, thus greatly reducing the potential hazard from this extremely hazardous reaction (KA).
- A continuous process has been developed for manufacturing phosgene on demand, thus eliminating the need for storage of liquid phosgene. Various important issues, such as quality control, understanding of transient reactor operation, and process control, were successfully resolved by a fundamental understanding of the chemical reaction. This enabled the design of a system that met all of the user's requirements in an inherently safer way (KA).
- It is understood that mixing and gas-liquid phase mass transfer controls the rate and efficiency of chlorination reactions. Replacing a stirred tank reactor with a loop reactor, specifically designed to optimize mixing and gas-liquid phase transfer has been shown to significantly reduce reactor size, processing time, and chlorine usage (KA).
- A 50-L loop reactor replaced a 5000-L batch reactor in a polymerization process (KA).

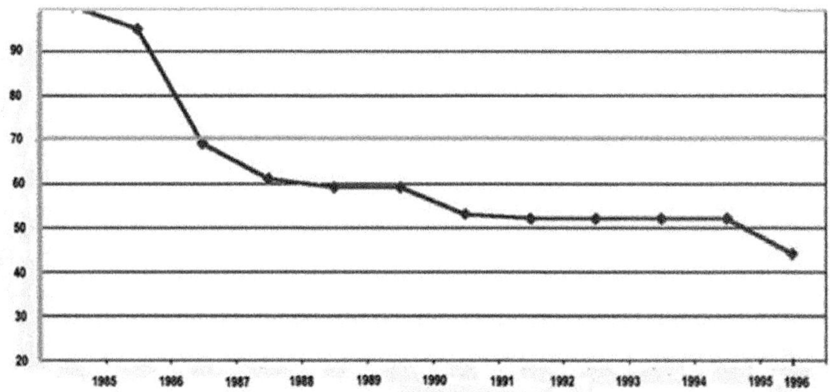

Figure 15.13: Dow LaPorte Phosgene Inventory Reduction

- MDI plants at Dow's LaPorte, Texas, manufacturing site began a campaign in 1986 to reduce phosgene inventory through improved process control and in situ production. Figure 15.13 shows that implementing a series of major capital projects reduced phosgene inventory by 50 % over 10 years while increasing MDI production—just one example of the impact of an evolutionary strategy for inherently safer technologies (OK).
- Dow's Union Carbide subsidiary redesigned the acrolein derivatives unit in Taft, Louisiana, during the mid-1980s to dramatically reduce acrolein inventory. In addition, by partnering with a key customer, another operating unit was added to convert acrolein to a less hazardous derivative at the Dow site prior to shipment to the customer site, thus reducing transportation risks (OK).
- Dow also uses continuous reaction processes instead of batch reaction processes to produce certain types of polyethylene and polypropylene plastic resins. This technology reduces the inventory of flammable material in these continuous reaction processes (OK).

- Direct chlorination using 95% chlorine vapor direct from chlorine cells rather than purified chlorine eliminates the need for liquid chlorine storage (Figure 15.14) (OK).

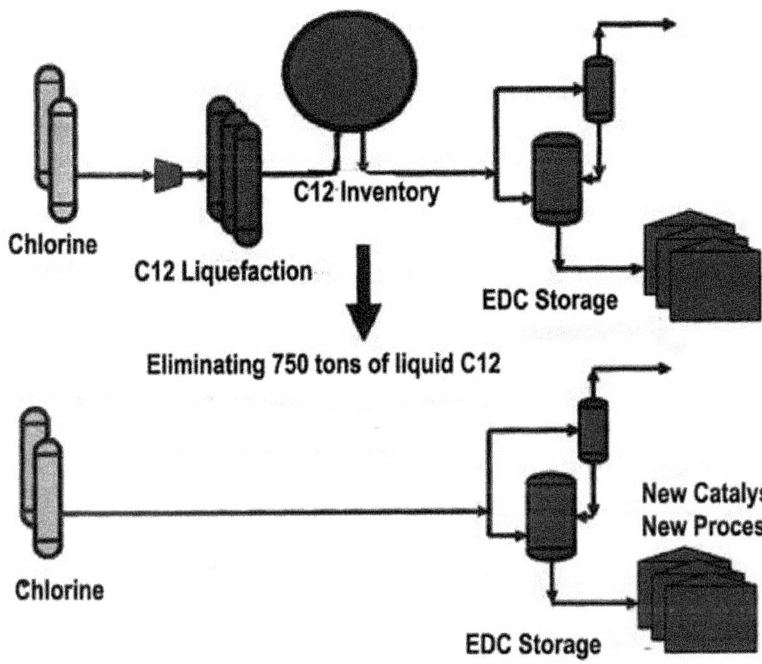

Figure 15.14: Use of Chlorine Vapor for Liquid Chlorine

- Dow's licensed EO METEORTM technology significantly reduces the portion of the process using concentrated EO (Figure 15.15).

15.8.2 Substitute

Replace a material with a less hazardous substance.

- Manufacture of acrylic esters by oxidation of propylene to produce acrylic acid, followed by esterification to manufacture the various esters, is inherently safer than the older Reppe process, which used acetylene, carbon monoxide, and nickel carbonyl (KA).

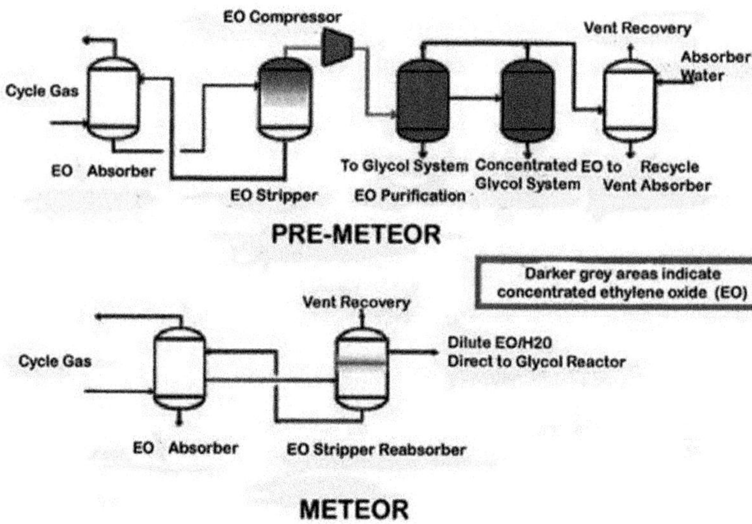

Figure 15.15: Streamlined METEORTM Ethylene Oxide Recovery System

- Disodium iminodiacetate (DSIDA), an intermediate used for agricultural chemicals, was traditionally manufactured using a process requiring ammonia, formaldehyde, hydrogen cyanide, and hydrogen chloride. Monsanto developed a new process for DSIDA that eliminates the use of hydrogen cyanide and formaldehyde, has a higher yield, is simpler, and produces a product of sufficiently high purity that purification is not required (Ref 15.8 Franczyk). Monsanto won a Presidential Green Chemistry Challenge Award for this process in 1996 (KA).
- Many cleaning and de-greasing operations have eliminated the use of organic solvents by substituting water-based systems (KA).

- A polymerization process conducted using a gradual addition batch process required a large quantity of organic solvent to keep the system viscosity low enough for effective mixing and heat transfer. In the event of an uncontrolled (runaway) polymerization, a large quantity of flammable and toxic material would be ejected through the reactor rupture disk. Instead of relying on an expensive and elaborate emergency relief discharge system to control the potential hazard, the basic process chemistry was re-considered. It was found that it was possible to make the product using a suspension polymerization process in water. Because of the significantly lower amount of solvent required in the new process, most of the material released in a runaway reaction would be water, with just a small amount of solvent and unreacted monomer. Reaction runaway was also less likely because of the higher heat capacity of water, allowing it to absorb more of the heat of reaction during a process upset (KA).
- Dow had been using benzene as an azeotroping agent in one of its oxide derivatives units but replaced the benzene with a less hazardous chemical (OK).
- Dow replaced chlorine gas with sodium hypochlorite (bleach) as a water purification chemical at selected non-chlorine producing sites, and replaced Varsol in a process with another, less hazardous solvent (OK).

15.8.3 Moderate

Use less hazardous conditions, a less hazardous form of a material, or facilities which minimize the impact of a release to hazardous material or energy.

- In the 1930s, ammonia plants typically operated at pressures up to 600 bar (9000 psia). Over the years, improved understanding of the chemistry has resulted in a downward trend in the operating pressure of ammonia plants; in the 1980s, plants operating in the range of 100-150 bar (1500 – 2250 psia) were common. The newer, low-pressure plants are inherently safer, cheaper, and more efficient than their high-pressure counterparts (KA).

- Dow's newest process for producing diphenyl oxide uses a catalyst to operate at temperatures and pressures that are significantly lower than the processes that it replaced (OK).
- Storing highly toxic or flammable materials (for example, chlorine, LNG) as a refrigerated liquid at atmospheric pressure, rather than at atmospheric temperature under pressure reduces the potential consequences of an accidental spill (OK).
- Using an aqueous hydrogen chloride (HCl) solution, rather than anhydrous HCl in selected processes also reduces the potential consequences of a spill (OK).
- One of Dow's facilities near a populated area replaced the use of aqueous 36% HCl with 17% HCl. Although this change did double the number of deliveries, and with that, increased transportation risks, and increased risk associated with connecting/disconnecting fill lines during deliveries, given that the partial pressure has reduced by a factor of 1,000 (i.e., reduced toxic hazard in the event of a spill), it was viewed as an overall reduction in total inherent risk (OK).
- Multiple processes within Dow use small diameter piping or restricting orifices to limit the flow rate in the event of an accidental release or operational error, making it almost impossible for a reaction to occur too quickly (OK).
- In some cases, Dow uses process equipment designed to contain the maximum foreseeable pressures from the process, eliminating the need for pressure relief (OK).
- Using high wall dikes and other types of secondary containment to severely limit the surface area available for evaporation of toxic liquid spills minimizes the impact, as shown in Figure 15.16 (OK).

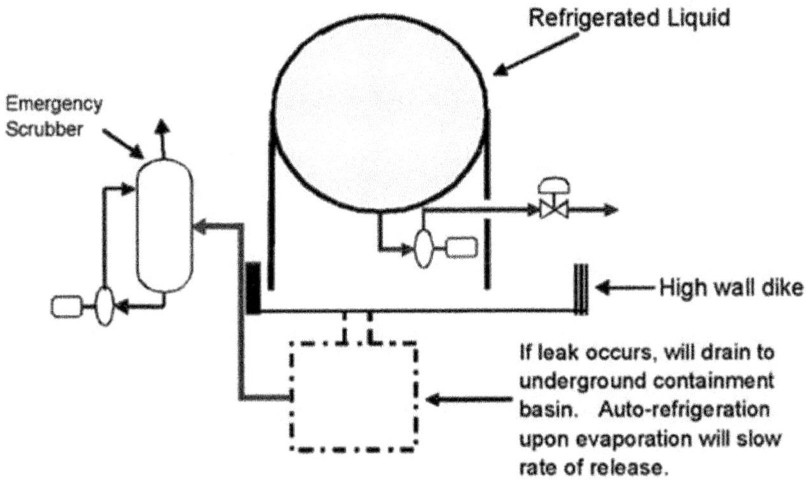

Figure 15.16: Dike and Secondary Containment System

15.8.4 Simplify

Design facilities that eliminate unnecessary complexity, make operating errors less likely, and are forgiving of errors that are made.

- An understanding of basic process chemistry allows process chemists and engineers to design simpler reaction systems and facilities. Processes where most or all of the unit operations are designed to take place in one vessel/reactor permit the generation of a hazardous intermediate in the vessel where it will be used, eliminating the need to store it, or to move it around the plant in piping systems. The inventory of hazardous material is thus limited to a maximum of one batch. This example illustrates the beneficial overlap that can occur with inherent safety principles, in this case, simplification and minimization (KA).

- Simplification sometimes involves a tradeoff between the complexity of an overall plant and complexity within one particular piece of equipment. For example, a reactive distillation process for producing methyl acetate requires only three columns and the associated support equipment. The older process required a reactor, an extractor, and eight other columns, along with the associated support equipment. The new process is simpler, safer, and more economical, but the successful operation of the reactive distillation component is itself more complex and knowledge-intensive (KA).
- Redesigned latex reactor cleaning equipment eliminates the potential for incorrect installation that could result in unwanted reactivity or cross contamination (OK).
- Dow eliminated the use of hoses in several hazardous services in favor of hard piped connections (OK).

15.9 ADDITIONAL LITERATURE GIVING EXAMPLES OF INHERENTLY SAFER OPERATIONS

- Amyotte, P.R., Goraya, A.U., Hendershot, D.C., and Khan, F.I. (2007). Incorporation of Inherent Safety principles in process safety management. Process Safety Progress, 26 (4), 333-346.
- CCPS (1993). "Inherently Safer Plants." *Guidelines for Engineering Design for Process Safety* (Chapter 2). New York: American Institute of Chemical Engineers.
- Commission of the European Community (1997). INSIDE Project and INSET Toolkit. Available for download at www.aeat-safety-and-risk.com/html/inset.html.
- Kletz, T.A. (1998). Process Plants - A Handbook for Inherently Safer Design, London, UK: Taylor and Francis.
- "Layer of Protection Analysis and Inherently Safer Processes." Process Safety Progress, 18, (4), 214-220, Winter 1999.
- 'Inherently Safer Approaches to Plant Design', DP Mansfield,

- Mansfield, D.P. and Cassidy, K. (1994). Inherently safer approaches to plant design: The benefits of an inherently safer approach and how this can be built into the design process. Presentation at IChemE Hazards XII—European Advances in Process Safety, April 19-21, 1994, UMIST, Manchester, UK. AEA/CS/HSE/R1016, HSE Books/HMSO, August 1994, ISBN 0853564159.
- Sanders, R.E. (2003). "Designs that lacked inherent safety: case histories." Journal of Hazardous Materials, 104, 1-3, 149-161.

15.10 REFERENCES

15.1 American Industrial Hygiene Association, *Emergency Response Planning Guidelines: Chlorine*. Akron, OH: American Industrial Hygiene Association, April 20, 1988.

15.2 Carrithers, G.W., Dowell, A.M., and Hendershot, D.C., *It's never too late for inherent safety.* In International Conference and Workshop on Process Safety Management and Inherently Safer Processes, October 8-11, 1996, Orlando, FL (pp. 227-241). New York: American Institute of Chemical Engineers, 1996.

15.3 Center for Chemical Process Safety (CCPS), *Guidelines for Engineering Design for Process Safety*. New York: American Institute of Chemical Engineers, 1993.

15.4 Crowl, D.A., and Louvar, J.F., *Chemical Process Safety Fundamentals With Applications* (pp. 14-15). Englewood Cliffs, NJ: Prentice Hall, 1990 and 2002.

15.5 Det Norske Veritas (DNV) - PHAST® (Process Hazard Analysis Software Tool) is a comprehensive hazard analysis software tool broadly used in the chemical and petroleum industries developed and marketed by DNV Software, the commercial software house of DNV.

15.6 Dow Chemical Company, *Dow's Chemical Exposure Index Guide, 1st Edition*. New York: American Institute of Chemical Engineers, 1994a.

15.7 Dow Chemical Company, *Dow's Fire and Explosion Index Hazard Classification Guide, 7th Edition*. New York: American Institute of Chemical Engineers, 1994b.

15.8 Franczyk, T. S., *The Catalytic Dehydrogenation of Diethanolamine.* In Paper Preprints from the 213th ACS National Meeting, April 13-17, 1997, San Francisco, CA. (342). Washington, DC: American Chemical Society Division of Environmental Chemistry, 1997.

15.9 Heikkila, A.M., *Inherent Safety in Process Plant Design An Index-Based Approach.* Technical Research Centre of Finland, VTT Publications 384. 129 p, 1999.

15.10 Khan F.I. and Amyotte, P.R., *How to make inherent safety practice a reality.* The Canadian Journal of Chemical Engineering, 81, 2-16, 2003.

15.11 Overton, T.A. and King G., *Inherently safer technology: An evolutionary approach.* Process Safety Progress 25 (2), 116-119, 2006.

15.12 Perry, R.H. and Green, D., *Perry's Chemical Engineer's Handbook, 6th Edition.* New York: McGraw-Hill, 1984.

15.13 Study, Karen, *In Search of an Inherently Safer Process Option*. In *Beyond Regulatory Compliance, Making Safety Second Nature.* Mary Kay O'Connor Process Safety Center 2005 Annual Symposium. College Station, TX: Texas A&M University, 2005.

15.14 United Stated Environmental Protection Agency (US EPA) (January 22, 1996). *Off-site Consequence Analysis Guidance* (Draft). Washington, D.C.: U.S., codified at Title 40 CFR Part 68.22, 1996..

16. Future Initiatives

Applying inherently safer design principles to new and existing processes will improve overall safety in the chemical process industry. The greatest gains in inherent safety will come from the invention of new chemistry and new processing technologies that eliminate the need for hazardous materials and operations. The key to spurring future invention in inherent safety is establishing a clear and broad appreciation of its potential benefits within each company, as well as among the entire research and chemical engineering community, including academia. This awareness is especially important among engineers, chemists and others working on the earliest development of process chemistry. Initiatives such as those presented below would lead to a quicker application of these concepts.

16.1 INCORPORATING INHERENTLY SAFER DESIGN INTO PROCESS SAFETY MANAGEMENT

Inherently safer design within each company should be incorporated into process safety management programs to make its application a "way of life." Seamlessly integrating IS concepts into all facets of process safety will encourage the development of inherently safer processes and procedures. To do so, the company must identify the best way to go about incorporating these concepts. Should inherent safety be considered separately and distinctly, or as part of the process hazards analysis programs? What are the advantages and disadvantages of each method of application?

Incident investigations are a part of process safety management. In investigations, lessons are learned that could be indicative as to how inherently safer design could have prevented or mitigated the outcomes. How can these learnings be disseminated such that future incidents in similar processes are avoided?

There are barriers impeding progress in applying inherently safer design. How might they be best identified and overcome?

16.2 ENCOURAGING INVENTION WITHIN THE CHEMICAL AND CHEMICAL ENGINEERING COMMUNITY

Publicizing the virtues of inherent safety beyond the *process safety* community and into the *broader chemistry and chemical engineering* community is essential to spur innovation. The authors of this book encourage readers to look for opportunities to "beat the drum" for inherent safety at every opportunity as they interact with this broader audience. Awareness can be raised in a variety of ways. Books by Trevor Kletz (Ref 16.12 Kletz 1984; Ref 16.13 Kletz 1991), CCPS (Ref 16.1 CCPS 1993), and Englund (Ref 16.7 Englund) have proven to be successful vehicles. The IChemE and IPSG Inherently Safer Processing training program (Ref 16.10 IChemE) is another successful format for promoting this topic. Other means need to be explored and exploited.

16.3 INCLUDING INHERENT SAFETY INTO THE EDUCATION OF CHEMISTS AND CHEMICAL ENGINEERS

Teaching inherently safer design concepts in undergraduate chemistry, chemical engineering, and related disciplines will provide great benefit as students move into industry after graduation.

16.4 DEVELOPING INHERENTLY SAFER DESIGN DATABASES AND LIBRARIES

The industry, and more importantly, individual companies, need to develop inherently safer design databases that are readily available, cataloged, cross-referenced, indexed, and shared across the broader company. These might take the form of libraries of information. Several examples of needed databases are:

- A continually updated database that describes and catalogs inherently safer design successes and failures.
- A collection of databases of chemicals, and of functional groups, ranked relative to their reactivity, stability, toxicity, and flammability categories. This would assist in the evaluation of the potential benefits of substituting one, somewhat safer, chemical for another.

- A database of the hazards associated with different types of equipment and unit operations, including the applicability of inherently safer design in each. As innovative solutions to hazards in equipment and process operations are discovered, these could be shared across the company through this database, reducing risk in similar equipment and processes. A summary of design approaches for a number of common types of chemical process equipment is published in CCPS (Ref 16.2 CCPS 1998).

16.5 DEVELOPING TOOLS TO APPLY INHERENTLY SAFER DESIGN

Inherent safety is still a young and evolving area of process risk assessment and decision-making. New tools and analytic approaches are needed to help process risk assessors and decision-makers identify inherent safety opportunities and evaluate them against other competing risks and opportunities.

16.5.1 The Broad View and Life Cycle Cost of Alternatives

One of the recognized barriers to the successful application of inherently safer design is the lack of appreciation of the benefits that can be derived from viewing a process broadly, rather than narrowly. Employing a cradle-to-grave and a feed-end-to-product-end view will lead to the development of processes which are as inherently safe as possible. Examples of myopic design could include the following.

- A chemist may think only of optimizing the route of synthesis to avoid a runaway reaction hazard and not consider the safety and design implications of flammable reaction by-products.
- A designer fails to identify and consider the environmental land use implications of dismantling a process at the end of its useful life.

At any given time, individuals are more concerned with a particular unit operation, or with a particular life cycle stage of the process. A broader approach, however, would require an analysis of the impact of the process on:

- upstream and downstream operations

- ancillary operations, such as waste stream management and the handling of byproducts
- later stages in the process life cycle

Tools to apply such a broader view of a process would pay inherently safer dividends. One tool might include instructions on how to estimate the life cycle cost of proposed alternative solutions. Organizations that are heavily invested in sustainability already possess the core of the tools. These tools can be tailored to address inherently safer design benefits just as they are already used to determine 'environmental footprint', 'energy impact', etc. Such a tool is not presently fully developed and available in the public domain. Training to assist in estimating life cycle cost is also needed.

16.5.2 Benefits of Reliability Analysis

Reliability relies on quality hardware, correct installation and commissioning, together with prudent inspection and test programs. The frequency and duration of failure are minimized, thus reducing the likelihood of such systems being unavailable when challenged by process hazards. Guidance in assessing such reliability could lead to simplifying SISs (Ref 16.3 CCPS 2007; Ref 16.8 Green 1995; Ref 16.9 Green 1996).

Reliability analysis of both the overall processes, and individual unit operations within those processes, is needed. High process reliability may reduce the overall need for large in-process inventories by minimizing the need for process intermediate storage facilities, or storage and blending facilities for "off-spec" product. It may also maximize the ability to produce product "as needed," thereby reducing on-plant product storage requirements. In addition, it should eliminate the run to failure philosophy for equipment and instrumentation in high risk processes. Highly reliable management of raw material purchasing and delivery systems may reduce the need for on-plant raw material storage.

16.5.3 Potential Energy

It is obvious that, all other things being equal, a process operated at high temperature and high pressure, containing exothermically reactive materials, and containing a large inventory, is inherently less safe than one at ambient temperature and pressure, containing a small inventory of thermally-stable materials. A measure of the magnitude of this difference in safety is the difference in the potential energy in the process. By comparing the potential energies of alternative process strategies, the lowest energy alternative can be identified. Methods and guidance are needed to estimate the total potential energy in different types of situations encountered in processes.

16.5.4 A Table of Distances and Consequence/Risk-Based Siting

Defining appropriate spacing of unit operations within a process, and appropriate distancing—from one process to another; from proximity of nonessential employees to day-to-day process operation; and from the process to the public—is an inherently safer approach. A definition of "appropriate spacing" would assist in evaluating the process location alternatives. One approach to this definition may take the form of a table of distances as a function of the type of hazard, inventory quantity, and other factors. Another approach, which is becoming more prevalent in industry, is to use consequence-based and risk-based modeling to determine the appropriate spacing. CCPS has provided guidance on spacing in a recent publication (Ref. 16.16 CCPS 2018).

16.5.5 Quantitative Measures of Inherent Safety

Several available quantitative risk indices can be used to measure the degree of application of inherent safety in a process. These include the Dow Chemical Exposure Index (Ref 16.4 Dow 1994a), the *Dow Fire and Explosion Index* (Ref 16.5 Dow 1994b), the Mond Index (Ref 16.11 ICI; Ref 16.15 Tyler), and the Prototype Index of Inherent Safety by Edwards, et al. (Ref 16.6 Edwards). Refinement of these quantitative measurement techniques and convergence to a single set of accepted indices would be beneficial. Inclusion of other criteria, such as the hazards posed by any by-products which are formed, should be considered. Accepted metrics would aid in comparing alternatives and in quantifying the degree of process improvement that the various applications of inherently safer design bring.

16.5.6 Other Suggestions

- Increase emphasis on training and promoting the benefits of IS in companies, universities, and through continuing education to practicing engineers and chemists.
- Develop tools to aid the engineer in conducting IS reviews through the life of a process.
- Make IS tools such as the INSET Toolkit (Ref 16.14 Mansfield) available to a broad community of engineers involved in process development and design.
- Increase efforts to promote IS to process development chemists, who work on processes early in the life cycle and have the greatest leverage for reducing or eliminating hazards.
- Promote IS to smaller companies and organizations, where the concept may be less well known and understood, as well as more challenging to implement in the facility life cycle.
- Develop and expand on the concepts of life cycle costing to better understand the potential economic advantages of inherently safer design.
- Continue research on process intensification through new and innovative technologies. The industry also needs to identify ways of demonstrating and promoting new and inherently safer technologies, perhaps through multi-company consortia, such as the Center for Waste Reduction Technology and AIChE's efforts through the Rapid Advancement in Process Intensification Deployment (RAPID) Institute.
- Publicize successes in implementing inherently safer processes, through symposia and literature articles. The requirements of regulations such as the US EPA RMP regulations, the New Jersey Toxic Catastrophe Prevention Act, and the Contra Costa County Industrial Safety Ordinance, as well as others across the globe, may provide additional incentive to companies to demonstrate

that they are committed to reducing the potential hazard zones in the area of a plant.

- Develop methods to measure various inherent safety process options, an essential first step to widespread implementation.
- Develop a method to measure inherent safety using "fuzzy logic" mathematics (a concept in which "ranges of truth" rather than discrete true or false values) - something that is now being researched.
- Government programs now support concepts research and development, such as green chemistry, solvent substitution, waste reduction, and sustainable growth, all of which are related to inherent safety. A similar approach involving industry, government and academia can enhance inherently safer chemical processes discovery, development and implementation.

16.6 REFERENCES

16.1 Center for Chemical Process Safety (CCPS 1993). *Guidelines for Engineering Design for Process Safety*. New York: American Institute of Chemical Engineers, 1993.

16.2 Center for Chemical Process Safety (CCPS 1998). *Guidelines for Design Solutions to Process Equipment Failures.* New York: American Institute of Chemical Engineers, 1998.

16.3 Center for Chemical Process Safety (CCPS 2007). *Guidelines for Safe and Reliable Instrumented Protective Systems.* New York: American Institute of Chemical Engineers, 2007.

16.4 Dow Chemical Company (1994a). *Dow's Chemical Exposure Index Guide, 1st Edition*. New York: American Institute of Chemical Engineers, 1994.

16.5 Dow Chemical Company (1994b). *Dow's Fire and Explosion Index Hazard Classification Guide, 7th Edition*. New York: American Institute of Chemical Engineers, 1994.

16.6 Edwards, D.W., Lawrence, D., and Rushton, A.G., *Quantifying the inherent safety of chemical process routes*. In 5th World Congress of Chemical Engineering, July 14-18, 1996, San Diego, CA (Paper 52d). New York: American Institute of Chemical Engineers, 1996.

16.7 Englund, S.M., *Process and design options for inherently safer plants.* In V. M. Fthenakis (ed.). Prevention and Control of Accidental Releases of Hazardous Gases (9-62). New York: Van Nostrand Reinhold, 1993.

16.8 Green, D.L., and Dowell, A.M., *How to design, verify, and validate emergency shutdown systems.* ISA Transactions 34, 261-272, 1995.

16.9 Green, D.L., and Dowell, A.M., *Cookbook safety shutdown system design.* In H. Cullingford (Ed.). 1996 Process Plant Safety Symposium, Volume 1, April 1-2, 1996, Houston, TX (pp.552-565). Houston, TX: South Texas Section of the American Institute of Chemical Engineers, 1996.

16.10 The Institution of Chemical Engineers and The International Process Safety Group, *Inherently Safer Process Design*. Rugby, England: The Institution of Chemical Engineers, 1995.

16.11 Imperial Chemical Industries (ICI), *The Mond Index, Second Edition*. Winnington, Northwich, Chesire, U. K.: Imperial Chemical Industries PLC, 1985.

16.12 Kletz, T.A., (1984) *Cheaper, Safer Plants, or Wealth and Safety at Work*. Rugby, Warwickshire, England: The Institution of Chemical Engineers, 1984.

16.13 Kletz, T.A. (1991). *Plant Design for Safety*. Rugby, Warwickshire, England: The Institution of Chemical Engineers, 1991.

16.14 Mansfield, D., Malmen, Y. and Suokas, E., *The Development of an Integrated Toolkit for Inherent SHE.* International Conference and Workshop on Process Safety Management and Inherently Safer Processes, October 8-11, 1996, Orlando, FL (pp. 103-117). New York: American Institute of Chemical Engineers, 1996.

16.15 Tyler, B.J., *Using the Mond Index to measure inherent hazards.* Plant/Operations Progress 4 (3), 172-75, 1985.

16.16 Center for Chemical Process Safety (CCPS 2018). *Guidelines for Siting and Layout of Facilities, Second Edition.* New York: American Institute of Chemical Engineers, 2018.

Appendix A. Inherently Safer Technology (IST) Checklist

A.1 IST Checklist Procedure

The following procedure can be used to answer each Inherently Safer Technology (IST) question listed in A.2 Checklist Questions:

1. Determine the applicability of the item to the process being reviewed. Indicate "Y" or "N" in the appropriate column to indicate applicability.
2. If an item is applicable to the process under review, determine whether there are any IST opportunities or applications available that could potentially reduce the risk (consequences and/or likelihood) of an accidental or intentional release. If an IST opportunity is identified, it should be described in the Opportunities/Applications column. If no new IST opportunities are identified, this is noted in the Opportunities/Applications column with supporting information, including general references to existing safeguards and IST already implemented, where applicable.
3. If an IST opportunity is identified, a screening evaluation of the alternative(s) should be performed by the review team and the results summarized in the Feasibility column. If the IST alternative is determined not to be feasible (due to cost, technology limitations, security, operability, safety, or other factors), this decision is noted as such along with supporting information, including general references to existing safeguards and IST already implemented, where applicable. The current status of the opportunity (e.g., "to be evaluated further," "evaluation in-progress," "implementation in-progress," etc.) is documented in the "Current Status" column.
4. If an IST opportunity is deemed potentially feasible, a recommendation for further evaluation or implementation is

made by the team and entered in the "Recommendations" column.

While proceeding through the IST checklist, existing safeguards are reviewed for each IST alternative determined to be not feasible. In some cases, the checklist suggests safeguards which can be active, passive, or procedural. Credible safeguards designed to prevent, detect, or mitigate a hazard may be as effective as implementing IST alternatives. For security issues, countermeasures can be detection, deterrence, delay, or response. Definitions and examples of safeguards reviewed include:

- *Active*. Using controls, alarms, safety instrumented functions (SIFs), and mitigation systems to detect and respond to process deviations from normal operation. Examples: High level, temperature, and flow SIFs, atmospheric sensors with automatic shutdowns, and pressure relief systems. Active safeguards require external inputs to function, such as electrical power.

- *Passive*. Minimize the hazard by process and equipment design features which reduce either the frequency or consequences of the hazard without the active functioning of any device. Examples: dikes, containment buildings. Passive safeguards do not require external inputs to function.

- *Procedural*. Using human response in the form of operating procedures, administrative checks, and emergency response. Examples: emergency shutdown or emergency response procedures, operator checks and corrective actions.

In general, supporting information for rejection of inherently safer design alternatives should include the presence and adequacy of existing safeguards combined with cost, operability, or other considerations that reduce the viability of implementing the inherently safer options.

You can access the IST Checklist based on these questions at the CCPS Website:

www.aiche.org/InherentlySafer/ISTChecklist

A.2 IST Checklist Questions

Table A.1 IST Checklist Substitute Questions

1	SUBSTITUTE Questions:
1.1	Is this (hazardous) process/product necessary?
1.2	Is it possible to completely eliminate hazardous raw materials, process intermediates, or by-products by using an alternative process or chemistry?
1.3	Is it possible to completely eliminate in-process solvents and flammable heat transfer media by changing chemistry or processing conditions?
1.4	Is an alternate process available for this product that eliminates or substantially reduces the need for hazardous raw materials or production of hazardous intermediates?
1.5	Is it possible to substitute less hazardous raw materials?
	• Noncombustible for flammable
	• Less volatile
	• Less reactive
	• More stable
	• Less toxic
	• Low pressure steam rather than flammable heat transfer fluid (i.e., operated above flash point)
1.6	Is it possible to substitute less hazardous final product solvents?
1.7	Is it possible to use a nonflammable refrigerant instead of a flammable one (or minimize inventory)?
1.8	Are there any other alternatives for substituting or eliminating the use of hazardous materials in this process?

Table A.2 IST Checklist Minimize Questions

2	MINIMIZE Questions:
2.1	Inventory Reduction
2.1.1	Can hazardous raw materials inventory be reduced?
	• Just-in-time deliveries based on production needs
	• Supplier management including strategic alliance
	• On-site generation of hazardous material (including in situ) from less hazardous raw materials
	• Hazardous raw material inventory management system based on production forecast
2.1.2	Can (hazardous) in-process storage and inventory be reduced?
	• Direct coupling of process elements
	• Eliminating or reducing size of in-process storage
	• Designing process equipment involving hazardous material with the smallest feasible inventory (see also Section 2.2)
2.1.3	Can hazardous finished product inventory be reduced?
	• Improving production scheduling/sales forecasting
	• Improving communication with transporters/material handlers
	• Hazardous finished product inventory management system based on sales forecast
2.2	Process Intensification Considerations
2.2.1	Can alternate equipment with reduced hazardous material inventory requirement be used?
	• Centrifugal extractors in place of extraction columns
	• Flash dryers in place of tray dryers
	• Continuous reactors in place of batch
	• Plug flow or loop reactors in place of continuous stirred tank reactors
	• Continuous in-line mixers (e.g., static mixer) in place of mixing vessels or reactors
	• Intensive mixers to minimize size of mixing vessel of reactor
	• High heat-transfer reactors (e.g., microreactor, HEX reactor)
	• Spinning-disk reactor (especially for high heat-flux or viscous liquids)

	• Compact heat exchangers (higher heat transfer area per unit volume, e.g., spiral, plate & frame, plate-fin) in place of shell-and-tube
	• More hazardous material on the tubeside in shell-and-tube exchangers
	• Use water or other non-flammable heat transfer medium, a vapor-phase medium, or a medium below its boiling point
	• Wiped film stills in place of continuous still pots (distillation columns)
	• Combine unit operations (such as reactive distillation or extraction in place of separate reactor with multi-column fractionation train or extractor; installing internal reboilers or heat exchangers) to reduce overall system volume
	• Use of acceleration fields (e.g., rotating packed bed for gas/liquid or liquid/liquid contacting for absorption, stripping, distillation, extraction, etc.)
	• Alternate energy sources (such as lasers, UV light, microwaves, or ultrasound) to control reaction or direct heat to the unit operation
2.2.2	Has the length of hazardous material piping runs been minimized?
2.2.3	Has hazardous material piping been designed for minimum pipe diameter?
2.2.4	Can pipeline inventory be reduced by using the hazardous material as a gas rather than a liquid?
2.2.5	Can process conditions be changed to reduce production of hazardous waste or by-products?
2.3	Are there any other alternatives for minimizing the inventory of hazardous materials in this process?

Table A.3 IST Checklist Moderate Questions

3	MODERATE Questions:
3.1	Is it possible to limit the supply pressure of (hazardous) raw materials to less than the maximum allowable working pressure of the vessels to which they are delivered?
3.2	Is it possible to make reaction conditions (for hazardous reactants or products) (temperature, pressure) less severe by using a catalyst, or a better catalyst (e.g., structured or monolithic vs. packed-bed)?
3.3	Can the process be operated at less severe conditions (for hazardous reactants or products) by considering:
	• Improved thermodynamics or kinetics to reduce operating temperatures or pressures
	• Changes in reaction phase (e.g., liquid/liquid, gas/liquid, or gas/gas)
	• Changes in the order in which raw materials are added
	• Raw material recycle to compensate for reduced yield or conversion
	• Operating at lower pressure to limit potential release rate
	• Operating at lower temperature to prevent runaway reactions or material failure
3.4	Is it possible to use less concentrated hazardous raw materials to reduce the hazard potential?
	• Aqueous ammonia and/or HCl instead of anhydrous
	• Sulfuric acid instead of oleum
	• Dilute nitric acid instead of concentrated fuming nitric acid
	• Wet benzoyl peroxide instead of dry
3.5	Is it possible to use larger particle size/reduced dust forming solids to minimize potential for dust explosions?
3.6	Are all process materials (e.g., heating/cooling media) compatible with process materials in event of inadvertent contamination (e.g., due to a tank coil or heat exchanger tube failure)?
3.7	Is it possible to add an ingredient to volatile hazardous materials that will reduce its vapor pressure?

3.8	For equipment containing materials that become unstable at elevated temperature or freeze at low temperature, is it possible to use heating/cooling media which limit the maximum and minimum temperatures attainable (i.e., self-limiting electric heat tracing or hot water at atmospheric pressure)?
3.9	Can process conditions be changed to avoid handling flammable liquids above their flash points?
3.10	Is equipment designed to totally contain the materials that might be present inside at ambient temperature or the maximum attainable process temperature (i.e., higher maximum allowable working temperature to accommodate loss of cooling, simplified reliance on external systems like refrigeration to control temperature such that vapor pressure is less than equipment design pressure)?
3.11	For processes handling flammable materials, is it possible to design the layout to minimize the number and size of confined areas and to limit the potential for serious overpressure in the event of a loss of containment and subsequent ignition?
3.12	Can process units (for hazardous materials) be designed to limit the magnitude of process deviations?
	• Selecting pumps with maximum capacity lower than safe rate of addition for the process
	• For gravity-fed systems, limiting maximum feed rate to be within safe limits by pipe size or fixed orifice
	• Minimum flow recirculation line for pumps/compressors (with orifice to control flow) to ensure minimum flow in event of deadheading
3.13	Can hazardous material liquid spills be prevented from entering drainage system/sewer (if potential for fire or hazardous reaction exists, e.g., water reactive material)?
3.14	For flammable materials, can spills be directed away from the storage vessel to reduce the risk of a boiling liquid expanding vapor explosion (BLEVE) in the event of a fire?
3.15	Can passive designs, such as the following, be implemented?
	• Secondary containment (e.g., dikes, curbing, buildings, enclosures)

	• Use of properly vented blowdown tank for dumping of runaway reaction mass
	• Permanent bonding and grounding systems for process equipment, tanks and vessels
	• Use of gas inerting systems for handling flammables and explosive dusts (e.g., nitrogen, CO2)
	• Use of diplegs with anti-siphon openings for feed to flammable liquid storage tanks
	• Fireproofing insulation vs. fixed/portable fire protection
3.16	Can gases be transported and stored at low or atmospheric pressure on a high capacity adsorbent instead of using pressurized gas cylinders?
3.17	Are there any other alternatives for moderating the use of hazardous materials in this process?

Table A.4 IST Checklist Simplify Questions

4	**SIMPLIFY Questions:**
4.1	Can equipment be designed such that it is difficult or impossible to create a potential hazardous situation due to an operating or maintenance error?
	• Easy access and operability of valves to prevent inadvertent errors
	• Elimination of all unnecessary cross-connections
	• Use of dedicated hoses and compatible couplings for reactants where hose connections are used
	• Designing temperature-limited heat transfer equipment to prevent exceeding maximum process or equipment design temperatures
	• Use of corrosion resistant materials for process equipment, piping and components
	• Operating at higher temperature to avoid cryogenic effects such as embrittlement failures
	• Using alternative agitation methods (e.g., external circulation using sealless pump which eliminates potential releases due to agitator seal failures)
	• Use of mixing feed nozzle instead of agitator for vessel mixing
	• Using underground or shielded tanks
	• Specifying fail-safe operation on utility failure (e.g., air, power)
	• Allocating redundant inputs and outputs to separate modules of the programmable electronic system to minimize common cause failures
	• Provide continuous pilots (independent, reliable source) for burner management systems
	• Using refrigerated storage vs. pressurized storage
	• Using independent power buses for redundant equipment to minimize consequences of partial power failures
	• Minimizing equipment wall area to minimize corrosion/fire exposure
	• Minimizing connections, paths and number of flanges in hazardous processes
	• Avoiding use of threaded connections in hazardous service

	• Using double-walled pipe
	• Minimizing number of bends in piping (potential erosion points)
	• Using expansion loops in piping rather than bellows for thermal expansion
	• Designing equipment isolation mechanisms for maintenance in the process
	• Limiting manual operations such as filter cleaning, manual sampling, hose handling for loading/unloading operations, etc.
	• Designing vessels for full vacuum to eliminate risk of vessel collapse
	• Designing both shell- and-tube side of heat exchangers to contain the maximum attainable pressure, eliminating the need for pressure relief (may still be needed to meet fire safety requirements)
	• Designing/selecting equipment which makes incorrect assembly impossible
	• Using equipment that clearly identifies status:
	• Check valves with easy to identify direction of flow
	• Gate valves with rising spindles to clearly indicate open or closed position
	• Spectacle (or figure 8) blinds instead of slip plates
	• Manual quarter-turn block valves with handles that clearly indicate position
	• For automated block valves, display actual valve position in addition to the output to the valve
	• Designing equipment with an MAWP to contain the maximum pressure generated without reliance on pressure relief systems, even if the "worst credible event" occurs?
	• Use open vent or overflow line to secondary containment for overpressure, overfill and vacuum protection
	• Eliminate utility connections above pressure rating of vessel
	• Carrying out several process steps in separate processing vessels rather than a single multi-purpose vessel (to reduce the complexity and number of raw materials, utilities, and auxiliary equipment connected to a specific vessel)

4.2	Can passive leak-limiting technology be used to limit potential loss of containment?
	• Blowout resistant gaskets (e.g., spiral wound)
	• Increasing wall strength of piping and equipment
	• Maximize use of all-welded pipe
	• Using fewer pipe seams and joints
	• Providing extra corrosion/erosion allowance (e.g., Sch. 80 vs. 40)
	• Reducing or eliminating vibration (e.g., through vibration dampening or equipment balancing)
	• Minimizing the use of open-ended (bleed or vent), quick-opening valves (for example, quarter-turn ball or plug valves)
	• Eliminating open-ended (bleed or vent), quick-opening valves (for example, quarter-turn ball or plug valves) in hazardous service
	• Using incompatible hose connections to prevent mis-connection (e.g., air/ nitrogen, raw materials)
	• Use of round valve handles for open-ended quarter-turn valves to minimize potential for bumping open
	• Improving valve seating reliability (e.g., using system pressure to seal valve seats where possible, using valve seat geometry, valve operations, and flow to eliminate or reduce seat damage)
	• Eliminating unnecessary expansion joints, hoses, and rupture disks
	• Use of articulated arms instead of hoses for loading/unloading of hazardous materials
	• Eliminating unnecessary sight glasses/glass rotameters; use high-pressure/armored sight glasses as needed
	• Eliminate use of glass, plastic or other brittle material as material of construction
	• Use of seal-less pumps (e.g., canned, magnetic drive)
	• Use of top-unloading vessels/storage tanks; minimize number of bottom connections/ fittings
	• Minimizing the number of different gaskets, nuts, bolts, etc. used to reduce potential for error
4.3	Has attention to control system human factors been addressed through:

	• Simplified control displays
	• Limited instrumentation complexity
	• Clearly displayed information about normal and abnormal process conditions
	• Logical arrangement of controls and displays that match operator expectations
	• Separate displays that present similar information in a consistent manner
	• Safety alarms that are easily distinguished from process alarms
	•Correction of nuisance alarms and elimination of redundant alarms as soon as practical to help prevent complacency
	• Control system displays that give adequate feedback for all operational actions
	• Layout of control system displays that are logical, consistent, and effective
	• Controls that are distinguishable, accessible, and easy to use
	• Controls which meet standard expectations (color, direction of movement, etc.)
	• Control arrangements which logically follow the normal sequence of operation
	• Operating procedure format and language which operators believe are easy to follow and understand and that include necessary information
4.4	Are there any other alternatives for simplifying operations involving hazardous materials in this process?

Table A.5 IST Checklist Location/Siting/Transportation Questions

5	LOCATION / SITING / TRANSPORTATION Questions:
5.1	Can the plant be located to minimize the need for transportation of hazardous materials? (e.g., co-located with supplier/customer, on-site production of hazardous raw materials)
5.2	Can hazardous process units be located to eliminate or minimize:
	• Adverse effects from adjacent hazardous installations
	• Off-site impacts
	• On-site impacts on employees and other plant facilities including control rooms, fire protection systems, emergency response and communication facilities, and maintenance and administrative facilities
5.3	Can a multi-step process, where the steps are done at separate sites, be divided up differently to eliminate the need to transport hazardous materials?
5.4	Can materials be transported:
	• In a less hazardous form (e.g., refrigerated liquid vs. pressurized)
	• In a safer transport method (e.g., via pipeline, top- vs. bottom-unloaded, rail vs. truck)
	• Along a safer route (e.g., avoiding high risk areas such as high population areas, tunnels, or high-accident-rate sections of roadway)?

Appendix B. Inherent Safety Analysis Approaches

Inherent safety (IS) can be analyzed in a number of ways, but in all cases, the intent is to formalize the consideration of inherent safety, rather than to include it by circumstance. By formally including inherent safety, in either a direct or indirect way, facilities can fully realize the potential benefits of inherent safety. In addition, all IS considerations will be fully documented.

Three analysis methods can be used to evaluate implementation of IS:

1. Inherent Safety Analysis: Guided Checklist Process Hazard Analysis (PHA)
2. Inherent Safety Analysis: Independent Process Hazard Analysis (PHA)
3. Inherent Safety Analysis: Integral to Process Hazard Analysis (PHA)

Method 1 employs a specialized checklist containing a number of practical inherent safety considerations organized around the four strategies of minimization, substitution, moderation, and simplification. The advantage of this approach is that it is very direct and asks pointed questions that have proven to be valuable in reducing hazards at past locations. The disadvantage is that, as with any checklist, it may be limiting in that other ideas may surface if the team was asked to more creatively determine applications for the inherent safety strategies given a safety objective. (Note that the checklist appearing here is only a representative subset to illustrate its use. See Appendix A for the complete checklist.)

For the second method, the team is asked to avoid a particular hazard at a designated part of the process. In this case, the team reviews a problem, determines which of the inherently safer strategies may apply, and then brainstorms possible ways the hazard can be reduced or eliminated.

The third method integrates Inherently Safer Design (ISD) into every PHA study the facility may conduct, including "What-If?," HAZOP, FMEA or other similar methodology. The concept is both to include questions (for "What-If?") or guidewords (for HAZOP) to introduce ISD to the discussion, and then to use the four strategies mentioned above as a possible means to mitigate each hazard identified. The input from those discussions are then evaluated alongside the other layers of protection strategies that may be used.

Each method is explained below. In each case, the analyst reviews the potential for applying inherent safety either at the process level or at the node level. A node is defined the same way as in Process Hazards Analysis studies, or existing studies. If the process is relatively simple and the ISD opportunities are limited, it is possible to do the analysis at the higher level required for PHA study. If the study is conducted at a detailed node level, it may result in additional considerations being given to smaller but important details. For example, at a macro level, the hazardous chemical in the process cannot be substituted, but on a micro level there may be opportunities to do so in given areas, process steps or with certain types of equipment.

In all cases, it is recommended to use a risk-ranking scheme that defines likelihood and consequences on a scale such as the one shown on Tables B.1-B.3. Inherent safety should be evaluated based on risks just as with other risk management strategies.

Table B.1. Generic Risk Matrix (R)

Likelihood	Very High (4)	Medium 2	High 3	Very High 4	Very High 4
	High (3)	Low 2	Medium 3	High 4	Very High 4
	Medium (2)	Low 2	Medium 3	Medium 3	High 4
	Low (1)	Low 1	Low 2	Low 3	Medium 3
		Low (1)	Medium (2)	High (3)	Very High (4)
		Severity			

Table B.2. Generic Severity (S) Rankings

Category	Low (1)	Medium (2)	High (3)	Very High (4)
Health & safety impacts	Minor injury or health effect	Moderate injury or health effect	Major injury or health effect; offsite public impacts	Fatality offsite, multiple onsite injuries or fatalities,
Asset damage (replacement cost)	Low	Medium	Moderate	High
Business interruption (days unavailable or $)	Low	Medium	Moderate	High
Environmental impact (remediation damages)	Low	Medium	Moderate	High

Table B.3. Generic Likelihood (L) Descriptors

Likelihood	Short descriptor	Description
1	Low	Not expected to occur in life of facility
2	Medium	Possible to occur in life of facility
3	High	Possible to occur in range of 1 year to 10 years
4	Very High	Possible to occur at least once a year

B.1 INHERENT SAFETY ANALYSIS – GUIDED CHECKLIST PROCESS HAZARD ANALYSIS (PHA)

Table B.4 offers an example of a guided checklist approach. The analyst asks the questions from the checklist (potential opportunities) and the team documents the potential consequences of any issue that may be applicable to the process or node under study. Considering the four ISD strategies, the team documents the potential recommendations that may address the concern ranked in the following order:

- First order ISD
- Second order ISD
- Layers of Protection

Table B.4 Case Summary: Inherent Safety Analysis – Checklist Process Hazard Analysis (PHA)

Location: Orange, New Jersey						**Risk Ranking**			**Unit:** Hydrofluoric Acid Alkylation Unit	**Analysis Date:** April 1, 2008
PFD No.: 1234-5678										
Node: Isobutane Storage										
Design Conditions/Parameters: Storage of isobutene in five bullets and two process vessels near the unit										
#	**QUESTION**	**POTENTIAL OPPORTUNITIES**	**FEASIBILITY**	**CONSEQUENCES**	**EXISTING SAFEGUARDS**	**S**	**L**	**R**	**RECOMMENDATIONS**	**COMMENTS/ STATUS**
1	Reduce hazardous raw materials inventory	Lower storage tank volume or eliminate some storage if possible.	Lowering tank volumes is already done. There may be one tank that could be eliminated.	Potential release from storage and exposure to south plant from unconfined vapor cloud explosion.	1. Administrative controls limit fill level of the five tanks.	4	1	3	1. Eliminate one of five flammable storage bullets to reduce potential releases from storage.[1]	In review.
2	Reducing in-process storage and inventory	Interim storage adds to inventory and could be eliminated.	Will require engineering analysis to evaluate.	Potential leak, fire and explosion.	1. High level alarms 2. Flammable gas detectors	4	1	3	2. Consider eliminating interim storage and providing a continuous flow operation.[2]	In review

[1] This uses the concept of ***Minimization*** to avoid a hazard. Originally, feed could be taken from any of five tanks, yet the process could be managed with only three tanks. A hazardous condition was reduced so this represents a **Second order Inherent Safety** change.

[2] Also uses the concept of ***Minimization***. A number of unit tanks could possibly be eliminated, thus eliminating a hazardous condition. This represents a **First order Inherent Safety** change.

Table B.4 Case Summary: Inherent Safety Analysis – Checklist Process Hazard Analysis (PHA)

#	QUESTION	POTENTIAL OPPORTUNITIES	FEASIBILITY	CONSEQUENCES	EXISTING SAFEGUARDS	S	L	R	RECOMMENDATIONS	COMMENTS/ STATUS
Location: Orange, New Jersey **PFD No.:** 1234-5678 **Node:** Isobutane Storage **Design Conditions/Parameters:** Storage of isobutene in five bullets and two process vessels near the unit						**Risk Ranking**			**Unit:** Hydrofluoric Acid Alkylation Unit	**Analysis Date:** April 1, 2008
3	Reducing finished product inventory	Not applicable (NA)[3]								
4	Reduce hazardous material by using alternate equipment		No alternatives available or feasible.[4]							
5	Minimize length of hazardous material piping runs	There are many piping runs that are no longer used that might be decommissioned.	Need to consider in engineering and operations evaluation but appears to be feasible.	Potential larger release		4	1	3	3. Consider moving the location of the planned storage vessels to a location not closer than 250 feet, but closer than the 1500 feet location presently planned.[5]	

[3] Some questions are not relevant to the particular process.

[4] Some questions may result in a conclusion that there are no alternatives or feasible inherent safety applications.

[5] See Footnote 3

Table B.4 Case Summary: Inherent Safety Analysis – Checklist Process Hazard Analysis (PHA)

Location: Orange, New Jersey **PFD No.:** 1234-5678 **Node:** Isobutane Storage **Design Conditions/Parameters:** Storage of isobutene in five bullets and two process vessels near the unit						**Risk Ranking**			**Unit** Hydrofluoric Acid Alkylation Unit	**Analysis Date:** April 1, 2008
#	**QUESTION**	**POTENTIAL OPPORTUNITIES**	**FEASIBILITY**	**CONSEQUENCES**	**EXISTING SAFEGUARDS**	**S**	**L**	**R**	**RECOMMENDATIONS**	**COMMENTS/ STATUS**
6	Smallest diameter piping	The line planned for the feed is oversized and could contain a larger inventory than is necessary.	Possible following engineering evaluation to reduce diameter.	Potential larger release		3	2	3	4. Reduce the planned feed line size from 6 to 4 inch.[6]	
7	Eliminate hazardous raw materials, process intermediates, or by-products by using an alternative process or chemistry		No alternative for isobutane in the process[7]							

[6] See Footnote 3.

[7] See Footnote 4.

Table B.4 Case Summary: Inherent Safety Analysis – Checklist Process Hazard Analysis (PHA)

Location: Orange, New Jersey						**Risk Ranking**			**Unit:** Hydrofluoric Acid Alkylation Unit	**Analysis Date:** April 1, 2008
PFD No.: 1234-5678										
Node: Isobutane Storage										
Design Conditions/Parameters: Storage of isobutene in five bullets and two process vessels near the unit										
#	**QUESTION**	**POTENTIAL OPPORTUNITIES**	**FEASIBILITY**	**CONSEQUENCES**	**EXISTING SAFEGUARDS**	**S**	**L**	**R**	**RECOMMENDATIONS**	**COMMENTS/ STATUS**
8	Eliminate in-process solvents and flammable heat transfer media.	Substitute the flammable solvent.	A nonflammable solvent may be commercially available.	Possible source of flammable release	Existing fire prevention controls and fire suppression systems.	3	3	4	5. Consider substituting the solvent used with a non-flammable solvent.[8]	

[8] This uses the concept of ***Substitution*** to avoid a hazard. A nonflammable alternative solvent was substituted for a planned flammable solvent. A hazardous condition was eliminated so this represents a **First order Inherent Safety** change.

B.2 INHERENT SAFETY ANALYSIS - INDEPENDENT PROCESS HAZARD ANALYSIS (PHA)

Table B.5 is an example of an IS Analysis approach, which is similar to a typical PHA, but focuses exclusively on inherent safety.

The analyst considers a hazard, such as runaway reaction caused by water reactivity in a reactor and sets a safety objective, such as "minimize potential for runaway reaction in the feed to the reactor." The team then documents each potential cause of the hazard being evaluated, reviews the consequences, existing safeguards, and potential means of eliminating it or reducing its risk through ISD strategies.

Considering the four ISD strategies, the team documents potential recommendations that may address the concern using the order of First Order ISD, Second Order ISD, followed by Layers of Protection. Each strategy is considered. Ideas that are feasible, practical, and best address the hazard are generated. The approach acknowledges that other risk management strategies besides ISD may be more effective.

Table B.5: Inherent Safety Analysis - Independent Process Hazard Analysis (PHA)

Node: 1. Feed system to reactor

Objective: 1. Minimize potential for runaway reaction in the feed to the reactor

CAUSES	CONSEQUENCES	EXISTING SAFEGUARDS	S	L	R	OPPORTUNITIES	FEASIBILITY	RECOMMENDATIONS	COMMENT//STATUS
1. High water content in feed tank due to settlement or water carryover from upstream process	1. Potential for operator error to leave water online or valve not fully closed, or failure of the valve, allowing leak of water into the feed line. Excess water in the reactor may cause shorter run life due to catalyst fouling; this has a possible safety hazard in more startups and shutdowns over the life of the process. Worst credible case: excessive water may cause a runaway reaction.	1. Control of unit operation to meet feed and operator monitoring of process conditions.	4	4	4	Evaluate way to positively eliminate water from entering the reactor rather than controls.	It may be feasible to switch to a 'clean' tank without the potential for water with minor piping changes.	1. Change from feeding from Tank 1 to only Tank 3 since Tank 1 has high water settlement potential. Tank 1 has water in upstream units that cannot be completely avoided whereas Tank 3 is clean feedstock.[1]	

[1] This uses the concept of ***Substitution*** to avoid a hazard. Previously the feed was taken from a tank with water contaminants that had to be controlled. The alternative tank did have this inherent condition. A hazardous condition was avoided altogether so this represents a **First order Inherent Safety** change.

Table B.5: Inherent Safety Analysis - Independent Process Hazard Analysis (PHA)

Node: 1. Feed system to reactor

Objective: 1. Minimize potential for runaway reaction in the feed to the reactor

CAUSES	CONSEQUENCES	EXISTING SAFEGUARDS	S	L	R	OPPORTUNITIES	FEASIBILITY	RECOMMENDATIONS	COMMENT//STATUS
2. Water into the feed from wrong valve opened in one of the water wash cross connections		1. Proper procedures for water washing	4	2	4	Evaluate ways to eliminate water contamination risk from human error	Operating procedures can be improved.	2. Improve operating procedures for water washing to ensure operators check the valve closure and water flow following a water wash.[2]	
		2. Operator training					There is an excess number of cross connections so some can be eliminated.	3. Reduce the number of water cross connections from the feed to the reactor from 3 to 1.[3]	
		3. Temperature instrumentation to monitor reaction rate.							

[2] This uses the concept of ***Applying Procedural Safeguards*** to avoid a hazard. Existing operating procedures had not explicitly called for checking on these aspects. The change in operating procedure may reduce the likelihood of water being left on mistakenly. The inherent condition of water usage and lead potential still exists, but the likelihood, and therefore the risk, may have been reduced.

[3] Note: This uses the concept of ***Minimization*** to avoid a hazard. The water wash was previously done from multiple connections. The alternative approach of eliminating unnecessary connections reduces the likelihood of the incident by eliminating equipment, representing a **Second order Inherent Safety** change.

B.3 INHERENT SAFETY ANALYSIS - INTEGRAL TO PROCESS HAZARD ANALYSIS (PHA)

Table B.6 is an example of a third IS Analysis approach. Essentially, it is the same as the second method presented above. However, they differ in the way in which the hazard is identified. In this example, a HAZOP method utilizes deviations from design intent while in the previous example the hazard was explicitly stated and directly addressed by ISD. The way in which ISD is considered remains the same. The ISD hierarchy of analysis of First Order ISD, then Second Order ISD, and finally Layers of Protection applies no matter what the analysis method unless the scope is limited to only identify ISD potential recommendations.

Table B.6: Inherently Safer Analysis – Integral to Process Hazards Analysis (PHA)

Node: 1. Feed system to reactor									
Intent: 1. Feed to the process									
Guideword: As Well As			**Parameter: Flow**					**Deviation: Contamination**	
CAUSES	**CONSEQUENCES**	**EXISTING SAFEGUARDS**	**S**	**L**	**R**	**OPPORTUNITIES**	**FEASIBILITY**	**RECOMMENDATIONS**	**COMMENT/STATUS**
1. Settlement or water carryover from upstream process.	1. Excess water in the feed and then the reactor may cause shorter run life due to catalyst fouling; this poses a possible safety hazard due to more frequent start-ups and shut-downs over the life of the process. Worst credible case: excessive water may cause a runaway reaction.	1. Control of unit operation to meet feed and operator monitoring of process conditions.	4	4	4	Evaluate ways to stop water from entering the reactor rather than controls.	It may be feasible to switch to a 'clean' tank without the potential for water with minor piping changes.	1. Change feeding from Tank 1 to only Tank 3. Tank 1 has high water settlement potential, and has water in upstream units that cannot be completely avoided. Tank 3 is clean feedstock.[1]	

[1] Note: This uses the concept of ***Substitution*** to avoid a hazard. The feed was previously taken from a tank that had water contaminants that had to be controlled. The alternative tank did not have this inherent condition. A hazardous condition was avoided altogether so this represents a **First order Inherent Safety** change.

Glossary

CCPS has developed a standard glossary that defines a number of common terms in process safety. By reference the current CCPS Process Safety Glossary at the time of publication is incorporated into this book and can be found at the CCPS Website:

www.aiche.org/ccps/resources/glossary

Additionally, there are some specific terms used in this book that are not currently included in the standard glossary. These terms are defined in the book as necessary when they are introduced.

References

Key Literature in Inherent Safety and Additional Reading

Agreda, V. H., Partin, L.R. and Heise, W.H. (1990). High-purity methyl acetate via reactive distillation. *Chemical Engineering Progress, 86* (2), 40-46.

Akay, G. and Azzopardi, B.J. (1995). In Proceedings of the First International Conference on Science, Engineering and Technology of Intensive Processing, September 18-20, 1995. Nottingham, U. K.: University of Nottingham.

Allen, D. (1992). The role of catalysis in industrial waste reduction. In D.T. Sawyer and A.E. Martell (Eds.). *Industrial Environmental Chemistry* (pp. 89-98). New York: Plenum Press.

Anastas, P.T. (1994). Benign by design chemistry: Pollution prevention through synthetic design. In *Preprints of Papers Presented at the 208th ACS National Meeting*, August 21-25, 1994, Washington, DC (pp. 204-205). Center for Great Lakes Studies, University of Wisconsin-Milwaukee, Milwaukee, WI: Division of Environmental Chemistry, American Chemical Society.

Arzoumanidis, G.G., and Karayannis, N.M. (1993). Fine-tuning polypropylene. *Chemtech*, 43-48.

Ashby, J. and Paton, D. (1993). The influence of chemical structure on the extent and sites of carcinogenesis for 522 rodent carcinogens and 55 different human carcinogen exposures. *Mutation Research, 286*, 3-74.

Ashford, N.A. (1993). The Encouragement of Technological Change for Preventing Chemical Accidents: Moving Firms From Secondary Prevention and Mitigation to Primary Prevention. Cambridge, MA: Center for Technology, Policy and Industrial Development, Massachusetts Institute of Technology.

Atherton, J.H. (March, 1993). Methods for study of reaction mechanisms in liquid/liquid and liquid/solid reaction systems and their relevance to the development of fine chemical processes. *Trans. IChemE, Part A, 71*, 111-118.

Bailey, J.D. (1992). Safer pesticide packaging and formulations for agricultural and residential applications. In *Reviews of Environmental Contamination and Toxicology* (pp. 17-27). New York: Springer-Verlag.

Balzani, V. (1994). "Greener Way to Solar Power." *New Scientist* 144, 1951 (12 November), 31-34.

Bashir, S., Chovan, T., Masri, B.J., Mukherjee, A., Pant, A., Sumit, S., et al. (1992). Thermal runaway limit of tubular reactors, defined at the inflection point of the temperature profile. *Ind. Eng. Chem. Res. 31* (9), 2164-2171.

Bell, A.T., Manzer, L.E., Chen, N.Y., Weekman, V.W., Hegedus, L.L., and Pereira, C.J. (1995). Protecting the environment through catalysis. *Chemical Engineering Progress, 91* (2), 26-34.

Benson, R.S., and Ponton, J.W. (March 1993). Process miniaturization - A route to total environmental acceptability? *Trans. IchemE, 71, Part A*, 160-168.

Berger, S.A. (1994). The pollution prevention hierarchy as an R&D management tool. In M. El-Halwagi and D.P. Petrides (Eds.). *Pollution Prevention Via Process and Product Modifications* (pp. 23-28). AIChE Symposium Series 303. New York: American Institute of Chemical Engineers.

Berger, S.A. (1995). "Estimating the environmental, safety, and health cost of processes in R&D." Presented at the American Industrial Hygienists Conference and Exhibition, May 26, 1995, Kansas City, MO.

Berger, S.A., Klawiter, I.L., Thomas, S.T., and Weber, V. (1994). "Estimate the environmental cost of new processes in R&D." Presented at the AIChE Spring National Meeting, April 1994, Atlanta, GA.

Berger, S.A., Cherry, J., Levengood, D., Weber, R., and Neman, D. (1996). A cost and time-efficient framework for inherent safety and pollution prevention during process development and engineering. In *International Conference and Workshop on Process Safety Management and Inherently Safer Processes, October 8-11, 1996, Orlando, FL* (pp. 416-428). New York: American Institute of Chemical Engineers.

Berger, S.A., and Lantzy, R.J. (1996). Reducing Inherent Risk Through Consequence Modeling. In H. Cullingford (Ed.). *1996 Process Plant Safety Symposium, Volume 1, April 1-2, 1996, Houston, TX* (pp. 15-23). Houston, TX: South Texas Section of the American Institute of Chemical Engineers.

Berglund, R.L. and Snyder, G.E. (December,1990). Waste minimization: The sooner, the better. *Chemtech*, 740-746.

Black, H. (1996). Supercritical carbon dioxide: the "greener" solvent. *Environmental Science and Technology* 30 (3), 124A-7A.

Blumenberg, B. (1992). Chemical reaction engineering in today's industrial environment. *Chemical Engineering Science, 47* (9-11,) 2149-2162.

Bodor, N. (October, 1995). Design of biologically safer chemicals. *Chemtech*, 22-32.

Borman, S. (November 30, 1992). Aromatic amine route is environmentally safer. *Chemical and Engineering News*, 26-27.

Bradley, D. (August 6, 1994). Solvents get the big squeeze. *New Scientist*, 32-35.

Bradley, D. (April 29, 1995). Incredible shrinking visions. *New Scientist*, 46-47.

Brennan, D.J. (May, 1993). Some challenges of cleaner production for process design. *Environmental Protection Bulletin* 024, 3-7.

Burch, W.M. (1986). Process modifications and new chemicals. *Chemical Engineering Progress, 82* (4) 5-8.

Callari, J. (November, 1992). Environmental pressures force widespread change. *Plastics World*, 40-43.

Calvert, C. (November, 1992). Environmentally-friendly catalysis using non-toxic supported reagents. *Environmental Protection Bulletin* 021, 3-9.

Calvert, C. (1992). Environmentally friendly catalysts using non-toxic supported reagents. *IChemE North Western Branch Symposium Papers, 3,* 4.1-4.16.

Carlsson, M., Habenicht, C., Kam, L.C., and Antal, M.J. (1994). Study of the sequential conversion of citric to itaconic to methacrylic acid in near-critical and supercritical water. *Ind. Eng. Chem. Res., 33,* 1989-1996.

Carrithers, G.W., Dowell, A.M., and Hendershot, D.C. (1996). It's never too late for inherent safety. In International Conference and Workshop on Process Safety Management and Inherently Safer Processes, October 8-11, 1996, Orlando, FL (pp. 227-241). New York: American Institute of Chemical Engineers.

Catanach, J.S., and Hampton, S.W. (1992). Solvent and surfactant influence on flash points of pesticide formulations. *ASTM Spec. Tech. Publ. 11*, 149-157.

Chementator: (August, 1993). A propylene process promises to slash operating costs. *Chemical* Engineering, 15.

Do away with solvents in polymer processing, 15.

Enzymes help make better phenolics, 17-18.

Chevalier, A.B.H. (1991). Passive safety features - A considered view of UK safety arguments. *Nuc. Energy, 30* (2), 79-83.

Clark, D.G. (2007). Applying the limitation of effects ISP strategy when siting and designing facilities. In *Proceedings of the 41st Annual Loss Prevention Symposium, April 22-26, 2007, Houston, TX* (Paper 1C). New York: American Institute of Chemical Engineers.

Cleary, J. A. (June, 1993). Overview of the new sealless pump standard *Hydrocarbon Processing*, 49-51.

Clifford,T., and Bartle, K. (17 June, 1996). Chemical reactions in supercritical fluids. *Chemistry & Industry*, 449-452.

Coeyman, M., and R. Mullin (October 13, 1993). Raw materials suppliers key in on performance enhancement. *Chemical Week*, 34-35.

Coghlan, A. (19 November, 1994). Dry ice leaves dirty cleaners out in the cold. *New Scientist*, 28.

Costa, R., Recasens, F. and Velo, E. (1995). Inherent thermal safety of stirred-tank batch reactors: A prognosis tool based on pattern recognition of hazardous states. In G.A. Melhem and H.G. Fisher (Eds.). *International Symposium on Runaway Reactions and Pressure Relief Design, August 2-4, 1995, Boston, MA* (pp. 690-709). New York: American Institute of Chemical Engineers.

Cottam, A. N. (1991). Risk assessment and control in biotechnology. In *IChemE Symposium Series,* No. 124, 341-w348.

Crabtree, E.W., and El-Halwagi, M.M. (1994). Synthesis of environmentally acceptable reactions. In M. El-Halwagi, and D.P. Petrides (Eds.).. *Pollution Prevention Via Process and Product Modifications* (pp. 117-127). AIChE Symposium Series, 303. New York: American Institute of Chemical Engineers.

Cusumano, J A. (August, 1992). New technology and the environment. *Chemtech,* 482-89.

Dartt, C.B., and Davis, M.E. (1994). Catalysis for environmentally benign processing. *Ind. Eng.Chem. Res. 33*, 2887-299.

Davis, G.A., Kincaid, L., Menke, D., Griffith, B., Jones, S., Brown, K., and Goergen, M. (1994). *The Product Side of Pollution Prevention: Evaluating the Potential for Safe Substitutes*. Cincinnati, Ohio: Risk Reduction Engineering Laboratory, Office of Research and Development, U. S. Environmental Protection Agency.

The design of inherently safer plants (1988). *Chemical Engineering Progress, 84* (9), 21.

DeSimone, J.M., Maury, E.E., Guan, Z., Combes, J.R., Menceloglu, Y.Z., Clark, M.R., et al. (1994). Homogeneous and heterogeneous polymerizations in environmentally-responsible carbon dioxide. In *Preprints of Papers Presented at the 208th ACS National Meeting, August 21-25, 1994, Washington, DC* (pp. 212-214). Center for Great Lakes Studies, University of Wisconsin-Milwaukee, Milwaukee, WI: Division of Environmental Chemistry, American Chemical Society.

DeVito, S. C. (November, 1996). Designing safer chemicals: Toxicological considerations. *Chemtech*, 34-47.

Doerr, W.W., and Hessian Jr., R.T. (1991). Control toxic emissions from batch operations. *Chemical Engineering Progress, 87* (9), 57-62.

Doherty, M., and Buzad, G. (27 August, 1992). Reactive distillation by design. *The Chemical Engineer*, s17-s19.

Donohue, M. D. and Geiger, J.L. (1994). Reduction of VOC emission during spray painting operations: A new process using supercritical carbon dioxide. In *Preprints of Papers Presented at the 208th ACS National Meeting, August 21-25, 1994, Washington, DC* (pp. 218-219). Center for Great Lakes Studies, University of Wisconsin-Milwaukee, Milwaukee, WI: Division of Environmental Chemistry, American Chemical Society.

Dowell, A.M. (1996). Vent systems: life cycle & inherently safer concepts. In International Conference and Workshop on Process Safety Management and Inherently Safer Processes, October 8-11, 1996, Orlando, FL (Workshop F: Case Studies on Inherent Safety: Cost Benefit Analysis; Life Cycle Cost).

Drexler, K. E. (1994). Molecular manufacturing for the environment. In *Preprints of Papers Presented at the 208th ACS National Meeting, August 21-25, 1994, Washington, DC* (pp. 263-265). Center for Great Lakes Studies, University of Wisconsin-Milwaukee, Milwaukee, WI: Division of Environmental Chemistry, American Chemical Society.

Drioli, E., and L. Giorno (1 January, 1996). Catalytic membrane reactors. *Chemistry and Industry*, 19-22.

Dutt, S. (1996). Safe design of direct steam injection heaters. In *The 5th World Congress of Chemical Engineering, July 14-18, 1996, San Diego, CA, Vol. II* (pp. 1107-1102). New York: American Institute of Chemical Engineers. (

Edwards, D.W. and Lawrence, D. (1993). Assessing the inherent safety of chemical process routes: Is there a relation between plant costs and inherent safety? *Trans. IChemE. 71*, Part B, 252-258.

Edwards, D.W. and Lawrence, D. (1995). Inherent safety assessment of chemical process routes. In *The 1995 ChemE Research*

Event: First European Conf. of Young Res. Chem. Eng, July 14-18, 1996 (pp.62-64). Rugby, UK: Institution of Chemical Engineers.

Edwards, D.W., Lawrence, D., and Rushton, A.G. (1996). Quantifying the inherent safety of chemical process routes. In *5th World Congress of Chemical Engineering, July 14-18, 1996, San Diego, CA* (Paper 52d). New York: American Institute of Chemical Engineers.

Eierman, R. G. (1995). Improving Inherent Safety With Sealless Pumps. In E.D. Wixom and R.P. Benedetti (Eds.). *Proceedings of the 29th Annual Loss Prevention Symposium, July 31-August 2, 1995, Boston, MA* (Paper 1e). New York: American Institute of Chemical Engineers.

Emsley, J. (12 March, 1994). A cleaner way to make nylon. *New Scientist*, 15.

Englehardt, J. D. (1993). Pollution prevention technologies: A review and classification. *Journal of Hazardous Materials, 35*, 119-50.

Englund, S.M. (1990). Opportunities in the design of inherently safer chemical plants. *Advances in Chemical Engineering, 15*, 69-135.

Englund, S. M. (1990). "The design and operation of inherently safer chemical plants." Presented at the American Institute of Chemical Engineers 1990 Summer National Meeting, August 20, 1990, San Diego, CA, Session 43.

Englund, S. M. (1991a). Design and operate plants for inherent safety - Part 1. *Chemical Engineering Progress, 87* (3), 85-91.

Englund, S.M. (1991b). Design and operate plants for inherent safety - Part 2. *Chemical Engineering Progress, 87* (5), 79-86.

Englund, S.M. (1993). Process and design options for inherently safer plants. In V. M. Fthenakis (ed.). *Prevention and Control of Accidental Releases of Hazardous Gases* (9-62). New York: Van Nostrand Reinhold.

Englund, S. M. (1994). "Inherently safer plants—Practical applications." Presented at the American Institute of Chemical Engineers 1994 Summer National Meeting, August 14-17, 1994, Denver, CO, Paper No. 47b.

US EPA urges chemical industry, universities to embrace "benign by design" production (August 27, 1993). *Chemical Regulation Reporter*, 989-990.

Finzel, W.A. (1991). Use low-VOC coatings. *Chemical Engineering Progress*, *87* (11), 50-53.

Flam, F. (9 September, 1994). US EPA campaigns for safer chemicals. *Science, 265*, 1519.

Flam, F. (14 October, 1994). Laser chemistry: The light choice. *Science, 266*, 215-217.

Forsberg, C.W., Moses, D.L., Lewis, E.B., Gibson, R., Pearson, R., Reich, W.J., et al. (1989). Proposed and Existing Passive and Inherent Safety-Related Structures, Systems, and Components (Building Blocks) for Advanced Light Water Reactors. Oak Ridge, TN: Oak Ridge National Laboratory.

Forsberg, C.W. (1990). Passive and inherent safety technologies for light-water nuclear reactors. Presented at the America Institute of Chemical Engineers 1990 Summer National Meeting, August 19-22, 1990, San Diego, CA, Session 43.

French, R. W., Williams, D.D. and Wixom, E.D. (1995). Inherent safety, health and environmental (SHE) reviews. In E. D. Wixom and R. P. Benedetti (Eds.). *Proceedings of the 29th Annual Loss Prevention Symposium, July 31-August 2, 1995, Boston, MA* (Paper 1c). New York: American Institute of Chemical Engineers.

Friedlander, S.K. (1989). The implications of environmental issues for engineering R&D and education. *Chemical Engineering Progress*, *85* (11), 22-28.

Gerritsen, H.G., and Van't Land, C.M. (1988). Intrinsic continuous process safeguarding. *IChemE Symposium Series* No. 110, 107-115.

Gibbs, W.W. (November, 1994). Ounce of prevention. *Scientific American*, 103-105.

Goldschmidt, G., and Filskov, P. (1990). Substitution—A way to obtain protection against harmful substances at work. *Staub - Reinhaltund Der Luft, 50*, 403-405.

Govardhan, C.P. and Margolin, A.L. (4 September,1995). Extremozymes for industry—From nature and by design. *Chemistry & Industry*, 689-693.

Gowland, R. T. (1995). Applying inherently safer concepts to an acquisition which handles phosgene. In E.D. Wixom and R.P. Benedetti (Eds.). *Proceedings of the 29th Annual Loss Prevention Symposium, July 31 - August 2, 1995, Boston, MA* (Paper No. 1d). New York: American Institute of Chemical Engineers.

Gowland, R.T. (1996). Putting numbers on inherent safety. *Chemical Engineering 103* (3), 82-86.

Green Chemistry Challenge (1995). *Chemical Health and Safety, 2* (6), 5.

Grewer, T., Klusacek, H., Loffler, U., Rogers, R.L. and Steinbach, J. (1989). Determination and assessment of the characteristic values for the evaluation of the thermal safety of chemical processes. *Journal of Loss Prevention in the Process Industries, 2*, 215-223.

Guettler, R. D., Jones, G.C., Posey, L. and Zare, R.N. (14 October, 1994). Partial control of an ion-molecule reaction by selection of the internal motion of the polyatomic reagent ion. *Science, 266*, 259-261.

Gygax, R.W. (1988). Chemical reaction engineering for safety. *Chemical Engineering Science, 43* (8), 1759-1771.

Haggin, J. (February 14,1994). Catalysis gains widening role in environmental protection. *Chemical and Engineering News*, 22-30.

Haggin, J. (April 18, 1994). Direct process converts methane to acetic acid. *Chemical and Engineering News*, 5.

Catalysis critical to benign process design. 22-25.

Haggin, J. (June 3, 1996). Chemical reaction engineering brings diversity of inputs to process designs. *Chemical and Engineering News*, 38-51.

Hall, N. (7 October, 1994). Chemists clean up synthesis with one-pot reactions. *Science, 266*, 32-34.

Hannon, J. (1992). Clean technology through process intensification. *IChemE North Western Branch Symposium Papers, 3*, 7.1-7.6.

Harris, C. (1993). Containment enclosures. In V. M. Fthenakis (Ed.). *Prevention and Control of Accidental Releases of Hazardous Gases* (pp. 404-410). New York: Van Nostrand Reinhold.

Hawksley, J. L., and. Preston, M.L (1996). Inherent SHE—20 Years of Evolution. In International Conference and Workshop on Process Safety Management and Inherently Safer Processes, October 8-11, 1996, Orlando, FL (pp.183-196). New York: American Institute of Chemical Engineers.

Heil, J.A. (1995). *Inherent Safety Characteristics of Innovative Nuclear Reactors*. Petten, Netherlands: Netherlands Energy Res. Foundation.

Hendershot, D. C. (1987). Safety considerations in the design of batch processing plants. In J.L. Woodward (Ed.). *Proceedings of the International Symposium on Preventing Major Chemical Accidents, February 3-5, 1987, Washington, D.C.* (pp. 3.2-3.16). New York: American Institute of Chemical Engineers.

Hendershot, D.C. (1988). Risk reduction alternatives for hazardous material storage. In *Proc. 1988 Hazardous Materials Spills Conference, May 16-19, 1988, Chicago, IL* (pp. 611-618). New York: American Institute of Chemical Engineers.

Hendershot, D.C. (1988). Alternatives for reducing the risks of hazardous material storage facilities. *Environmental Progress 7* (3), 180-184.

Hendershot, D.C. (1991). "Design of inherently safer chemical processing facilities." Presented at Texas Chemical Council Safety Seminar, June 11, 1991, Galveston, TX, Session D.

Hendershot, D.C. (1993). Inherently safer plants. In *Guidelines for Engineering Design for Process Safety* (pp. 5-52). New York: American Institute of Chemical Engineers.

Hendershot, D.C. (1994). Chemistry—The Key to Inherently Safer Manufacturing Processes. In *208th American Chemical Society National Meeting, August 21-25, 1994, Washington, DC* (Paper No. ENVR-135).

Hendershot, D.C. (1995). Conflicts and decisions in the search for inherently safer process options. *Process Safety Progress, 14* (1), 52-56.

Hendershot, D.C. (1995). Some thoughts on the difference between inherent safety and safety. *Process Safety Progress, 14* (4), 227-28.

Hendershot, D.C. (1996). Conflicts and decisions in the search for inherently safer process options. In G.F. Nalven (Ed.). *Plant Safety* (pp. 58-62). New York: American Institute of Chemical Engineers.

Hendershot, D.C. (1996). "The philosophy of inherently safer chemical process design." Presented at the 89th Annual Meeting and Exhibition of the Air & Waste Management Association, June 23-28, 1996, Nashville, TN (Session 27).

Hendershot, D.C. (1996). Case studies of inherently safe design. In *The 5th World Congress of Chemical Engineering: Technologies Critical to a Changing World July 14-18, 1996, San Diego, CA* (pp. 98-99). New York: American Institute of Chemical Engineers.

Hendershot, D.C. (1996). How do you measure inherent safety early in process development? In International Conference and Workshop on Process Safety Management and Inherently Safer Processes, October 8-11, 1996, Orlando, FL (Workshop F).

Hill, R.G. (July, 1995). Inherent safety. *Safety & Health News*, 4.

Hill, R.G. (April, 1995). Ongoing work—Inherent safety. *Safety and Health News*, 1.

Hodel, A.E. (March, 1993). Butyl acetate replaces toluene to remove phenol from water. *Chemical Processing*, 53-56.

Hugo, P., and Steinbach, J. (1986). A comparison of the limits of safe operation of a SBR and a CSTR. *Chemical Engineering Science, 41* (4), 1081-1087.

Humphrey, J L., and Seibert, A.F. (December, 1992). New horizons in distillation. *Chemical Engineering*, 86-98.

Illman, D.L. (September 6,1993). "Green" technology presents challenge to chemists. *Chemical and Engineering News*, 26-30.

Illman, D.L. (September 5, 1994). Environmentally benign chemistry aims for processes that don't pollute. *Chemical and Engineering News*, 22-27.

Inherent Safety (July, 1995). *Safety & Health News*, 3.

Inherently Safe Technology Advances Despite Doubts (1994). *Chemical Process Safety Report, 5* (1), 1.

International Conference and Workshop on Process Safety Management and Inherently Safer Processes (1996). New York: American Institute of Chemical Engineers.

Irwin, A. (March, 1993). Changing the unchangeable. *Process Engineering*, 29-30.

Ishizaki, A., Tanaka, K, Takeshita, T., Kanemaru, T., Shimoji, T., and Kawano, T. (1993). Equipment and operation for fermentative PHB production using gaseous substrate to guarantee safety from explosion. *J. Chem. Eng. Jpn. 26* (2), 225-227.

Jacobson, A. and Willingham, G.L. (1994). Designing an Environmentally Safe Marine Antifoulant." In *Preprints of Papers Presented at the 208th ACS National Meeting, August 21-25, 1994, Washington, DC* (pp. 358-388). Center for Great Lakes Studies, University of Wisconsin-Milwaukee, Milwaukee, WI: Division of Environmental Chemistry, American Chemical Society.

Jarabek, A.M., et. al. (1994). Mechanistic insights aid the search for CFC substitutes: Risk Assessment of HCFC-123 as example. *Risk Analysis, 14* (3) 231-250.

Jessop, P.G., Ikariya, T. and Noyori, R. (17 March, 1994). Homogeneous catalytic hydrogenation of supercritical carbon dioxide. *Nature,* 368, 231-233.

Joback, K.G. (1994). Solvent substitution for pollution prevention. In M. El-Halwagi, and D.P. Petrides (Eds.). *Pollution*

Prevention Via Process and Product Modifications (pp. 98-103). AIChE Symposium Series, 303. New York: American Institute of Chemical Engineers.

Johnson, R.W., Unwin, S.D., and McSweeney, T.I. (1996). Inherent safety: how to measure it and why we need it. In International Conference and Workshop on Process Safety Management and Inherently Safer Processes, October 8-11, 1996, Orlando, FL (pp. 118-127). New York: American Institute of Chemical Engineers.

Johnston, K.P. (17 March 1994). Safer solutions for chemists. *Nature 368*, 187-88.

Jones, D.O. (1996). *Process Intensification of Batch, Exothermic Reactors*. Sudbury, Suffolk, UK: HSE Books.

Kauffman, G.B. (21 February,1994). CFCs, TEL and "Midge." *Chemistry and Industry*, 143.

Keeports, G.L. (1996). Major accident prevention program. In International Conference and Workshop on Process Safety Management and Inherently Safer Processes, October 8-11, 1996, Orlando, FL (Workshop E). New York: American Institute of Chemical Engineers.

Kelleher, T. and Fair, J.R. (1996). Distillation studies in a high-gravity contactor. *Ind. Eng. Chem. Res., 35*, 4646-4655.

Keller, A., Heinxle, E. and K. Hungerbuhler (1996). Development and assessment of inherently safe processes in the fine chemical industry." In *International Conference and Workshop on Process Safety Management and Inherently Safer Processes, October 8-11, 1996, Orlando, FL* (213-223). New York: American Institute of Chemical Engineers.

Kelley, K P. (April, 1992). Pollution prevention pays off. *Control*, 27-34.

Kharbanda, O.P. and Stallworthy, E.A. (1980 and 1988). *Safety in the Chemical Industry*. London: Heinemann Professional Publishing, Ltd.

Kirby, G.N. (1985). Explosion pressure shock resistance. *Chemical Engineering Progress, 81* (11), 48-50.

Kletz, T.A. (6 May, 1978). What you don't have, can't leak. *Chemistry and Industry*, 287-292.

Kletz, T.A. (March, 1979). Is there a simpler solution? *The Chemical Engineer*, 161-64.

Kletz, T.A. (June, 1983). Inherently safer plant—The concept, its scope and benefits. *Loss Prevention Bulletin, 051*, 1-8.

Kletz, T.A. (1984). *Cheaper, Safer Plants, or Wealth and Safety at Work*. Rugby, Warwickshire, England: The Institution of Chemical Engineers.

Kletz, T.A. (1985). Inherently safer plants. *Plant/Operations Progress 4* (3), 164-167.

Kletz, T.A. (1990). Plants should be friendly. In *Safety and Loss Prevention in the Chemical and Oil Processing Industries, October 23-27, 1989, Singapore* (pp. 423-435). IChemE Symposium Series, no. 120. Rugby, Warwickshire, U. K.: The Institution of Chemical Engineers.

Kletz, T. A. (1991). Present trends in process safety. Speculations in Science and Technology, Developments in Chemical Engineering, 15 (2), 83-90.

Kletz, T.A. (25 April, 1991). Billiard balls and polo mints. *The Chemical Engineer, 495*, 21-22.

Kletz, T.A. (June 1991). Inherently safer plants: An update. *Plant/Operations Progress, 10* (2), 81-84.

Kletz, T.A. (1991). Inherently safer plants: Recent progress. *IChemE Symposium Series*, 124, 225-33.

Kletz, T.A. (1992). Titles can mislead. Journal of Loss Prevention in the Process Industries, 5 (5), 259.

Kletz, T.A. (1992). Managing risk in chemical manufacture. *Spec. Pub.- R. Soc. Chem*, 101, 92-105.

Kletz, T.A. (1993). What can we do to encourage the development of inherently safer designs. *Safety and Health News*, 3, 1.

Kletz, T.A. (1993). Accident prevention: Lessons Learned. In V. M. Fthenakis (ed.). *Prevention and Control of Accidental Releases of Hazardous Gases* (pp. 83-93). New York: Van Nostrand Reinhold.

Kletz, T.A. (1995). Inherently safer design: The growth of an idea. In E.D. Wixom and R.P. Benedetti (Eds.). *Proceedings of the 29th Annual Loss Prevention Symposium, July 31-August 2, 1995, Boston, MA (Paper 1a)*. New York: American Institute of Chemical Engineers.

Kletz, T.A. (1996). Inherently safer design: Achievements and prospects. In International Conference and Workshop on Process Safety Management and Inherently Safer Processes, October 8-11, 1996, Orlando, FL (pp. 197-206). New York: American Institute of Chemical Engineers.

Koch, T. A., Krause, K.R. and Mehdizadeh, M. (1996). Improved safety through distributed manufacturing of hazardous chemicals. In *The 5th World Congress of Chemical Engineering, July 14-18, 1996, San Diego, CA, Vol. II* (pp. 1095-1098). New York: American Institute of Chemical Engineers.

Kollman, C.J. (1993). "Achieving environmental excellence in process development." Presented at HazMat International '93 Environmental Management and Technology Conference, June 9-11, 1993, Atlantic City, NJ (Session 20).

Komarek, R. (December, 1993). Briquets used to reduce environmental hazards. *Pollution Engineering*, 47-48.

Laing, I.G. (21 September, 1992). Waste minimisation: The role of process development. *Chemistry and Industry*, 682-686.

Laird, T. (1992). "Is your chemical process operating at its optimum?" Presented at Chemicals and the Environment Symposium, in Association With Chemical Specialities USA'92 Exhibition. Bramhall, Stockport England: Spring Innovations Limited.

Lammertink, E. (1992). "Safe storage of propylene oxide, a new approach." Presented at the International Process Safety Group Technical Meeting, April 29-May 1, 1992, Taormina, Italy (Paper 1.2)

Lawrence, D., and Edwards, D.W. (1994). Inherent safety assessment of chemical process routes by expert judgement. In *The 1994 IChemE Research Event* (886-888).

Lawrence, D., Edwards, D.W., and Rushton, A.G. (1993). Quantifying inherent safety. *Proc. Inst. Mech. Eng, 4,* 1-8.

Lazari, L., S. J. Burley, and S. T. S. Al-Hassani (1991). Containment of explosive blast in a spherical vessel with nozzles. *IChemE Symposium Series No. 124*, 135-153.

Lerou, J. J., and K. M. Ng (1996). Chemical reaction engineering: A multiscale approach to a multiobjective task. *Chemical Engineering Science, 51* (10), 1595-614.

Lim, P. K. (1994). Environmentally benign oil-water interfacial synthesis. *Preprints of Papers Presented at the 208th ACS National Meeting, August 21-25, 1994, Washington, DC* (pp. 238). Center for Great Lakes Studies, University of Wisconsin-Milwaukee, Milwaukee, WI: Division of Environmental Chemistry, American Chemical Society.

Lin, D., Mittelman, A., Halpin, V. and Cannon, D. (1994). *Inherently Safer Chemistry: A Guide to Current Industrial Processes to Address High Risk Chemicals*. Washington, DC: Office of Pollution Prevention and Toxics, U. S. Environmental Protection Agency.

Lin, M., and Ayusman, S. (14 April, 1994). Direct catalytic conversion of methane to acetic acid in aqueous medium. *Nature,* 238, 613-615.

Lodal, P. N. (1996). "Inherently Safer Processes Using the Low Cost Plant Approach." Presented at International Conference and Workshop on Process Safety Management and Inherently Safer Processes, October 8-11, 1996, Orlando, FL, Workshop E.

Lorentzen, G. (1995). The use of natural refrigerants: A complete solution to the CFC/HCFC predicament. *Int. J. Refrig. 18* (3), 190-197.

Lubineau, A. (19 February, 1996). Making a splash in synthesis: Water as a solvent. *Chemistry and Industry*, 123-26.

Lutz, W.K. (1995). Take chemistry and physics into consideration in all phases of chemical plant design. *Process Safety Progress 14* (3), 153-162.

Lutz, W.K. (1995). Putting safety into chemical plant design. *Chemical Health and Safety, 2* (6), 12-15.

Lutz, W.K. (1996). Advancing inherent safety into methodology. In *The 5th World Congress of Chemical Engineering, July 14-18, 1996, San Diego, CA, Vol. II* (pps.1015-1019). New York: American Institute of Chemical Engineers.

Lutz, W.K. (1996). "Advancing inherent safety into methodology." Presented at the International Conference and Workshop on Process Safety Management and Inherently Safer Processes, October 8-11, 1996, Orlando, FL, Workshop E.

Mandich, N.V.and Krulik, G.A. (1992). Substitution of nonhazardous for hazardous process chemicals in the printed circuit industry. *Me. Finish. 90* (11), 49-51.

Mansfield, D.P. (1994). *Inherently Safer Approaches to Plant Design*. Warrington, Cheshire, UK: United Kingdom Atomic Energy Authority.

Mansfield, D.P. (1996). Viewpoints on implementing inherent safety. *Chemical Engineering, 103* (3), 78-80.

Mansfield, D.P., and Cassidy, K. (1994). Inherently safer approaches to plant design. *IChemE Symposium Series, 134,* 285-99.

Mansfield, D.P., Turney, R.D., Rogers, R.L., Verwoerd, M. and Bots, P. (1995). How to integrate inherent SHE in process development and plant design. Presented at *IChemE Conference, Major Hazards Onshore and Offshore*. UMIST'95, Manchester, UK.

Mansfield, D., Malmen, Y. and Suokas, E. (1996). "The Development of an Integrated Toolkit for Inherent SHE." International Conference and Workshop on Process Safety Management and Inherently Safer Processes, October 8-11, 1996, Orlando, FL (pp. 103-117). New York: American Institute of Chemical Engineers.

Manzer, L.E. (1993). Environmentally safer processes: Opportunities for catalysis and process R&D. *Chemistry for Sustainable Development, 1*, 147-151.

Manzer, L.E. (1993). Toward catalysis in the 21st century chemical industry. *Catalysis Today, 18*, 199-207.

Manzer, L.E. (1994). Chemistry and catalysis. In P.T. Anastas and C.A. Farris (Eds.). *Benign by Design: Alternative Synthetic Design for Pollution Prevention* (pp. 144-154). Washington, D. C.: American Chemical Society.

Maroldo, S. G., and M. T. Vandersall (1993). Carbonaceous adsorbents for the prevention of bed fires. Presented at the Air & Waste Management Association, 86th Annual Meeting and Exhibition, June 13-18, 1993, Denver, CO (93-TP-31B.04).

Marshall, V.C. (1990). The social acceptability of the chemical and process industries. *Trans. IChemE, Part B, 68*, 83-93.

Marshall, V.C. (1992). The management of hazard and risk. *Applied Energy, 42*, 63-85.

Marshall, J., Mundt, A., Hult, M., McKealvy, T.C., Myers, P. and Sawyer, J. (1995). The relative risk of pressurized and refrigerated storage for six chemicals. *Process Safety Progress, 14* (3), 200-211.

McCarthy, A.J., Ditz, J.M. and Geren, P.M (1996). Inherently safer design review of a new alkylation process. *5th World Congress of Chemical Engineering, July 14-18, 1996, San Diego, CA (Paper 52b).* New York: American Institute of Chemical Engineers.

McQuaid, J. (1991). Know your enemy: The science of accident prevention. *Trans. IchemE, Part B. 69*, 9-19.

Mekitarian, A.E., and Keller, Jr., J.A (August, 1992). Product safety and modern grease formulations. *NLGI Spokesman*, 181-190.

Melhem, G.A. (1993). Hazard reduction benefits from reduced storage temperature of pressurized liquids. In V.M. Fthenakis (Ed.). *Prevention and Control of Accidental Releases of Hazardous Gases* (pps. 411-437). New York: Van Nostrand Reinhold.

Misono, M., and Okuhara T. (1993). Solid superacid catalysts. *Chemtech, 23* (11), 23-29.

Mittelman, A., and Lin, D. (4 September,1995). Inherently safer chemistry. *Chemistry & Industry*, 694-696.

Mittelman, A., Lin, D., Cannon, D., and Ho, C K. (1994). Inherently Safe Chemistry Guide." *Preprints of Papers Presented at the 208th ACS National Meeting, August 21-25, 1994, Washington, DC* (pps. 271-272). Center for Great Lakes Studies, University of Wisconsin-Milwaukee, Milwaukee, WI: Division of Environmental Chemistry, American Chemical Society.

Mizerek, P. (January/February 1996). Disinfection techniques for water and wastewater. *The National Environmental Journal*, 22-28.

Moore, S., Samdani, S., Ondrey, G., and Parkinson, G. (March, 1994). New roles for supercritical fluids. *Chemical Engineering*, 32-35.

More Poke, Less Pounds (10 December, 1992). *The Chemical Engineer*, 18.

Mullin, R., and Plishner, E. (April 21, 1993). Big ticket changes in store for North American pulp and paper. *Chemical Week*, 32-34.

Nakazato, K. (1992). Anti-explosion system at recycling plants. *Inorganic Transformations and Ash Deposition During Combustion*, 915-920

Nasr, A. (1992). Prevent failures of mag drive pumps. *Chemical Engineering Progress, 88* (11), 31-34.

Nasr, A. (July, 1994). Sealless pumps: Limitations and developments. *Hydrocarbon Processing*, 63-66.

Negron, R.M. (March/April 1994). Using ultraviolet disinfection in place of chlorination. *The National Environmental Journal*, 48-50.

Negron, R.M. (March/April 1994). Using ultraviolet disinfection in place of chlorination. *The National Environmental Journal*, 48-50.

New look fridges take off without HFCs (21 February 1994). *Chemistry and Industry*, 130.

New Refrigerant Gives More for Less (1994). *Environmental Protection Bulletin*, 032 (September), 33.

Newby, T., and Forth, D. (1991). Glandless pumps and valves—A technical update. *IChemE Symposium* Series No. 124, 119-134.

Newman, A. (1994). CFC phase-out moving quickly. *Environ. Sci. Technol.*, 28 (1), 35-37.

Newman, A. (1994). Designer chemistry. *Environ. Sci. Technol.* 28, 11, 463.

Ondrey, G. (March, 1995). Microreactor engineering: Birth of a new discipline? *Chemical Engineering*, 52.

Oosterkamp, W.J. (1993). Inherent and passive safety aspects in the design of advanced natural circulation boiling water reactors. In *Proc. ASME/JSME Nucl. Eng. Jt. Conf., Vol. 1* (pp. 713-728). New York: ASME.

Orrell, W., and Cryan, J. (August, 1987). Getting rid of the hazard. *The Chemical Engineer*, 14-15.

Paint Removers: New products eliminate old hazards (May 1991). *Consumer Reports*, 340-343.

Pan, H.Y., Minet, R.G., Benson, S.W. and Tsotsis, T.T. (1994). Process for converting hydrogen chloride to chlorine. *Ind. Eng. Chem. Res., 33*, 2996-3003.

Papayannokos, N.G., Koufopanos, C.A., and Karetsou, A. (1993). Studies on parametric sensitivity and safe operation criteria of batch processes. *Chem. Eng. Technol. 16*, 318-324.

Paquet Jr., D.A., and Ray, W.H. (1994). Tubular reactors for emulsion polymerization: I. Experimental investigation. *AIChE Journal, 40* (1), 73-87.

Paquet Jr., D. A., and Ray, W.H. (1994). Tubular reactors for emulsion polymerization: II. Model Comparisons With Experiments. *AIChE Journal, 40* (1), 88-96.

Park, D. E. (1996). "Inherently Safer Chemical Processes - A Life Cycle Approach." International Conference and Workshop on Process Safety Management and Inherently Safer Processes, October 8-11,

1996, Orlando, FL (pp. 316-319). New York: American Institute of Chemical Engineers.

Parthasarathy, R.V., and C.R. Martin (26 May, 1994). Synthesis of polymeric microcapsule arrays and their use for enzyme immobilization. *Nature, 369*, 298-301.

Paul, B.O. (February, 1994). Mag-drive pumps eliminate VOC emissions. *Chemical Processing*, 66-68.

Paul, E.L. (1988). Design of reaction systems for specialty organic chemicals. *Chemical Engineering Science, 43* (8), 1773-1782.

Pease, R. (10 June, 1995). Nanoworlds are made of this. *New Scientist*, 26-29.

Penteado, F.D., and. Ciric, A.R (1996). An MINLP approach for safe process plant layout. *Ind. Eng. Chem. Res., 35*, 1354-1361.

Perrin, D. A. (1996). Inherent safety in processes: Softer aspects of process safety. In *International Conference and Workshop on Process Safety Management and Inherently Safer Processes*, October 8-11, 1996, Orlando, FL (pp. 207-212). New York: American Institute of Chemical Engineers.

Perry, R.G. (1995). The paradox of inherent safety. Presented at the International Symposium on Runaway Reactions & Pressure Relief Design, August 2-4, 1995, Boston, MA.

Ponton, J. (1996). The disposable batch plant. In *5th World Congress of Chemical Engineering, July 14-18, 1996, San Diego, CA, Vol. II* (pp. 1119-1123). New York: American Institute of Chemical Engineers.

Popoff, F.P. (January 11, 1993). Full-cost accounting. *Chemical and Engineering News*, 8-10.

Poudrier, J.K. (September,1996). The chemical industry wins the green chemistry challenge. *Today's Chemist at Work*, 64-70.

Preston, M.L., and Turney, R.D. (1991). The process systems contribution to reliability engineering and risk assessment. In L. Puigjaner and A.L. Espuna, *Computer-Oriented Process Engineering* (pp.249-257). Amsterdam: Elsevier Science Publishers.

Preventing Pollution (March 23, 1995). *Right-to-Know Planning Guide,* 8 (14), 3.

Prugh, R.W. (1992). Hazardous fluid releases: Prevention and protection by design and operation. *Journal of Loss Prevention in the Process Industries, 5* (2), 67-72.

Puranik, S.A., Hathi, K.K., and Sengupta, R. (1990). Prevention of hazards through technological alternatives. In *Safety and Loss Prevention in the Chemical and Oil Processing Industries, October 23-27, 1989, Singapore* (pp. 581-587). IChemE Symposium Series, No. 120. Rugby, Warwickshire, U.K.: The Institution of Chemical Engineers.

Raghaven, K.V. (1992). Temperature runaway in fixed bed reactors: Online and offline checks for intrinsic safety. *Journal of Loss Prevention in the Process Industries, 5* (3), 153-159.

Report Updates Accident Patterns, Emphasizes Prevention (1997). *Chemical Process Safety Report* 7, 4 (February), 1.

Riezel, Y. (1996). Inherently safer process: an integral part of PSM implementation. In *International Conference and Workshop on Process Safety Management and Inherently Safer* Processes, October 8-11, 1996, Orlando, FL (pp. 320-328). New York: American Institute of Chemical Engineers..

Rogers, R.L., Hallam, and S. (1991). A chemical approach to inherent safety. *IChemE Symposium Series* No. 124, 235-241.

Rogers, R.L., Mansfield, D.P., Malmen, Y., Turney, R.D., and Verwoerd, M. (1995). The INSIDE Project: Integrating inherent safety in chemical process development and plant design. In G.A. Melhem and H.G. Fisher (Eds.). *International Symposium on Runaway Reactions and Pressure Relief Design, August 2-4, 1995, Boston, MA* (pp. 668-689). New York: American Institute of Chemical Engineers.

Rogers, W.G. (December, 1996). Working knowledge: Pop tops." *Scientific American*, 132.

Ror, B.O. (1992). Secondary containment examples in norsk hydro. Presented at the International Process Safety Group Technical Meeting, April 29-May 1, 1992, Taormina, Italy (Paper 1.1).

Rotman, D. (October 14, 1992). ICI claims breakthrough. *Chemical Week*, 24.

Rotman, D. (September 22, 1993). Chemists map greener synthesis pathways. *Chemical Week*, 56-57.

Rotman, D. (July 27, 1994). Supercritical CO2 points to greener polymer synthesis. *Chemical Week*, 12.

Rotman, D. (December 7, 1994). EPA, industry eye ways to safer chemistry. *Chemical Week*, 27.

Rotman, D. (June 14, 1995). California chemists make nanocatalyst. *Chemical Week*, 12.

Rotman, D., and Wood, A. (December 9, 1992). Successes emerge in search for cleaner processes. *Chemical Week*, 66.

Rushton, A.G., Edwards, D.W., and Lawrence, D. (May, 1994). Inherent safety and computer aided design. *Trans. IChemE 72, Part B*, 83-87.

Safe as a soda can (January, 1995). *ChemEcology*, 3-4.

Samdani, G.S. (October, 1994). "Miniaturization Reaches the CPI." *Chemical Engineering*, 41-43.

Savage, P.E., Gopalan, S., Mizan, T.I., Martino, C.J., and Brock, E.E. (1995). Reactions at supercritical conditions: Applications and fundamentals. *AIChE Journal, 41* (7), 1723-1778.

Scheffler, N.E. (1995). Inherently safer latex plants. In E.D. Wixom and R.P. Benedetti (Eds.), *Proceedings of the 29th Annual Loss Prevention Symposium, July 31 - August 2, 1995, Boston, MA.*. New York: American Institute of Chemical Engineers, Paper 1b

Schwartz, A.T. (March,1996). Sir Humphry Davy and the safety lamp. *Chemistry Review*, 12-15.

Seldon, R. A. (6 January, 1997). Catalysis and pollution prevention. *Chemistry & Industry*, 12-15.

Sherringron, D.C. (7 January, 1991). Polymer supported systems: Towards clean chemistry? *Chemistry and Industry*, 15-19.

Shinnar, R., Doyle, F.J., Budman, H.M., and Morari, M. (1992). Design considerations for tubular reactors with highly exothermic reactions. *AIChE Journal, 38* (11), 1729-1743.

Shonnard, D.R., Herlevich, J., Parikh, P., Stark, D., Barna, B.A., Rogers, T.N., et al. (1996). Pollution assessment software as chemical industry process simulator enhancements. In *The 5th World Congress of Chemical Engineering, July 14-18, 1996, San Diego, CA, Vol. III* (pp. 329-337). New York: American Institute of Chemical Engineers.

Siirola, J.J. (1995). An industrial perspective on process synthesis. In *AIChE Symposium Series*, 91, 222-233.

Singh, J. (February 5, 1993). Assessing semi-batch reaction hazards. *The Chemical Engineer, 537* (21), 23-25.

Singh, J. (1993). Safe disposal of reactive chemicals following emergency venting. *IChemE Symposium Series, 130,* 249-274.

Singh, M.M., Pike, R.M., and Szafran, Z. (1994). Pollution prevention in the organic and inorganic chemistry laboratory: Microscale approach. In *Preprints of Papers Presented at the 208th ACS National Meeting, August 21-25, 1994, Washington, DC* (pp.194-197). Center for Great Lakes Studies, University of Wisconsin-Milwaukee,

Snyder, P.G. (1996). Inherently safe(r) plant design. In H. Cullingford (Ed.). *1996 Process Plant Safety Symposium, Volume 1, April 1-2, 1996, Houston, TX* (pp.203-215). Houston, TX: South Texas Section of the American Institute of Chemical Engineers.

Snyder, P. G. (July, 1996). Is your plant inherently safer. *Hydrocarbon Processing*, 77-82.

Somerville, R. L. (1993). Control of liquefied toxic gas releases. In V. M. Fthenakis *Prevention and Control of Accidental Releases of Hazardous Gases* (pp. 438-447). New York: Van Nostrand Reinhold.

Sorensen, F. and Petersen, H.J.S. (1992). Substitution of organic solvents. *Staub - Reinhaltund Der Luft, 52*, 113-18.

Speechly, D., Thornton, R.E. and Woods, W.A. (1979). Principles of total containment system design. *North Western Branch Papers No. 2, Institution of Chemical Engineers*, 7.1-7.21.

Steensma, M., and Westererp, K.R. (1988). Thermally safe operation of a cooled semibatch reactor: Slow liquid-liquid reactions. *Chemical Engineering Science, 43* (8), 2125-2132.

Steensma, M., and Westererp, K.R (1990). Thermally safe operation of a cooled semibatch reactor: Slow reactions. *Ind. Eng. Chem. Res., 29*, 1259-1270.

Summary and Highlights: Industry/Department of Energy Workshop on Chemical Safety (1993). Washington, D.C.: U. S. Department of Energy, Assistant Secretary for Environment, Safety and Health, Office of Safety and Quality Assurance.

Sundell, M.J. and Nasman, J.H. (1993). Anchoring catalytic functionality on a polymer. *Chemtech, 23* (12), 16-23.

Tavener, S.J., and Clark, J.H. (6 January, 1997). Recent highlights in phase transfer catalysis. *Chemistry & Industry*, 22-24.

Techniques for evaluating chemical reaction hazards. (1993). In J.A. Barton and R.L. Rogers (Eds.). *Chemical Reaction Hazards-A Guide* (19-26). Rugby, Warwickshire, UK: Institution of Chemical Engineers.

Technology for a greener world (March 28, 1991). *The Chemical Engineer*, 13.

Ten Alps Publishing (15 May, 1995). A block of gas. *Chemistry and Industry*, 371.

Tickner, J. (3 October, 1994). The case for inherent safety. *Chemistry and Industry*, 796.

Tietze, L.F. (19 June, 1995). Domino reactions in organic synthesis. *Chemistry & Industry*, 453-457.

Tillman, J.W. (1991). Achievements in Source Reduction and Recycling for Ten Industries in the United States. Cincinnati, Ohio: U. S. Environmental Protection Agency.

Timberlake, D.L., and Govind, R. (1994). Expert system for solvent substitution. In *Preprints of Papers Presented at the 208th ACS National Meeting, August 21-25, 1994, Washington, DC* (pp. 215-217). Center for Great Lakes Studies, University of Wisconsin-Milwaukee,

Milwaukee, WI: Division of Environmental Chemistry, American Chemical Society.

Tobin, P.S. (1994). Risk management through inherently safe chemistry. In *Preprints of Papers Presented at the 208th ACS National Meeting, August 21-25, 1994, Washington, DC* (pp. 268-270). Center for Great Lakes Studies, University of Wisconsin-Milwaukee, Milwaukee, WI: Division of Environmental Chemistry, American Chemical Society.

Tyler, B.J. (1985). Using the Mond Index to measure inherent hazards. *Plant/Operations Progress 4* (3), 172-75.

van Steen, J. F.J. (1996). The Inside Project, Phase 1, Task 1f Report: An Assessment of Hurdles and Solutions Concerning the Adoption of Inherently Safer Approaches in Industry. Apeldoom, Netherlands: TNO.

Velo, E., C. M. Bosch, and F. Recasens (1996). Thermal safety of batch reactors and storage tanks: Development and validation of runaway boundaries. *Ind. Eng. Chem. Res., 35*, 1288-1299.

Verwijs, J. W., van den Berg, H., and Westerterp, K.R. (1996). Startup strategy design and safeguarding of industrial adiabatic tubular reactor systems. *AIChE Journal, 42* (2), 503-515.

Villermaux, J. (1996). New horizons in chemical engineering. In The 5th World Congress of Chemical Engineering: Technologies Critical to a Changing World, July 14-18, 1996, San Diego, CA, Summary Proceedings Volume (pp. 16-23). New York: American Institute of Chemical Engineers.

Wallington, T.J., et al. (1994). The environmental impact of CFC replacements – HFCs and HCFCs. *Environmental Science and Technology, 28* (7), 320-325.

Walsh, F., and Mills, G. (2 August, 1993). Electrochemical techniques for a cleaner environment. *Chemistry and Industry*, 576-579.

Weirick, M.L., Farquhar, S.M., and B.P. Chismar (1994). Spill containment and destruction of a reactive, volatile chemical. *Process Safety Progress, 13* (2), 69-71.

Wells. G.L. and Rose, L.M. (1986). *The Art of Chemical Process Design, 266*. Amsterdam: Elsevier.

Welter, T.R. (February 4, 1991). The quest for safe substitutes. *Industry Week*, 38-43.

Whiting, M.J.L. (March,1992). The benefits of process intensification for caro's acid production. *Trans. IChemE, 70*, 195-196.

Wilday, A.J. (1991). The safe design of chemical plants with no need for pressure relief systems. In *IChemE Symposium Series No. 124*, 243-53.

Wilkinson, M., and Geddes, K. (December 1993). An award winning process. *Chemistry in Britain*, 1050-1052.

Wilson, S. (4 April,1994). Peroxygen technology in the chemical industry. *Chemistry and Industry*, 255-258.

Windhorst, J.C.A. (1995). Application of inherently safe design concepts, fitness for use and risk driven design process safety standards to an lpg project. In J.J. Mewis, H.J. Pasman, and E.E. DeRademacker (Eds.). *Loss Prevention and Safety Promotion in the Process Industries* (543-554). Amsterdam: Elsevier Science B. V.

Wiss, J., Fleury, C. and Fuchs, V. (1995). Modelling and optimization of semi-batch and continuous nitration of chlorobenzene from safety and technical viewpoints. *Journal of Loss Prevention in the Process Industries, 8* (4), 205-213.

Wood, A. (October 14, 1992). DuPont develops waste-free route to terathane. *Chemical Week*, 9.

Wood, A. (August 17, 1994). Mobil cultivates new zeolites to grow process licensing. *Chemical Week*, 34-35.

Wright, M.E., Toplikar, E.G., and Svejda, S.A. (1991). Details concerning the chloromethylation of soluble high molecular weight polystyrene using dimethoxymethane, thionyl chloride and a Lewis acid: A full analysis. *Macromolecules, 24,* 5879-5880.

Index